PARTICULATE CARBON
Atmospheric Life Cycle

PUBLISHED SYMPOSIA
Held at
General Motors Research Laboratories
Warren, Michigan

1980   G. T. Wolff, R. L. Klimisch, eds., *Particulate carbon: Atmospheric life cycle,* Plenum Press, New York, 1982.

1980   D. C. Siegla, G. W. Smith, eds., *Particulate carbon: Formation during combustion,* Plenum Press, New York, 1981.

1979   R. C. Schwing, W. A. Albers, Jr., eds., *Societal risk assessment: How safe is safe enough?* Plenum Press, New York, 1980.

1978   J. N. Mattavi, C. A. Amann, eds., *Combustion modeling in reciprocating engines,* Plenum Press, New York, 1980.

1978   G. G. Dodd, L. Rossol, eds., *Computer vision and sensor-based robots,* Plenum Press, New York, 1979.

1977   D. P. Koistinen, N.-M. Wang, eds., *Mechanics of sheet metal forming: Material behavior and deformation analysis,* Plenum Press, New York, 1978.

1976   G. Sovran, T. A. Morel, W. T. Mason, eds., *Aerodynamic drag mechanisms of bluff bodies and road vehicles,* Plenum Press, New York, 1978.

1975   J. M. Colucci, N. E. Gallopoulos, eds., *Future automotive fuels: Prospects, performance, perspective,* Plenum Press, New York, 1977.

1974   R. L. Klimisch, J. G. Larson, eds., *The catalytic chemistry of nitrogen oxides,* Plenum Press, New York, 1975.

1973   D. F. Hays, A. L. Browne, eds., *The physics of tire traction,* Plenum Press, New York, 1974.

1972   W. F. King, H. J. Mertz, eds., *Human impact response,* Plenum Press, New York, 1973.

1971   W. Cornelius, W. G. Agnew, eds., *Emissions from continuous combustion systems,* Plenum Press, New York, 1972.

1970   W. A. Albers, ed., *The physics of opto-electronic materials,* Plenum Press, New York, 1971.

1969   C. S. Tuesday, ed., *Chemical reactions in urban atmospheres,* American Elsevier, New York, 1971.

1968   E. L. Jacks, ed., *Associative information techniques,* American Elsevier, New York, 1971.

1967   P. Weiss, G. D. Cheever, eds., *Interface conversion for polymer coatings,* American Elsevier, New York, 1968.

1966   E. F. Weller, ed., *Ferroelectricity,* Elsevier, New York, 1967.

1965   G. Sovran, ed., *Fluid mechanics of internal flow,* Elsevier, New York, 1967.

1964   H. L. Garabedian, ed., *Approximation of functions,* Elsevier, New York, 1965.

1963   T. J. Hughel, ed., *Liquids: Structure, properties, solid interactions,* Elsevier, New York, 1965.

1962   R. Davies, ed., *Cavitation in real liquids,* Elsevier, New York, 1964.

1961   P. Weiss, ed., *Adhesion and cohesion,* Elsevier, New York, 1962.

1960   J. B. Bidwell, ed., *Rolling contact phenomena,* Elsevier, New York, 1962.

1959   R. C. Herman, ed., *Theory of traffic flow,* Elsevier, New York, 1961.

1958   G. M. Rassweiler, W. L. Grube, eds., *Internal stresses and fatigue in metal,* Elsevier, New York, 1959.

1957   R. Davies, ed., *Friction and wear,* Elsevier, New York, 1959.

PARTICULATE CARBON
Atmospheric Life Cycle

Edited by
GEORGE T. WOLFF and RICHARD L. KLIMISCH

General Motors Research Laboratories

PLENUM PRESS • NEW YORK – LONDON • 1982

Library of Congress Cataloging in Publication Data

Main entry under title:

Particulate carbon, atmospheric life cycle.

(General Motors symposium series)
"Proceedings of an international symposium...held October 13-14, 1980, at the General Motors Research Laboratories, Warren, Michigan" — Copr. p.
Includes bibliographical references and indexes.
1. Air — Pollution — Congresses. 2. Soot — Congresses. I. Wolff, George T. II. Klimisch, Richard L., 1938-   . III. General Motors Corporation. Research Laboratories. IV. Series.

TD884.5.P37	628.5'3	81-21017
ISBN-13: 978-1-4684-4156-7	e-ISBN-13: 978-1-4684-4154-3	₹2
DOI: 10.1007/978-1-4684-4154-3		

Proceedings of an international symposium on Particulate Carbon:
Atmospheric Life Cycle, held October 13-14, 1980, at the
General Motors Research Laboratories, Warren, Michigan

© 1982 Plenum Press, New York
Softcover reprint of the hardcover 1st edition   1982
A Division of Plenum Publishing Corporation
233 Spring Street, New York, N.Y. 10013

PREFACE

This book contains the papers and discussions from the symposium, "PARTICU-
LATE CARBON: Atmospheric Life Cycle," held at the General Motors Research
Laboratories on October 13-14, 1980. This symposium, which focused on atmospheric
particulate elemental carbon, or soot, was the twenty-fifth in this series sponsored by
the General Motors Research Laboratories. The present symposium volume contains
discussions of the following aspects of particulate elemental carbon (EC): the atmospheric life cycle of EC including sources, sinks, and transport processes, the role of EC
in atmospheric chemistry and optics, the possible role of EC in altering climate, and
measurement techniques as well as ambient concentrations in urban, rural, and
remote areas.

Previous symposia have covered a wide range of scientific and engineering subjects.
Topics are selected because they are new or represent rapidly changing fields and are of
significant technical importance. It is ironic that the study of particulate elemental
carbon or soot should meet the above criteria for selection because soot, especially
from coal and wood combustion, has been a recognized air pollutant for centuries.
However, since the 1950s, when intense efforts to study air pollution were initiated, to
until a few years ago, the role of elemental carbon in the atmosphere was largely
ignored. The major reason for this was the lack of a suitable measurement technique.
Recently, this situation has changed, and presently there are about 20 different
measurement techniques being employed by various research groups. Unfortunately,
however, the various techniques appear to give different results, and before the
symposium there had been no coordinated effort to compare the various methods.
Such an effort was initiated at the meeting and is currently in progress under the
coordination of our group.

In addition to the measurement difficulties, this field is further complicated by
inconsistencies, redundancies, and contradictions in nomenclature. For example,
nearly every measurement method results in unique operational terminology for

elemental carbon. The other descriptors include: apparent elemental carbon, soot, dry soot, black carbon, nonvolatile carbon, nonsoluble carbon, absorbing carbon, residual carbon, and total noncarbonate/nonvolatile carbon. No attempt was made to change the nomenclature used by the various authors in this book so the reader should be aware of these terminology difficulties. Hopefully, a more universal set of nomenclature can be adopted in the near future when the relationships between the results from the various analytical methods become known.

In preparing for the symposium, we discovered that there were a surprisingly large number of researchers engaged in various aspects of research on particulate elemental carbon. We attempted to gather the leading investigators in each area so that a holistic view of the subject could be obtained. We believe that the symposium was successful in accomplishing this and we feel that the papers represent important original contributions to the field.

The efforts of a number of people were responsible for the success of this symposium. The advice and suggestions of Dr. Robert J. Charlson of the University of Washington and Dr. Tihomir Novakov of Lawrence Berkeley Laboratories were especially appreciated. At General Motors Research Laboratories, we would like to thank R. Thomas Beaman for making the symposium arrangements, David N. Havelock for overseeing the manuscript layout and art work, Denise M. Pierson for her assistance in the indexing, and Cheryl Clark for her concientiousness and skills as a secretary and discussion transcriber.

George T. Wolff and Richard L. Klimisch

CONTENTS

SESSION I
THE IMPORTANCE OF
PARTICULATE ELEMENTAL CARBON

Session Chairman
G. M. HIDY

Environmental Research and Technology, Inc.
Westlake Village, California

THE ATMOSPHERIC CYCLE OF ELEMENTAL CARBON

R. J. CHARLSON and J. A. OGREN

*University of Washington
Seattle, Washington*

ABSTRACT

Four sets of factors determine the overall nature of the cycling of elemental carbon through the atmosphere and thereby determine the concentration fields, and fluxes in and out of the atmosphere. The source factor controls mass emission rates, initial microphysical properties such as size distribution, initial chemical composition, and location of injection into the atmosphere. Aerosol mechanics determine the rate of coagulation of the elemental carbon particles with themselves and with other aerosol particles, the rate of diffusive removal to surface sinks, and sedimentation.Chemical factors, largely the physical and chemical properties of impure graphitic carbon, subsequently govern the refractive indices along with the chemical interaction of the particles with other gas and aerosol constituents and with liquid water. Finally, meteorological factors include mixing in the planetary boundary layer, advection, incorporation into clouds and/or into cloud droplets, chemical processes inside of cloud drops, cloud evaporation and removal by precipitation. These factors may be linked together in a system flow diagram to explain the observed presence and behavior of carbon particles in air.

INTRODUCTION

The presence in the atmosphere of particulate elemental carbon (PEC) is a generally accepted fact. This presence can be deduced from any of several points of view. First, sources of PEC exist chiefly in the form of combustion of carbon-based materials and fuels. Second, under atmospheric conditions, PEC is inert to oxidation and modification of its usual graphitic molecular structure. As a result, once it is injected into the atmosphere it must necessarily reside there for some time until aerosol scavenging processes can remove it. Third, and perhaps most commonly, it is observed that filter samples of air are grey or black in color. This observation of

References pp. 15-16.

blackness has been used for over 70 years as a gross indicator of the amount of air pollution (see e.g., Hill [1] and Waller [2]).

The rationale for studying and understanding the presence of PEC in the atmosphere stems from a desire to understand and predict its effects. In general, these effects can be organized into four categories.

1. Effects which are functions of *concentration,* such as atmospheric heating rate due to absorption of sunlight or such as the amount of adsorbed, cogenerated organic matter.
2. Effects which are functions of *dosage* or a product of concentration and time of exposure. The accumulation of carbon in human lungs might serve as an example.
3. Effects depending on a *column burden,* such as the influence on visibility along a sight path or on the optical depth of the atmosphere.
4. Effects depending on a *flux density* such as the rate at which carbon is deposited on windows.

In order to explore and understand the relationship of effects such as these to the sources of PEC, a number of questions may be asked. Among the more important ones are:

- What are typical concentrations and how do they vary with time, over a region, and vertically?
- How do these concentrations compare (magnitude, time and space variation) with other important aerosol constituents?
- What are the major sources and what factors control the source strengths?
- What fraction of the atmospheric burden is due to natural processes and what is due to human activities?
- What are the dominant removal mechanisms, their magnitudes, and controlling factors?
- What are the residence times and how do they compare with those for other aerosol constituents?
- What are the magnitudes of the effects (e.g., on climate, visibility, and the chemistry of other aerosol constituents)?

APPROACH — THE CYCLE CONCEPT

An integrating framework for addressing this family of questions is found in the concept of the atmospheric cycle of PEC. A cycle involving sources, transport, physical transformation and removal can be represented in a variety of different flow schemes or box models.

A simple representation of the cycle is a gross, one-box model of the atmospheric cycle of graphitic carbon aerosols, as illustrated in Fig. 1. This model, which is defined over a specific region, consists of a spatially uniform source, sink, and burden (volume integral of concentration).Such a model is clearly a simplistic representation of cycling through the atmosphere, but it is nevertheless useful because it forces study of the cycle as a whole rather than just one or a few of its aspects. This approach is based on the principle of conservation of mass, which

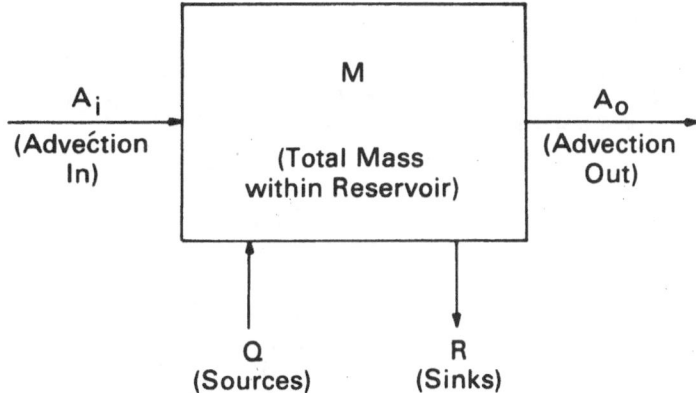

Fig. 1. Gross box model of the atmospheric cycle of particulate elemental carbon (PEC).

requires that the magnitudes and time variation of the sources, sinks, and burden must be consistent. If an internally consistent model can be constructed, then it is likely that all the major fluxes (chemical and physical) have been identified. The important point is that an overall model of the cycle, no matter how simple, is needed in order to interpret measurements or predict effects of atmospheric graphitic carbon.

A major weakness of the gross model is that it does not include any mechanisms for physical and chemical transformations, such as coagulation with other particles or adsorption of trace gases. One way to include these processes is to incorporate a system flow diagram (mechanistic model) within the gross model, as illustrated in Fig. 2. Each of the "mechanistic" reservoirs are characterized by specific transformations of the original particles, and these transformations (fluxes) and reservoirs can be studied independently of the gross model. Eventually, the gross and mechanistic models need to be combined into a complete description of the life cycle, although it is not clear that the measurements needed to effect such a merger are presently available, or even possible.

Gross Model Description — The gross model consists of source and sink strengths, average concentration (and thus burden), and advective transport into the region. One derived quantity of particular interest is the average residence time (or turnover time), defined as the burden divided by sink strength. This quantity can be defined for a particular sink or for the sum of all sinks. Earlier studies [3] of the atmospheric cycle of sulfur aerosols indicated that a spatial scale of the order of several thousand kilometers is appropriate for this type of model. This scale is based on a desire to have source and sink strengths that are large compared with advective fluxes into and out of the region. An alternate approach is to choose a smaller region and include advection terms in the model.

As its name implies, this model can be used to study some of the gross features of the atmospheric cycle of graphitic carbon aerosols, such as the relative importance of advective transport versus local sources and sinks, the average lifetime in the

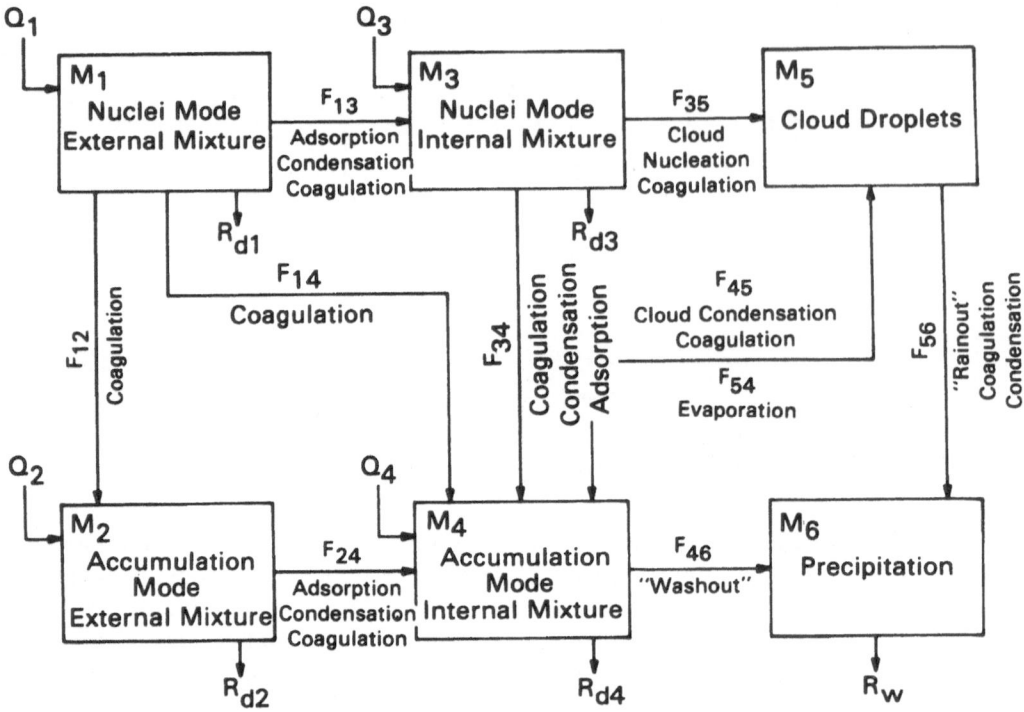

Fig. 2. Mechanistic model of the atmospheric cycle of particulate elmental carbon (Q denotes sources; F is a flux between reservoirs; R_W and R_D are wet and dry removal, respectively; and M denotes the mass or burden of elemental carbon in each reservoir).

atmosphere, and the area influenced by a given source region. By comparison with other aerosol constituents (notably sulfates), the relative importance of graphitic carbon aerosols to atmospheric visibility and radiative climate can be calculated. Comparison of residence times for different aerosol constituents can be used to determine if they have similar removal mechanisms.

Mechanistic Model Description — A total of six reservoirs, differentiated by size and chemical composition of the graphitic carbon-containing particles, are included in the mechanistic model (Fig. 2). Defining the terms used to identify these reservoirs, *external mixture* means that PEC is physically isolated from the other aerosol constituents as individual particles. Thus the properties of the graphitic carbon-containing particles with respect to water are determined by PEC. An *internal mixture* is one where other compounds are physically attached or coated on PEC and hence may dominate the physical and chemical properties of the graphitic carbon-containing particles. *Nuclei mode* and *accumulation mode* refer to the size of the particles, with the nuclei mode including particles smaller than about 0.05-0.1 micron diameter and the accumulation mode covering the particle diamter interval from about 0.1 to 1 micron. The *cloud droplets* reservoir includes graphitic carbon contained in cloud droplets (or ice particles), while the *precipitation* reservoir contains atmospheric graphitic carbon on its way to the surface in rain, snow, or hail (wet removal).

The reasons for selection of these particular reservoirs lie with the physical and chemical properties of graphitic carbon aerosols. Nuclei mode aerosols have higher diffusion coefficients than those in the accumulation mode, resulting in higher rates of coagulation and of collision with the surface (dry removal). However, pure graphitic carbon is insoluble and hygrophobic, implying that wet removal is not very efficient for externally mixed graphitic carbon aerosols. In contrast, accumulation mode sulfate aerosols are hygrophillic and thus are readily incorporated into cloud droplets, making removal by precipitation a dominant mechanism for them. By defining the reservoirs in terms of the physical and chemical properties, the mechanisms which transfer graphitic carbon among the reservoirs are highlighted. Some of the mechanisms thought to be most important are included in Fig. 2, although there are probably other important fluxes that are not shown. If these fluxes are thought of as first-order processes, then there are corresponding rate constants (or reciprocal lifetimes) for each process. For any particular source-to-sink pathway, these lifetimes can be combined to yield an overall lifetime, and compared with the lifetimes observed in the atmosphere (e.g., via the gross model). Ultimately, calculation of equilibrium concentration or burden and response times may be accomplished if the coefficients are known via an approach similar to that of Yuen et al. [4].

CONTROLLING FACTORS

Consideration of the processes involved in the cycling of PEC through the atmosphere results in the definition of four sets of controlling factors:
1. Source characteristics,
2. Aerosol mechanics,
3. Chemical properties of PEC,
4. Meteorological factors.

Source Characteristics — Sources of PEC are ubiquitous in both natural and polluted settings but almost all involve combustion. Small amounts of coarse particle PEC can be generated by the physical weathering of graphite-containing sedimentary rocks or exposed charcoal (e.g., in a forest after a fire has stopped). If we limit our focus to fine (sub μm radius) particles, all PEC is generated by combustion of carbon containing materials.

There are basically two mechanisms by which PEC can be generated, both involving pyrolysis. Gas phase reactions exist by which hydrocarbons are dehydrogenated and the carbon eventually ends up reassembled into a graphitic structure. Such reactions are known for both aromatic [5] and aliphatic compounds [6]. Commercial production of carbon black often utilizes methane, and acetylene flames can be very sooty, demonstrating that even small hydrocarbon molecules can produce graphitic carbon. It is speculated that in the case of such low carbon numbers, droplets of aromatic material may be formed as an intermediate preceding the final pyrolysis to graphitic structures [7]. Another gas-phase production mechanism involves the equilibrium:

$$CO + CO \leftrightarrows C + CO_2$$

which is shifted to the right at $T \sim 600°C$ [8]. The other mechanism involves the pyrolysis of a droplet or particle of a carbon containing material. An example of this process may be the production of soot in oil burners or diesel engines.

Both of these processes tend to make primary particles in the radius range from 0.02 to 0.1 μm. Larger carbon particles can be emitted from sources if the residence time and/or concentration of primary particles is sufficiently large to permit coagulation. Fig. 3 shows a few typical size distributions for sources, and includes a freeway distribution [9, 10] showing a probable contribution of direct injection to the atmosphere of primary particles. It thus seems clear that sources control the initial size distribution of PEC and that the initial sizes are considerably, perhaps a factor of ten, smaller than the bulk of the mass of the fine particle, accumulation mode aerosols. This difference in size dictates that the PEC has a factor of ca. 10^{-3} smaller particle mass, decreased Stokes number, and about a factor of 6 increase in Brownian displacement in comparison to the accumulation mode.

The chief consequences of the small initial size of PEC lie in the realm of aerosol mechanics [11], in the morphology of the particles [12] and in the likelihood that light absorption is proportional to the mass concentration [13]. The small initial particle size encourages coagulation, but at the same time the physical rigidity and inertness of a solid phase dictates that the surface area per unit mass of PEC is maintained as the particles agglomerate. Soots may have surface/mass ratios up to 1000 m²/gm [6]. Depending on the source, this surface area may be covered with adsorbed cogenerated materials, or it may be exposed to the atmosphere for interaction with other substances.

Besides these size-dictated quantities, the source factors determine the initial chemical composition. Some sources produce relatively pure PEC, while others may produce a soot which is only 50% elemental carbon [15], the rest often being organic matter. In turn, the initial chemical composition determines whether the soot is hygroscopic (such as PEC coated with H_2SO_4) or perhaps most often it determines that the PEC is hygrophobic.

Source factors also govern the mass of PEC injected into the atmosphere. The amount of PEC produced per unit of fuel burned varies with the type of source and individual source operating conditions. Table 1 shows some examples to illustrate this variability.

Finally, source factors determine location of injection into the atmosphere. Some, perhaps most, sources are widely distributed and close to the ground and may be called area sources. Others are elevated and isolated point sources, while jet aircraft might be considered as elevated line sources.

Aerosol Mechanics — Due mainly to small initial particle size, PEC has a strong tendency to coagulate, both within the source and in the atmosphere. In the latter case, coagulation ultimately brings the PEC into physical contact with the other

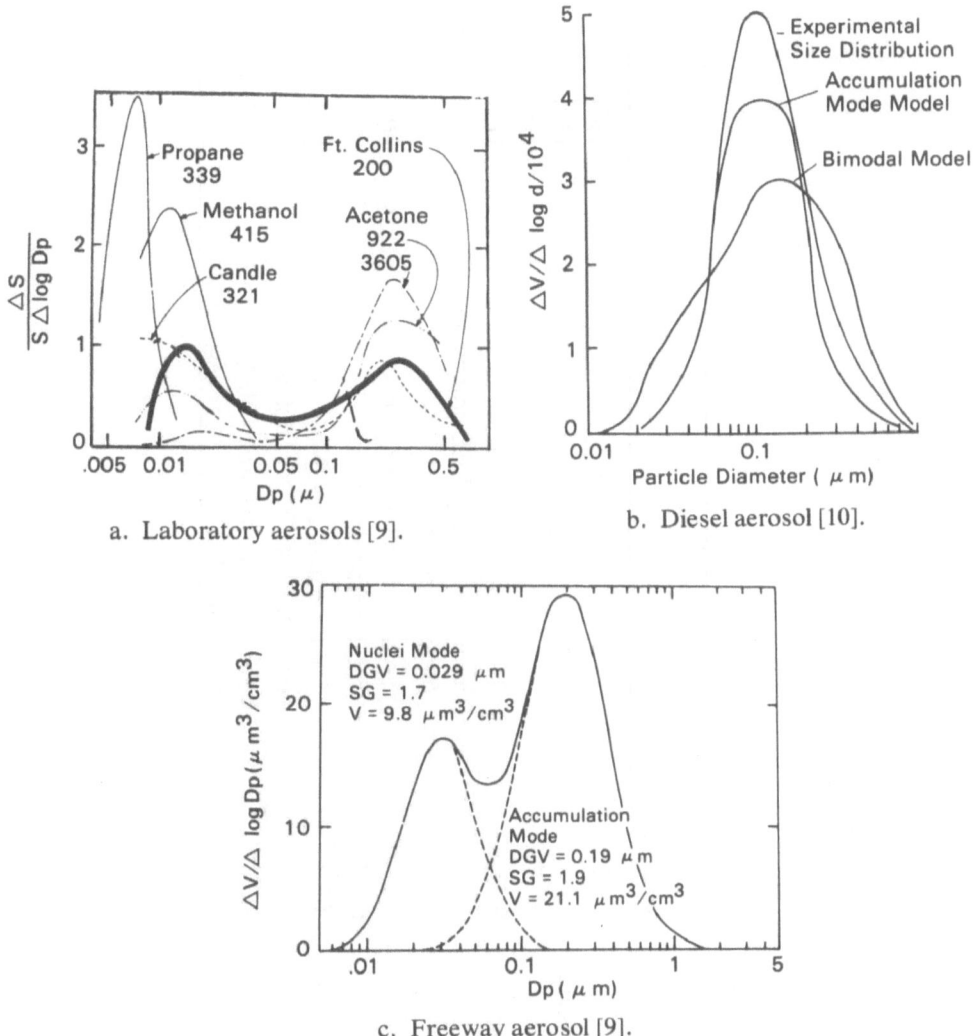

a. Laboratory aerosols [9].

b. Diesel aerosol [10].

c. Freeway aerosol [9].

Fig. 3. Source size distributions.

constituents of the accumulation mode. The morphology of PEC usually includes some degree of coagulation of primary particles. Fig. 4 shows photographs of laboratory generated PEC demonstrating various degrees of coagulation. The atmospheric lifetime of individual, primary PEC is governed by the concentrations of both primary particles and the pre-existing fine particles. In urban areas this lifetime appears to be of the order of one hour or less while in the unpolluted upper troposphere it could be much longer, perhaps as much as several days [19].

Besides coagulation, the small initial particle sizes dictate that some diffusion occurs toward surfaces. Although this diffusion is distinctly larger than that for accumulation mode aerosol particles, it is still too small to represent a major sink for PEC. Nonetheless, diffusive removal may represent a finite buildup of PEC on

References pp. 15-16.

TABLE 1
Source Emission Factors

Fuel	Source	g PEC/ kg Fuel	Comments	Reference
Natural Gas	Steam Generator	3×10^{-4}		[14]
	Domestic Water Heater	0.1		[14]
	Heating Boiler	.01-.07		[14]
Gasoline	Automobile Engine	0.1	Tuscarora tunnel (1975) assumes 10 mpg, PEC comprises 60% of total carbon	[15]
Diesel	Automobile Engine	2-4	Total carbon less organic soluble fraction	[16]
	Truck/bus Engine	0.6-1		
Jet A	Aircraft Turbine	.05-3	Total particulates. No elements with Z > 11 detected.	[17]
Fuel Oil (No. 2)	Utility Turbine (20 MWe)	.08	Total carbon. Assumes 40% thermal efficiency, 3% excess O_2	[18]

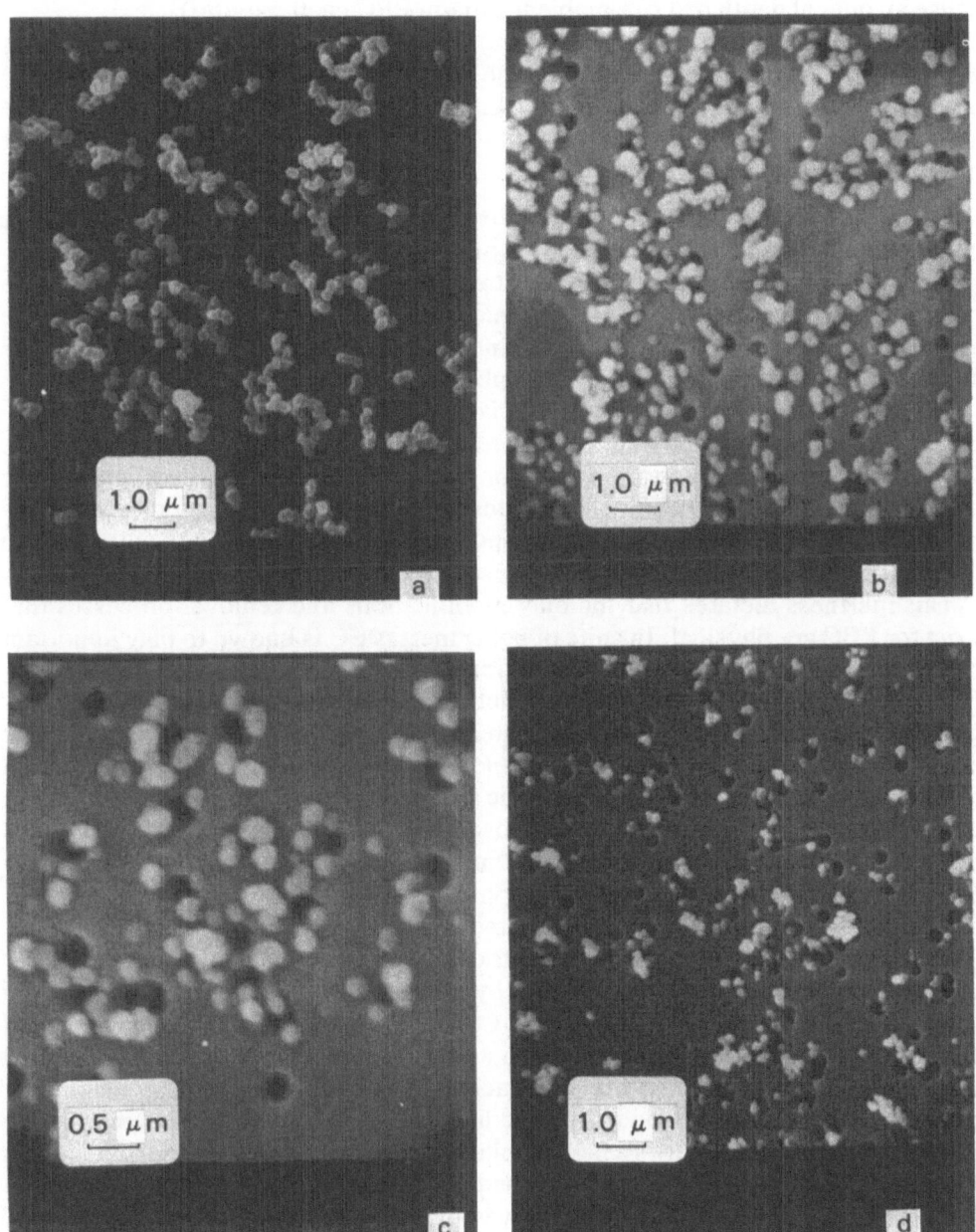

Fig. 4. Scanning electron microscope photograph of laboratory-generated PEC.
   a. Methane pyrolysis; 1120°C; pore size=0.2 μm; magnification 8,700x.
   b. Benzene pyrolysis; 1080°C; pore size=0.4 μm; magnification 9500x.
   c. Benzene pyrolysis; 1080°C; pore size=0.4 μm; magnification 19,400x.
   d. Tunnel sample; pore size=0.4 μm; magnification 9,300x.

surfaces, e.g., on windows. Limited urban measurements reported by Ogren [20] suggest that about 35 years are required for the deposition of sufficient PEC to cause an optical depth of 1 (for ambient sub μm PEC ca. 0.5 μg/m^3).

Sedimentation of fine particle PEC is not important as a sink. If we estimate that the density of a 0.02 μm spherical particle is 2 g/cm^3, its fall speed is about 6 x 10^{-5} cm/s or ca. 10 cm/day. A 0.1 μm particle has a fall speed of only 4 x 10^{-4} cm/s. These velocities are extremely small compared to ordinary wind velocities and hence the motion of PEC is dictated by the wind.

Chemical Factors — Although the overall chemical compositions of soots are complex and variable, all have in common a significant fraction of elemental carbon. Although they are often referred to as amorphous soot, charcoal, or lamp black, are all impure carbon containing microcrystalline forms of graphite [21]. The chemical properties of PEC are determined largely by the nature of graphite, the surface characteristics and co-existing substances.

Graphite is an exceedingly inert material at ordinary temperatures, is hygrophobic, and insoluble in any solvent. It can be oxidized in air at temperatures above ca. 600°C, will react in liquid systems of strong oxidizing agents (such as fuming nitric acid, $KClO_4$ or $KMnO_4$) and reacts explosively with F_2. However, it seems completely safe to say that these sorts of conditions cannot be met in the atmosphere and hence PEC is, for all practical purposes, inert.

This inertness dictates that the only modifications and removal processes that exist for PEC are physical. In spite of being inert, PEC is known to play important chemical roles. The large surface mentioned above along with impurities in the PEC itself, probably as terminating functional groups at the edges of the planar graphite macromolecules, make the material very active as an adsorber of other materials. Either co-generated organic and inorganic species or material adsorbed from the atmosphere may be found on the surfaces of PEC.

While pure graphite can exist in macroscopic crystalline form (hexagonal), PEC as found in the atmosphere shows only microcrystalline structure and hence the particle morphology does not reflect graphitic character. Rather, the primary particles tend to be spheres, often agglomerated into chains or clumps.

The molecular form of graphite consists of carbon atoms bound together in planar units with each carbon atom surrounded by three others. Besides a σ bond between each C atom, the extra electron on each C atom is available for a π bond such that the system of bonds resonates between a structure as in Fig. 5 or equivalent arrangements. Each carbon-carbon bond achieves one third of double bond character, unlike a strictly aromatic or olefinic linkage. However, the resonance among various configurations allows a high mobility of the π electrons and hence accounts for electrical conductivity as well as broad-band light absorption. The large imaginary refractive index of soot thus derives directly from the molecular nature of graphite.

Meteorological Factors — After injection into the atmosphere, PEC is subjected to the influences of the physical motion of the atmosphere and of humidity increases, cloud formation, cloud evaporation and precipitation. Because in most instances PEC has an expected lifetime of order days (assumed to be similar to

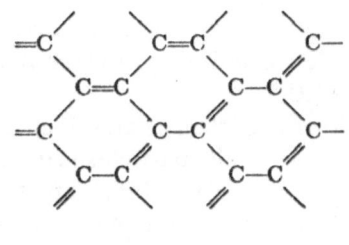

Molecule

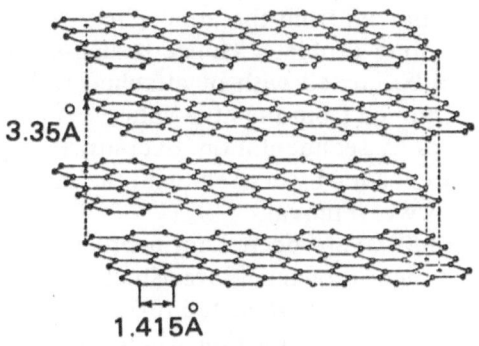

3.35Å

1.415Å

Platelet

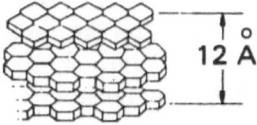

12 Å

Platelets

Particle

Fig. 5. Structure of particulate elemental carbon. Adapted from Cotten and Wilkinson [21], Pauling [22], and Lipkea *et al*. [23].

other fine particle aerosols), it should tend toward being well mixed into the planetary boundary layer. It also should be distributed fairly widely, with horizontal spatial scales of 10^3 km or so, as is the case for sulfates [24]. Since many of the sources are small and widely dispersed into an area source, there is probably a slowly varying background of PEC on which are superimposed plumes from a few discrete, elevated, sources. Other than mixing fresh PEC with pre-existing aerosols, atmospheric motion (turbulence and advection) should do nothing at all to the physical or chemical nature of PEC.

However, atmospheric motions eventually carry PEC into high humidity locations and clouds. Here, the behavior of the particles is dictated by the hygroscopic nature of their surfaces. If the PEC is hygrophobic, it must remain inactivated in the cloud and will not grow at high RH. On the other hand, if it is hygroscopic it may or may not be activated into a cloud droplet, depending on a complex set of considerations determined by the amount of water soluble material on the particles [25] and on any surface active compounds that might be present [26]. In order for the particles to be efficiently removed from the atmosphere, it seems necessary for them to incorporated into cloud droplets. Hence it also seems to be necessary for most of the particles to be coated with or attached to hygroscopic, soluble particles before they are removed from the atmosphere. A small fraction may be deposited by dry processes or by sedimentation. In contrast to the time of 35 years for unit optical depth to be deposited by dry removal, the same conditions resulted in a time of only 2.3 years for wet removal.

Since most clouds or cloud layers do not form precipitation (rain or snow), particles that act as cloud condensation nuclei are usually evaporated. This cycle of cloud formation and evaporation may be repeated 10-20 times on the average before precipitation occurs. Meanwhile, during the time when a small ($\sim 0.05 \mu$m) particle of insoluble, graphitic carbon is imbedded in a cloud droplet of 5-10 μm radius, chemical reactions proceed within the liquid portion of the droplet and on the surface of the insoluble particle. Some of these reactions undoubtedly produce non-volatile material from gaseous reaction products, such as $SO_2 SO_4^{-2}$ [27, 28].
Hence the repeated cycling of PEC into and out of clouds probably builds up the mass of soluble but non-volatile material coated onto or attached to the carbon particles.

Finally, the meteorological processes of rain or snow carry the PEC to the ground where it remains or is carried in surface runoff to rivers, the ocean and to its final resting place in sedimentary formations.

CONCLUSION

Returning to the cycle diagrams, and with the four sets of controlling factors in mind, we conclude by stating some of the quantities that might be measured in order to quantify the cycle. In the simplest case of the one-box model, and with the 10^3 km horizontal spatial scale in mind, a comparison of sources and sinks can be made, i.e., a budget can be prepared. In addition, studies of the quantities of PEC advected into and out of the test region will allow a closer definition of the mass budgets. One outstanding question arises at this simplest level: Does the deposition of PEC correlate with other materials such as SO_4^{-2}?

Other, local and small scale measurements seem appropriate for understanding even the qualitative nature of the mechanistic model. One set of questions pertains to the degree of internal or external mixing of the PEC aerosol as it moves from the source location into the larger spatial scales. Similar questions exist for the incorporation of PEC into activated cloud droplets and subsequently into rain or snow. Laboratory studies with controlled aerosols may be useful in estimating lifetimes of PEC before it coagulates with the accumulation mode and/or is incorporated into cloud droplets.

Chemical measurements already have shown that the graphitic structure is present [29] leaving detailed, systematic studies of the associated impurities in the source aerosol. Other chemical studies in the laboratory have been done, but still seem appropriate for elucidating surface chemical and catalytic properties of PEC, particularly regarding those processes which determine its lifetime and degree of internal mixing.

REFERENCES

1. *A. S. G. Hill, Trans. Farad. Soc., Vol. 32 (1936), p. 1126.*
2. *R. E. Waller, J. Air. Poll. Control Assoc., Vol. 14 (1964), p. 323.*
3. *H. Rodhe, Atmos. Environ. Vol. 12 (1978), p. 671.*
4. *T. Y. Yuen, H. Harrison, R. J. Charlson and M. B. Baker, Atmos. Environ., Vol. 13 (1979), p. 1351.*
5. *J. Lahaye, G. Prado and J. B. Donnet, Carbon, Vol. 12 (1974), p. 27.*
6. *J. Donnet and A. Voet, Carbon Black: Physics, Chemistry, and Elastomer Reinforcement, Marcel Dekker, Inc., New York, 1976.*
7. *H. B. Palmer and C. F. Cullis, "The Formation of Carbon From Gases" in Chemistry and Physics of Carbon, P. L. Walker, Jr., editor, Marcel Dekker, Inc., New York, (1964), pp. 265-325.*
8. *S. A. Pursley, "Kinetics of Carbon Dioxide and Carbon Formation from Carbon Monoxide," Ph.D. Dissertation, Purdue University, 1965.*
9. *K. Whitby, "Size distribution and physical properties of combustion aerosols" in Proceedings of the Carbonaceous Particles in the Atmosphere, Lawrence Berkeley Laboratory, Berkeley, California, 1979.*
10. *D. F. Dolan and D. B. Kittelson, SAE Paper 780110, SAE Transactions, (1979), p. 462.*
11. *R. Husar, "Coagulation of Knudsen Aerosols," Ph.D. Dissertation, University of Minnesota, 1971.*
12. *P. A Russel, "Carbonaceous particulates in the atmosphere: Illumination by electron microscopy" Proc. Conf. On Carbonaceous Particles in the Atmosphere, Lawrence Berkeley Laboratory, Berkeley, California, 1979.*
13. *A. P. Waggoner, M. B. Baker and R. J. Charlson, App. Opt., Vol. 12 (1973), p. 895.*
14. *A. D. A. Hansen, W. H. Benner and T. Novakov, "A carbon and lead emission inventory for the Greater San Francisco Bay Area," in Atmospheric Aerosol Research Annual Report, 1977-1978, Lawrence Berkeley Laboratory, Berkeley, California, 1978.*
15. *W. R. Pierson, "Particulate organic matter and total carbon from vehicles on the road," Proc. Conf. On Carbonaceous Particles in the Atmosphere, Lawrence Berkeley Laboratory, Berkeley, California, 1979.*
16. *K. J. Springer, "Exhaust particulate – the diesel's achilles' heel," Paper No. 78-14.3, 71st Annual Meeting of the Air Pollution Control Assoc., Houston, Texas, 1978.*
17. *J. D. Stockham, D. L. Fenton, R. H. Johnson and P. P. Campbell, "Turbine Engine Particulate Emission Characterization," U.S. Dept. of Transportation Report No. FAA-RD-79-15, 1979.*
18. *S. Hersh, J. F. Hurley and R. C. Carr, "The Effects of Smoke Suppressant Additives on the Particulate Emissions from a Utility Gas Turbine," Paper 76-8.1, 69th Annual Meeting of the Air Pollution Control Assoc., Portland, OR., 1976.*

19. C. E. Junge and N. Abel, *Modification of aerosol size distribution in the atmosphere and development of an ion counter of high sensitivity. Final Techn. Rep. No. DA 91-591-EUC-3483, DDCNo. AD469376.* Johannes Gutenberg University, Mainz, Germany, 1965.
20. J. A. Ogren, These Proceedings, 1981.
21. F. A. Cotten and G. Wilkinson, *Advanced Inorganic Chemistry*, Interscience Publishers, John Wiley & Sons, (1962), p. 217.
22. L. Pauling, *The Nature of the Chemical Bond*, Cornell Univ. Press, Ithaca, New York, (1960), p. 235.
23. W. H. Lipkea, J. H. Johnson and C. T. Vuk, SAE Transactions, Vol. 87 (1979), p. 405.
24. R. E. Weiss, A. P. Waggonner, R. J. Charlson and N. C. Ahlquist, Science, Vol. 195 (1977), p. 979.
25. B. J. Mason, *The Physics of Clouds*, Oxford Press, 1957.
26. W. P. Giddings and M. B. Baker, J. Atmos. Sci., Vol. 34 (1977), p. 1957.
27. D. A. Hegg and P. V. Hobbs, Atmos. Environ., Vol. 12 (1978), p. 241.
28. S. G. Chang, R. Brodzinsky, R. Toossi, R. R. Markowitz and T. Novakov, "Catalytic oxidation of SO_2 on carbon in aqueous solution," Proc. Conf. On Carbonaceous Particles in the Atmosphere, Lawrence Berkeley Laboratory, Berkeley, California, 1979.
29. H. Rosen and T. Novakov, Nature, Vol. 266 (1977), p. 708.

DISCUSSION

J. Heicklen, *(Pennsylvania State University)*

In your model you seemed to indicate that most of the removal would be through wet deposition and you did not mention the possibility of dry deposition. Do you think dry deposition is unimportant?

R. Charlson

Dry deposition is certainly important and it's measurable. You need only to go to a window that has not been cleaned for a long period of time and wipe your finger on it and it comes out black. Its a matter of the relative amounts of wet removal versus dry removal. Calculations which John Ogren* will report later in the conference, indicate that the dry removal is very much slower than the wet removal. The quantitative estimates that I recall are that wet removal is more than ten times as efficient as dry removal. The reason being that, unlike gases, the brownian movement of the 0.1 micron particles is very slow. So their diffusive removal from the atmosphere is slow. Removal by impaction is slow because the particles are too small to readily impact on surfaces at ordinary wind speeds.

R. Williams, *(General Motors Research Laboratories)*

What kind of break do we need in terms of methodologies to get at the removal process? You talked briefly about dry deposition and wet deposition, but we are missing some important technology in order to get at the issue.

*J. Ogren, These Proceedings.

R. Charlson

Again, John Ogren will be talking about this in the last paper in the Symposium. But, the technology for studying removal is quite new. I know of no papers on the subject that have been published. The technologies that we are using are very ordinary; we simply collect rain in buckets and study the black material that is present. The chemical methods we are using are very ordinary and include the usual list of optical and combustion techniques. For dry removal, we let the material diffuse to the surface and measure its presence there after a long period of time.

V. Mohnen, *(State University, Albany, New York)*

First, dry deposition is indeed influenced by brownian motion, by impaction and by inertia and these are the same parameters that influence the removal of particles by cloud droplets. So obviously what we call the end-point aerosol, the aerosol between 0.1 and one micron is precisely the size range where you showed most of the carbonaceous material in the atmosphere is concentrated. I believe it is also the one that Dr. Rahn* will tell us shows up in the Arctic. This means that this particular type of aerosol can survive those rigorous cleansing processes by cloud and precipitation elements. Thus the removal efficiency might be so low that it is in the realm of dry deposition. Second, on dry deposition, the problem is that nobody has measured it precisely or even within an order of magnitude for 0.1 to one micron size particles. At a 1979 workshop sponsored by EPA at Battelle Northwest, it was pointed out that the only true way of measuring dry deposition in that size range would be by the correlation coefficient method. In this method, one measures the mean verticle velocity fluctuation and multiplies it by the mean concentration measurement (C'W'). This instantaneous flux, measured over a period of an hour, ten times a second, is the only acceptable way. Now, for particles in that size range, we do have the technique of measuring W' (the variation of the vehicle wind velocity) ten times a second or a hundred times a second with sonic anomometers. We also can measure the particle concentration in the 0.1 to one micron size range, five times a second with the condensation nuclei techniques with appropriate pre-filters. So the technology is ready but nobody has picked it up. The third point is that when we look at the overall process that you showed with boxes, I think that Seinfeld† and more recently Friedlander†† have gone far beyond the simple box model approach. They have developed a condensation coagulation-nucleation model. All that it requires is proper data and out comes the answers to the type of questions that you have raised. Do you have any objections to the Seinfeld model?

R. Charlson

No. I think those are very important developments that follow after one qualitatively identifies the pathways. The primary purpose of utilizing these simple models is to be able to identify where we should be looking first. I certainly agree with both Seinfeld and Friedlander in their approach. Getting back to your second

---

*Rahn, K., Brosset, C. & Ottar, B., These Proceedings.

†Seinfeld, J. H. Air Pollution, Physical & Chemical Fundamentals, McGraw Hill, N.Y., N.Y., 1975.
††Friedlander, S. K., Smoke, Dust & Haze, John Witey, N.Y., N.Y., 1977.

point, it is critically important to realize that the lifetime of a 0.1 or 0.05 micron soot particle will change dramatically if it collides with and sticks to a 0.5 micron sulfate particle. And that makes a big difference as to the subsequent behavior of that soot particle. That qualitative question is one of the things we really need to investigate soon. If a soot particle gets injected in the high troposphere where there is almost no accumulation mode, it has very little chance of coagulating with a hygroscopic particle and its lifetime could be very long. If it was injected into Los Angeles on a smoggy day, the probability of it being captured by a hygroscopic particle would be orders of magnitude higher. Those kinds of differences need to be focused on as we begin to look at the overall regional and large scale cycle of the soot. As to the possiblity of measuring $W'C'$ for soot, I doubt that methods exist for measuring fluctuating concentrations adequately.

D. Stedman, *(University of Michigan)*

You showed the beautiful structure of graphite with its six angstrom interplane space. This is why activated charcoal is a very efficient adsorber of gases. Then you suggested that adsorption was not likely to be an important feature of the chemistry. Why do you suggest that?

R. Charlson

I did not mean to imply that. I think absorption and adsorption are both very important features. The large surface area to mass ratio of a material with polar functional groups on it would certainly participate in absorption processes that occur in the atmosphere or in sources.

C. Brosset, *(Swedish Water & Air Pollution Research Institute)*

Do you mean that the conversion of a soot particle or carbon particle to a hydrophilic state is mainly due to the collision with a sulfate particle?

R. Charlson

That is one possibility.

C. Brosset

I think that a more important process is the gas-phase reaction on the surface.

R. Charlson

That is certainly another possibility. I just mentioned that one because it came to mind first. By no means do I mean to suggest that any one of these processes have proven to be the case. The point I wanted to make was that the life cycle of an individual particle will dramatically change when it goes from a hydrophobic state to the hygroscopic or hydrophilic state.

SOOT IN THE ATMOSPHERE

T. NOVAKOV

Lawrence Berkeley Laboratory
University of California •
Berkeley, California

ABSTRACT

Carbonaceous particles in the atmosphere consist of two major components — graphitic or black carbon and organic material. The organic component can either be directly emitted from sources (primary organics) or be produced by atmospheric reactions from gaseous precursors (secondary organics). We define soot as the total primary carbonaceous material, i.e., the sum of black carbon and primary organics. The complex set of questions concerning the origin and the chemical and physical characterization of carbonaceous particles has been central to the research of the Atmospheric Aerosol Research group at Lawrence Berkeley Laboratory since the group's beginning in 1972. This paper will present an overview of our efforts to quantitate the amount of soot in a variety of urban locations. The results will demonstrate that soot is a major component of the carbon aerosol in all locations and can be the dominant component in many.

INTRODUCTION

Carbon-, sulfur-, and nitrogen-containing particles account for most of the anthropogenically generated particulate burden in urban areas. Considerable attention has been given to understanding the origin and speciation of the sulfur and nitrogen components, but until recently relatively little effort has been directed toward the carbonaceous aerosol which is often the single most important contributor to the submicron aerosol mass.

Carbonaceous particles in the atmosphere consist of two major components — graphitic or black carbon (sometimes referred to as elemental or free carbon) and organic material. The latter can either be directly emitted from sources (primary organics) or produced by atmospheric reactions from gaseous precursors (secondary organics). Black carbon, however, can be produced only in a combustion

References pp. 36-37.

process and is therefore definitely primary. For the sake of clarity, we define soot as the total primary carbonaceous material, i.e., the sum of graphitic carbon and primary organics.

According to the view prevailing at the time our research was initiated (1972), it was postulated that most of the particulate material was produced by certain atmospheric photochemical reactions from gaseous hydrocarbons [1]. The products of these reactions were believed to be certain highly oxygenated hydrocarbons which condense into carbonaceous particles. Such highly oxygenated carbonaceous species, if present in sufficient concentrations, should be detected easily by X-ray photoelectron spectroscopy (ESCA).

We have used ESCA extensively in attempts to chemically characterize particulate carbon [2-4]. In most instances the carbon (1s) peak of ambient particulates appears essentially as a single peak with a binding energy compatible with either elemental carbon or condensed hydrocarbons or both. As seen in Fig. 1, where the carbon (1s) spectrum of an ambient air particulate sample is shown, chemically shifted carbon peaks, due to suspected oxygen bonding, are of low intensity compared with the intense neutral chemical state peak. This and other similar experiments suggested that the oxygenated (i.e., secondary) species are not the major constituents of carbonaceous particles, even when the samples were collected under conditions of high photochemical activity.

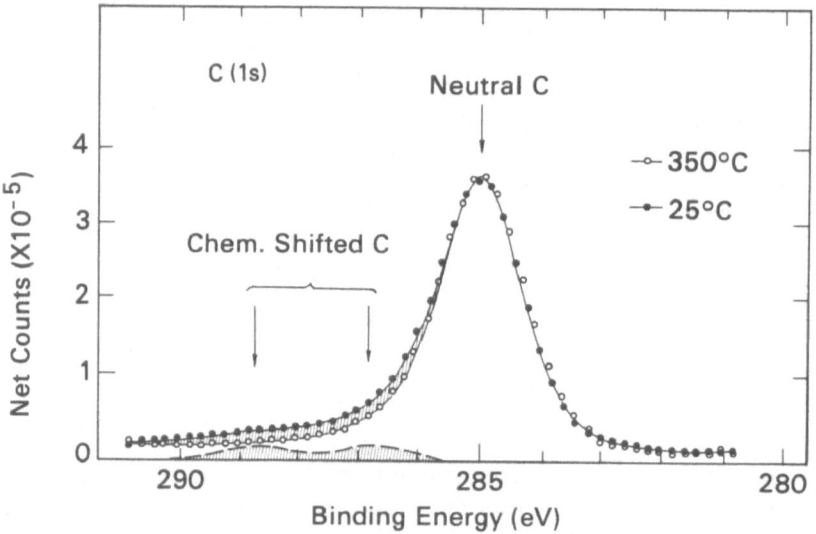

Fig. 1. Carbon (1s) photoelectron spectrum of an ambient (West Covina, California) sample as measured at 25 and 350°C. The chemically shifted peaks are of low intensity compared to the main peak. Most of the carbon appears to be nonvolatile. The shaded area represents the difference between low- and high-temperature spectra (from Ref. 2).

We have also employed other supplementary measurements to estimate the relative abundance of primary particulate carbon [2-4]. This was attempted by comparing the carbon (1s) peak obtained from the sample at 25°C with the carbon

(1s) peak from the sample at 350°C. The difference between the low-temperature and the high-temperature runs should give the fraction of volatile carbon. Fig. 1 shows the result of one such experiment for a sample collected in 1975 from West Covina. This experiment suggests that most of the ambient particulate carbon is nonvolatile in vacuum at 350°C. Assuming that the secondary hydrocarbons will have a substantial vapor pressure at 350°C, we have suggested that a substantial fraction of the total particulate carbon is primary (sootlike) material.

The preliminary results on the origin and nature of carbonaceous particles were first reported at the First Annual NSF Trace Contaminants Conference at Oak Ridge National Laboratory in August 1973 [3], just one year after the start of our research. Additional results strengthening this hypothesis were reported in several publications from 1973 to 1975 [2, 4]. From that time the complex set of questions concerning the origin and the chemical and physical characterization of carbonaceous particulates has been central to the program of the Atmospheric Aerosol Research group at Lawrence Berkeley Laboratory (LBL). In this paper we will briefly review some of our recent work and attempt to estimate the concentrations and relative amounts of soot in several urban atmospheres.

METHODOLOGY

The principal goal of the research described here is to quantitatively assess the relative amounts of primary and secondary carbonaceous material in atmospheric aerosols and to differentiate between secondary carbonaceous species produced by photochemical and nonphotochemical reactions. The approach we have used most extensively involves the use of an optical attenuation technique, combined with total particulate carbon determination; it is based on the following rationale. The black component of soot, which is an unambiguous tracer for primary emissions, can be conveniently monitored because of its large and uniform optical absorptivity [5-8]. The black carbon content of the particles can easily be determined by an optical attenuation method developed in our laboratory [6, 7]. Determination of total particulate carbon mass enables us to study the relations between the black and the total carbon content.

An important characterization of a particulate sample is given by the fraction of black carbon to total carbon. Measurements of this ratio for numbers of source samples give insights into the relative black-to-total-carbon ratio of primary emissions and the source variabilities. Secondary material will not contain the black component, but it will increase the total mass of carbon and will therefore reduce the black-to-total-carbon fraction.

Photochemical gas-to-particle conversion reactions should be most pronounced in the summer in the Los Angeles air basin, while in the winter these reactions should play a much smaller role and the primary component should be much more important. These different photochemical conditions should manifest themselves in the ratio of the black carbon to total carbon of these particles. That is, under high photochemical conditions one would expect this ratio to be significantly smaller than under conditions obviously heavily influenced by sources.

This approach to the identification and quantitation of primary and secondary carbonaceous aerosols involves a systematic comparison of particulates collected from a wide range of ambient sites as well as combustion sources. Ambient particulates are sampled at sites that differ significantly in meteorology, photochemical activity, and source composition. Source samples have been obtained at a tunnel and a parking garage, and from direct source samplings.

In our field experiments the samples are collected in parallel on prefired quartz fiber and Millipore filter membranes. The Millipore filter is used for X-ray fluorescence (XRF) elemental analysis and for the LBL laser transmission technique described in detail elsewhere [6]. The latter technique gives a measurement that is proportional to the amount of light-absorbing (black) carbon present on the filter. The quartz filter is used for total carbon determination.

The LBL optical method [6] compares the transmission of a 633 nm He-Ne laser beam through a loaded filter relative to that of a blank filter. The relationship between the optical attenuation and the black carbon content can be written as:

$$[C_{black}] = (1/K) \times ATN \; , \tag{1}$$

where $ATN = -100 \ln (I/I_0)$. I and I_0 are the transmitted light intensities for the loaded filter and for the filter blank.

Besides the black carbon, particulate material also contains organic material which is not optically absorbing. The total amount of particulate carbon is then:

$$[C_{tot}] = [C_{black}] + [C_{org}] \; . \tag{2}$$

A fundamental characterization of a particulate sample can be given by its attenuation per unit mass of total carbon, i.e., its specific attenuation, σ:

$$\sigma \equiv \frac{ATN}{[C_{tot}]} = K \times [C_{black}]/[C_{tot}] \; . \tag{3}$$

The determination of specific attenuation therefore gives an estimate of black carbon as a fraction of total carbon.

The proportionality constant K, which is equal to the specific attenuation of black carbon alone, was recently shown to have an average value of 20 [9]. This value was obtained by determining the optical attenuation of 25 samples for which the absolute concentration of black carbon was known. In principle the percentage of soot (i.e., primary carbonaceous material) in ambient particles can be determined from the ratio of ambient specific attenuation and an average specific attenuation of major primary sources:

$$[Soot]/C = \sigma_{ambient}/\sigma_{source} \quad . \tag{4}$$

The approach for estimating the soot content depends on the validity of using the optical attenuation as a measure of the black carbon content. That this procedure is valid can be illustrated by the following example involving a direct analysis for total carbon and black carbon by thermal analysis [10, 11]. Thermal analysis is used to obtain total carbon, black carbon, organic carbon, and carbonate carbon. A schematic representation of the thermal analysis apparatus used in our studies is shown in Fig. 2. The main components of this apparatus are a quartz tube and a temperature-programmed furnace. The tube is mounted axially inside the furnace. The particulate sample, collected on a prefired quartz filter, is placed in the quartz tube so its surface is perpendicular to the tube axis. The tube is constantly supplied with pure oxygen. The excess oxygen escapes through an axial opening at the end of the tube, while the remainder of the oxygen (and other gases evolved during analysis) passes through a nondispersive infrared analyzer at a constant flow. In addition to the variable temperature furnace, the apparatus also contains a constant temperature furnace, usually kept at about 850°C. The segment of quartz tube inside the constant temperature furnace is filled with a copper oxide catalyst. The purpose of the catalyst is to ensure that carbon-containing gases evolved from the samples are completely converted to CO_2. This is especially important at relatively low temperatures when complete oxidation to O_2 does not occur.

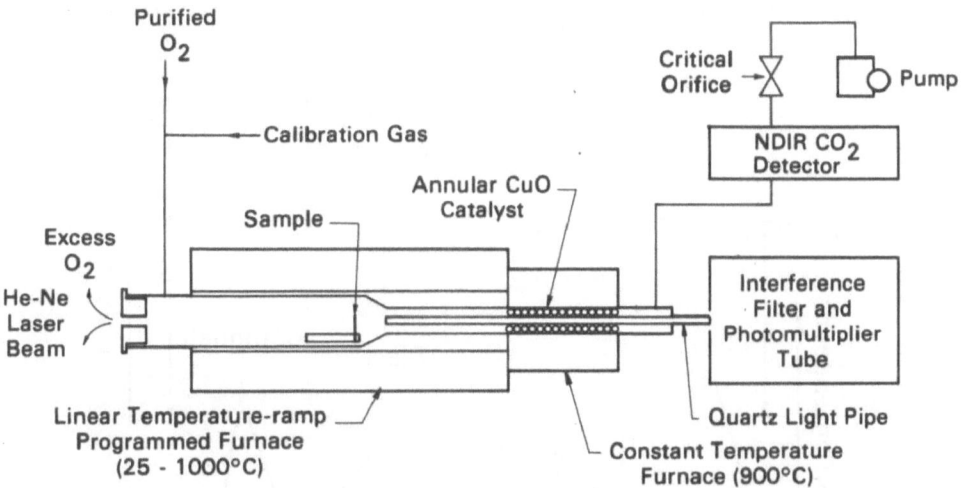

Fig. 2. Schematic representation of the thermal analysis apparatus.

The actual measurement consists in monitoring the CO_2 concentration as a function of the sample temperature. The result is a "thermogram — a plot of the CO_2 concentration vs. temperature. The area under the thermogram is proportional to the carbon content of the sample. The carbon content is quantitated by calibrating with a calibration gas (CO_2 in oxygen) and by measuring the flow rate through the system. This calibration is crosschecked by analyzing samples of known carbon content. The thermograms of ambient and source aerosol samples reveal distinct features in the form of peaks or groups of peaks that correspond to volatilization, pyrolysis, oxidation, and decomposition of the carbonaceous material.

To determine which of the thermogram peaks corresponds to black graphitic carbon, the intensity of the light beam produced by a He-Ne laser is monitored by a photomultiplier and displayed by the second pen of the chart recorder, simultaneously with the measurement of the CO_2 concentration [11]. In actual experiments the light penetrating the filter is collected by a quartz light guide and filtered by a narrow band interference filter to eliminate the effect of the glow of the furnaces. An examination of the CO_2 and light intensity traces enables the assignment of the peak or peaks in the thermograms corresponding to the black carbon because they appear concomitantly with the decrease in sample absorptivity.

In Fig. 3, a complete thermogram of an ambient sample is shown. The lower trace represents the CO_2 concentration vs. the sample temperature, while the upper curve corresponds to the light intensity of the laser light beam that reaches the detector during the temperature scan. Inspection of the thermogram shows that a sudden change in the light intensity occurs concomitantly with the evolution of the CO_2 peak at about 470°C. The light intensity I_0, after the 470°C peak has evolved, corresponds to that of a blank filter. This demonstrates that the light-absorbing species in the sample are combustible and carbonaceous. We refer to these species as black carbon. The carbonate peak evolves at about 600°C; and as carbonate is not light absorbing, it does not change the optical attenuation of the sample. In addition to black carbon and carbonate, the thermogram in Fig. 3 also shows several distinct groups of peaks at temperatures below~400°C that correspond to various organics.

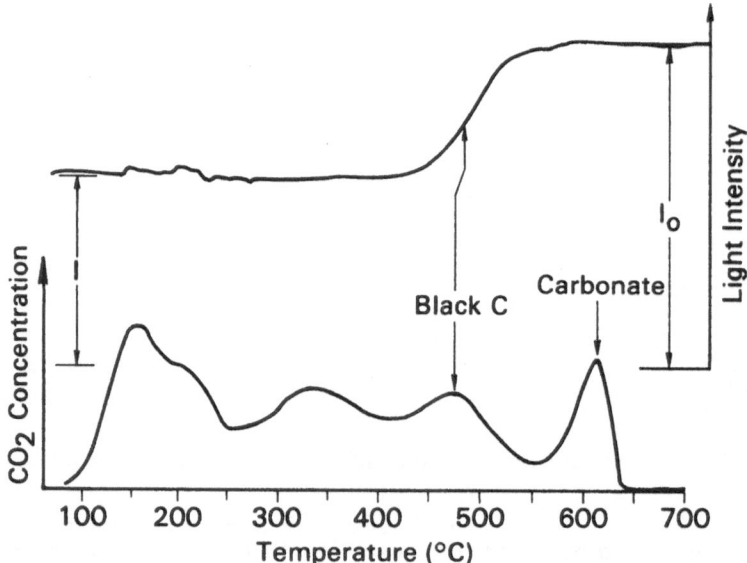

Fig. 3. Thermogram of an ambient sample showing carbonate, black carbon, and several forms of organic material.

The thermogram in Fig. 3 was obtained with a 1.46 cm diameter disc cut out of a sample collected on prefired quartz fiber filters. The temperature ramp rate was 10°C/minute. The integrated area under the CO_2 trace is proportional to the total carbon concentration. For this sample the total carbon concentration, determined by thermal analysis, was 17.9 μg (C)/cm^2. The black carbon, determined from the thermogram, composes 14% of the total carbon. This value can be crosschecked by using the optical attenuation and total carbon data. The specific attenuation for this sample, determined in a separate measurement, is $\sigma \equiv ATN/C = 3.00$. The estimated percentage of black carbon (as a percent of total C), determined from measurement of optical attenuation and total carbon only, is 100 x 3.0/20.0 = 15%. This value is in excellent agreement with the percentage of black carbon determined directly from the CO_2 thermogram.

RESULTS AND DISCUSSION

The data presented in this paper consist of information obtained from analyses of 24-hr. samples (collected weekdays) and multi-day samples (collected over weekends [9]. Table 1 lists the routine sampling sites with the beginning date of sampling.

TABLE 1
LBL Aerosol Sampling Sites

Site	Location	Date of first sample
Lawrence Berkeley Laboratory	Berkeley, California	1 June 1977
BAAQMD monitoring station	Fremont, California	15 July 1977
SCAQMD monitoring station	Anaheim, California	19 August 1977
Argonne National Laboratory	Argonne, Illinois	22 January 1979
DOE Environmental		
Measurments Laboratory	Manhattan, New York	22 November 1978
National Bureau of Standards	Gaithersburg, Maryland	23 January 1979
Denver Research Institute	Denver, Colorado	15 November 1978

Total Carbon and Optical Attenuation — Fig. 4 shows the variations of 24-hr total carbon (weekends excluded) at the Fremont, California, site. These data cover the period from July 1977 to January 1980. The 24-hr histogram superimposed on the bar diagram represents the monthly averages.

It is evident from Fig. 4 that there are significant day-to-day variations in total carbon. The maximum and minimum daily concentrations differ by an order of magnitude. The monthly averages are at peak values during the November-December periods of each year. The variations in optical attenuation for the same samples are represented in Fig. 5. The pattern of ATN values resembles that of total carbon and shows similar seasonal variations. The specific attenuation (ATN/C) variations represented in Fig. 6 are much less pronounced and show no clear seasonal variations. Similar features of total C, ATN, and ATN/C are also observed at the Berkeley and Anaheim sites.

References pp. 36-37.

At the New York, Gaithersburg, and Argonne sites, daily and monthly variations of total C and ATN are much less pronounced than at the three west coast sites.

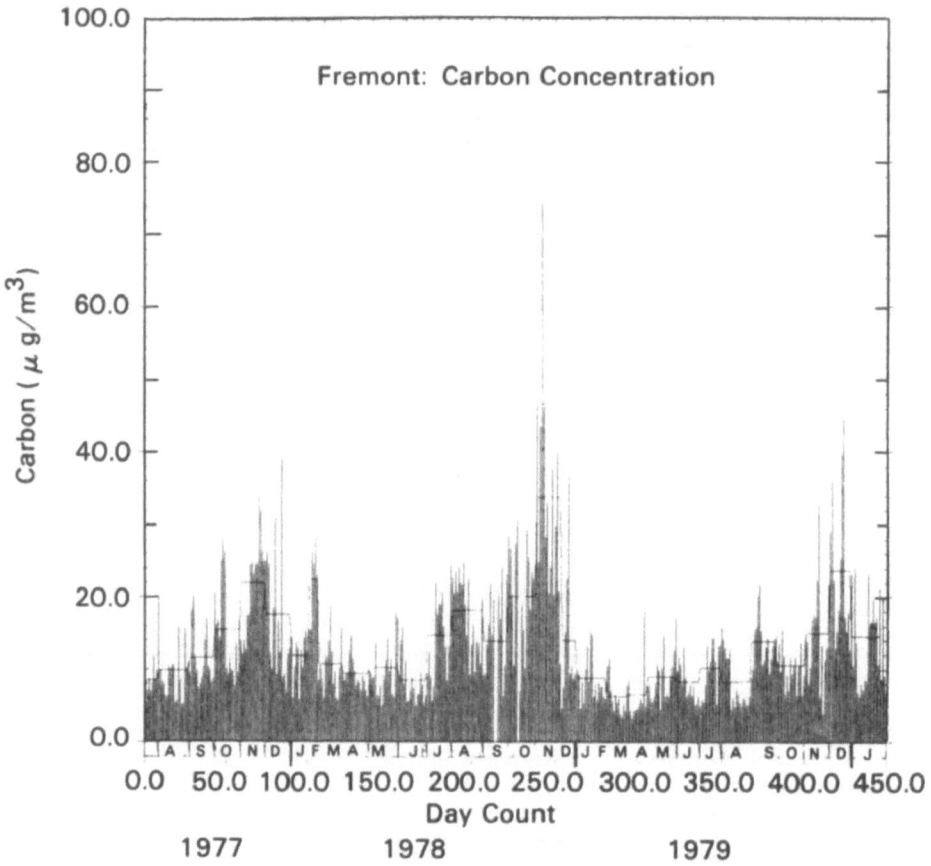

Fig. 4. Variations in the daily total carbon concentration at the Fremont, California, site (from Ref. 9).

Correlations between Total Carbon and ATN — Statistical analysis of the data shows that there is a strong correlation ($r > 0.85$) between optical attenuation and total particulate carbon at every site studied [12, 13]. Furthermore, a study of a number of source samples shows that there is also a strong correlation between optical attenuation and total carbon for these samples. The correlations between optical attenuation and total carbon for the three California sites, Argonne, and source samples are shown in Fig. 7 (a-e) [13].

Results obtained from ambient samples imply that the fraction of graphitic soot to total particulate carbon is approximately constant under the wide range of conditions occurring at a given site. On specific days, however, there can be large variations in the ratio, reflecting the variations in the relative amounts of organic and black carbon. The least squares fit of the data shows regional differences which

are related to the fraction of black carbon due to primary emissions. These differences would suggest an increase in the relative importance of the primary component for samples collected respectively at Berkeley, Fremont, Anaheim, and
Argonne.

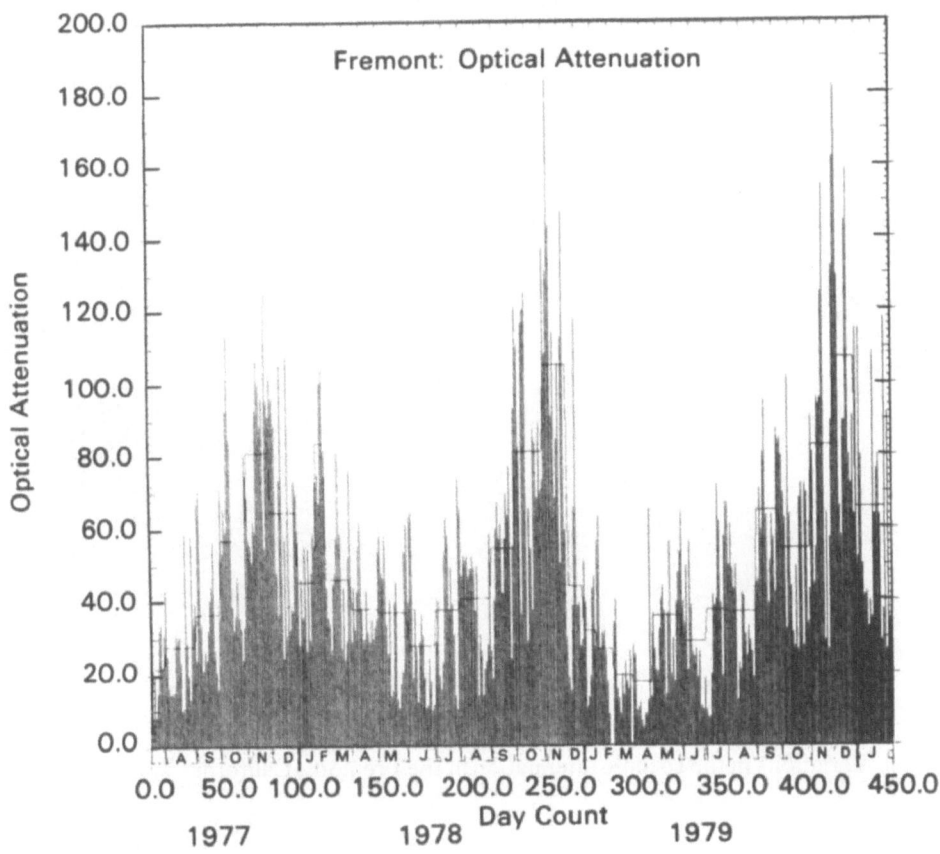

Fig. 5. Variations in the optical attenuation at the Fremont, California, site (from Ref. 9).

Concentrations of Black Carbon — Determination of specific attenuation,
$\sigma \equiv ATN/C$, enables a straightforward estimation of black carbon. From relation (3)
one can calculate black carbon as a percentage of total carbon, and the concentration of black carbon in $\mu g/m^3$. Table 2 lists the average specific attenuation and
black carbon percentages for all samples (including multi-day samples) analyzed to
date. In addition to the average values, the highest and lowest values are given.
Based on this estimate, on the average 20% of the total carbon is black carbon. This
fraction can on occasion be as high as 56% or as low as 6%. The latter occurs as a
rule when total carbon concentrations are low. Table 3 lists the concentrations of
total particulate carbon and estimated concentrations of black carbon for the same
samples as Table 2.

References pp. 36-37.

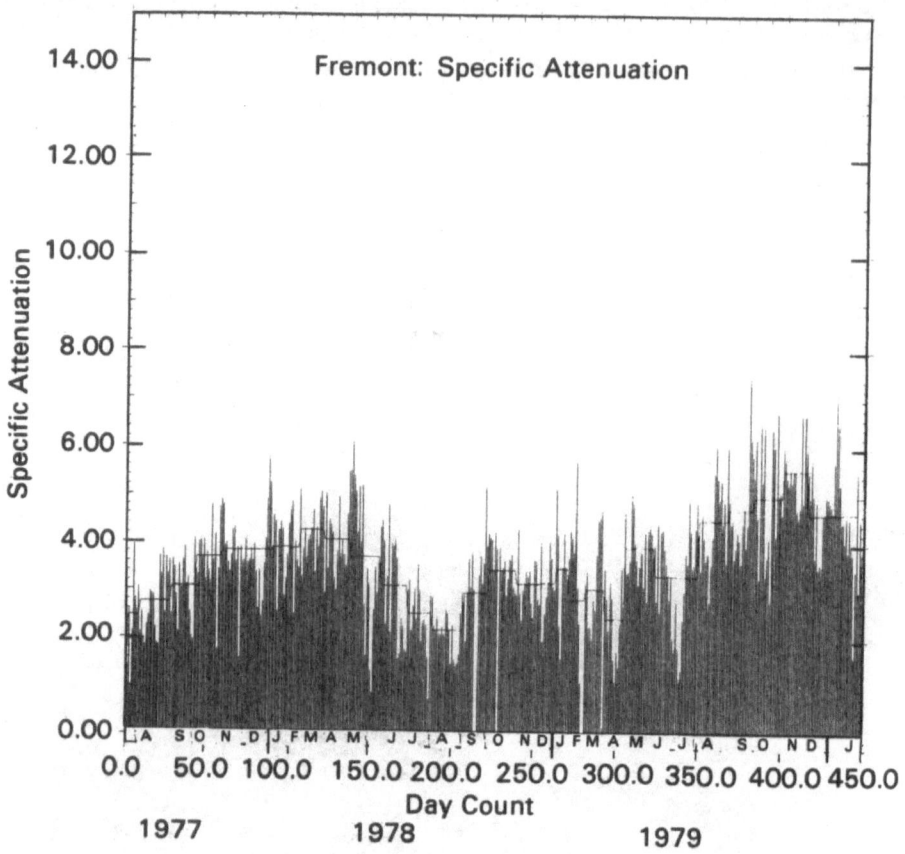

Fig. 6. Variations in the specific attenuation at the Fremont, California, site (from Ref. 9).

TABLE 2
Specific Attenuation (σ) and Black Carbon (BC) (% of Total C)
from Ambient Samples

Site	Date on file	# Samples	Average		Highest		Lowest	
			σ	%BC	σ	%BC	σ	%BC
New York	Nov 78—Apr 80	439	5.44	27%	11.1	56%	2.8	14%
Argonne	Jan 79—Mar 80	438	4.30	22%	9.1	46%	1.1	6%
Gaithersburg	Jan 79—Mar 80	381	4.33	22%	8.0	40%	1.8	9%
Denver	Nov 78—May 79	141	3.23	16%	5.7	29%	1.4	7%
Anaheim	Aug 77—Jan 80	852	3.70	19%	9.6	48%	0.8	4%
Fremont	Jul 77—Mar 80	924	3.55	18%	8.3	42%	1.6	8%
Berkeley	Jun 77—Apr 80	998	4.09	20%	9.2	46%	1.2	6%

TABLE 3
Carbon Concentrations (μg/m^3)

Site	Average		Highest		Lowest	
	C	BC	C	BC	C	BC
New York	15.2	4.2	53.1	12.6	3.4	0.6
Argonne	8.1	1.7	25.1	5.2	3.1	0.2
Gaithersburg	6.1	1.4	17.6	5.6	2.3	0.3
Denver	9.8	1.6	30.8	5.3	4.1	0.2
Anaheim	16.6	3.1	112.9	17.4	3.1	0.3
Fremont	12.0	2.1	75.6	9.2	3.4	0.3
Berkeley	6.7	1.3	31.7	5.2	3.0	0.3

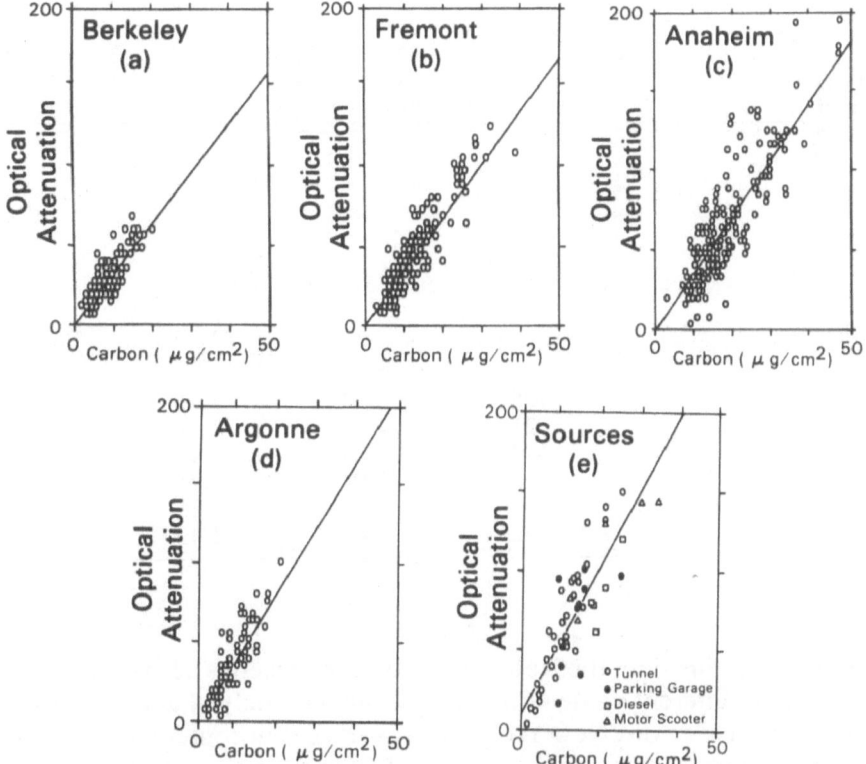

Fig. 7. Plots of optical attenuation versus carbon loading in μg/cm^2 for particulate samples collected at Berkeley, Fremont, Anaheim, and Argonne, and from various combustion sources. The solid line represents the least squares fit of the data points (from Ref. 13).

Concentrations of Soot — Soot contains not only black carbon but also various organic material. Because the organic soot component does not absorb light, the specific attenuation of soot is much less than 20, the σ value of pure black carbon.

Table 4 lists the average and extreme values of specific attenuation and the black carbon fraction of a number of source samples.

The percentage of soot in ambient carbonaceous particulates can be estimated by comparing the σ of sources with that of ambient samples. The fraction of soot is given in equation (4). Table 5 lists the mean specific attenuation of ambient samples (weekends excluded) in order of decreasing σ and soot fractions obtained by using equation (4) and $\sigma_{source} = 5.85$.

TABLE 4
Specific Attenuation (σ) and Black Carbon (BC) Content of Source Samples

Source	# Samples	Average		Highest		Lowest	
		σ	%BC	σ	%BC	σ	%BC
Parking garage	12	5.4	27%	7.7	39%	2.25	11%
Diesel	6	5.6	28%	5.7	29%	3.5	18%
Scooter	9	5.1	26%	6.1	31%	4.2	21%
Tunnel	63	6.3	32%	12.5	63%	3.7	19%
Natural gas	6	2.6	13%	3.3	17%	1.9	10%
Garage and tunnel		5.9	29%				

TABLE 5
Mean Specific Attenuation of Ambient Samples

Site	# samples	σ	SDEV	Soot (%)
New York	211	5.69	1.34	97
Gaithersburg	155	4.72	1.51	81
Argonne	221	4.35	1.64	74
Berkeley	513	4.28	1.47	73
Anaheim	444	3.99	1.71	68
Fremont	461	3.74	1.25	64
Denver	42	3.47	1.49	59

Based on this estimation, the New York City carbonaceous aerosol is essentially primary soot. A different value of σ_{source} would certainly change the estimated soot percentage. However, New York City's average soot content would neverthe-less remain the highest, irrespective of the actual numerical value of σ_{source}. It is logical that samples from this location have the highest soot content because the site is representative of a heavily traveled street canyon. The Fremont and Anaheim samples have on the average the smallest soot content, as may be expected, be-cause both sites represent receptor sites. According to the above estimate, Denver has the smallest specific attenuation value. It is possible that high-altitude combus-tion results in increased emissions of primary organics; however, we note that the number of samples from this location is small compared to that from other sites, so this conclusion should be taken with caution.

It is instructive to present the specific attenuation data in the form of histograms representing their frequency of occurrence. Histograms for New York and Fremont (Fig. 8) show that the occurrence of high specific attenuation samples is much greater for New York than for Fremont. In Fig. 9 the histogram of specific attenuations of a number of source samples is shown together with those for New York and Fremont. The distribution for sources looks similar to the distribution for New York. This supports the inference that the New York samples, on the average, consist almost entire of primary carbonaceous material. Histograms for other sampling sites are shown in Figs. 10 and 11.

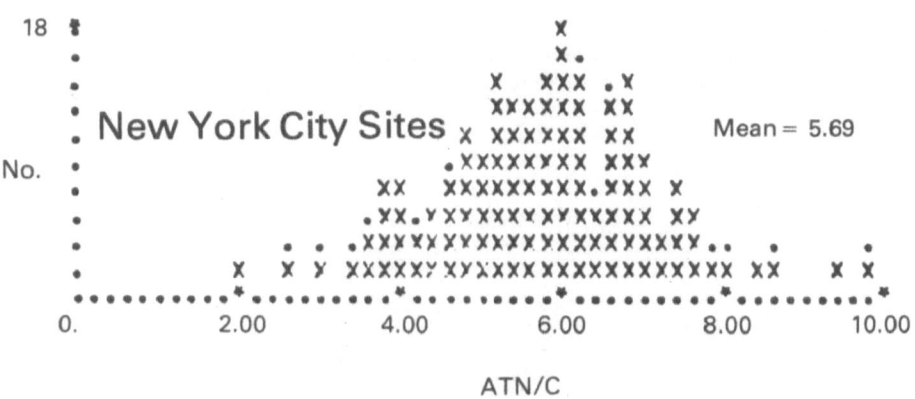

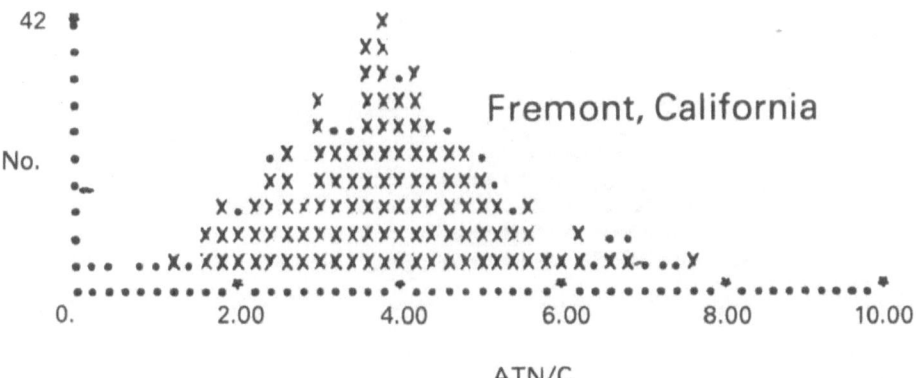

Fig. 8. Distribution of specific attenuation for the New York (Nov., 1978-April, 1980) and Fremont (July, 1977-March, 1980) sites (from Ref. 9). are included. No weekend samples are included.

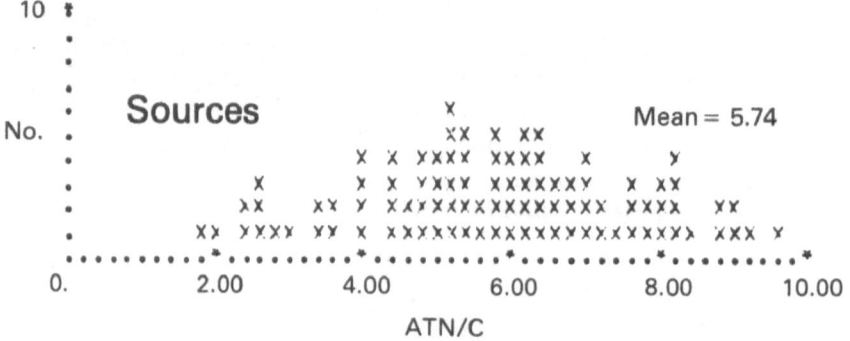

Fig. 9. Distribution of specific attenuation for various sources including samples from a garage, tunnel, motor scooter, diesel vehicle and a natural gas burner.

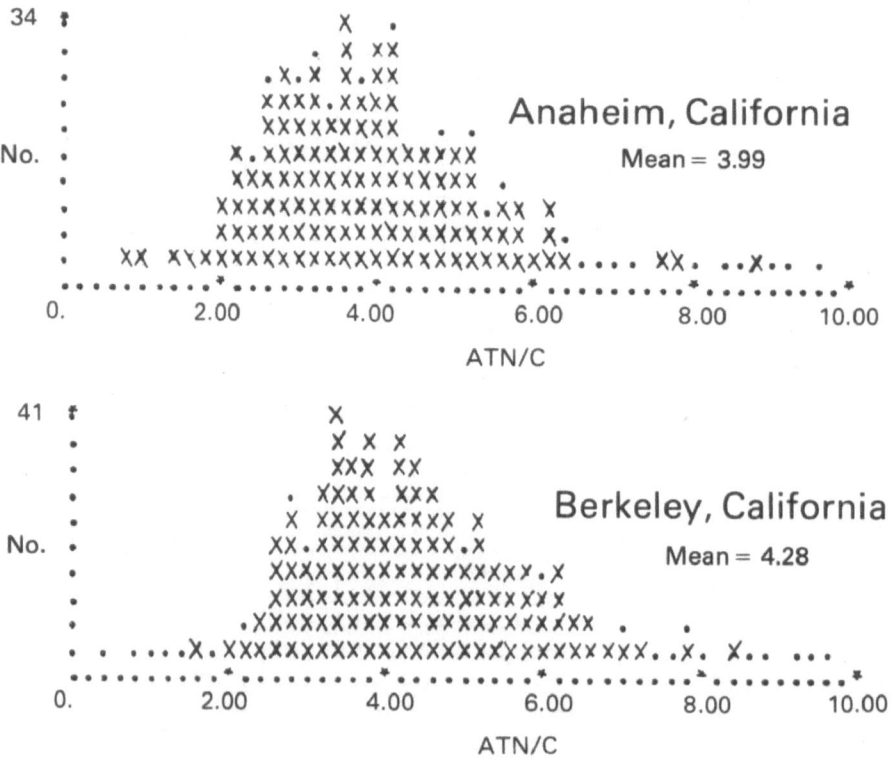

Fig. 10. Distribution of specific attenuation for Anaheim, (Aug., 1977-Jan., 1980) and Berkeley, California (June, 1977-April, 1980), sites (from Ref. 9). No weekend samples are included.

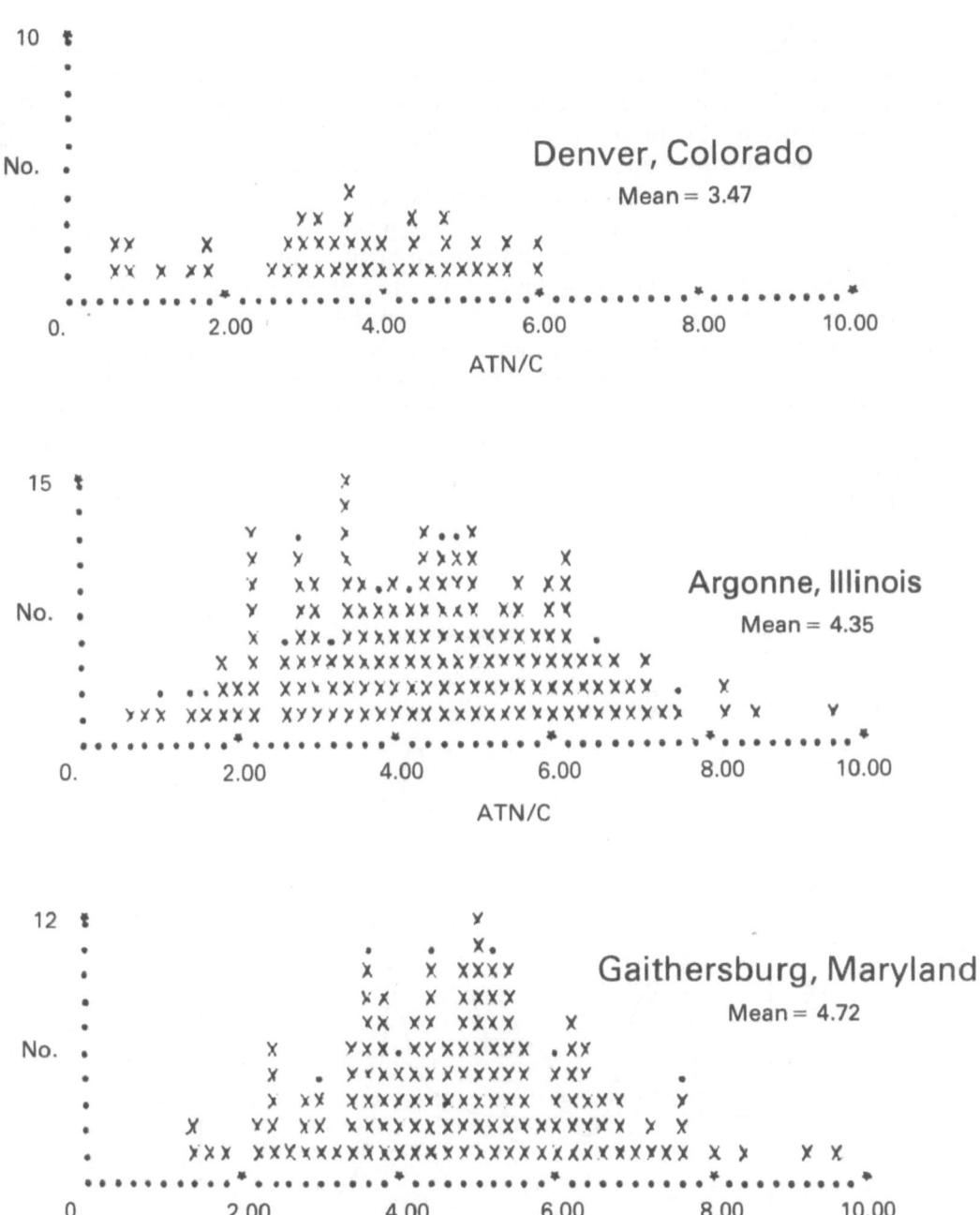

Fig. 11. Distribution of specific attenuation for Denver, Colorado (Nov., 1978-May, 1979) Argonne, Illinois (March, 1978-March, 1980) , and Gaithersburg, Maryland (Jan., 1979-March, 1980), sites (from Ref. 9). No weekend samples are included.

References pp. 36-37.

Secondary Organics — Results in Table 5 suggest that the West Coast sites have an organic component that occurs in excess of source-dominated organics. This excess should be equal to the secondary organic material, which can be conveniently identified by the thermal analysis method.

We have already described how thermal analysis can be used to obtain the total carbon, black carbon, organic carbon, and carbonate carbon. The greatest strength of this method, however, is its ability to "fingerprint" source-produced carbonaceous particles and their contribution to the ambient aerosols. As an illustration, Figs. 12 and 13 show the thermograms of a sample collected in Manhattan (high σ) and one collected in Berkeley (low σ). The two thermograms are substantially different. Common features of both samples are the black carbon and the group of peaks below 250°C, corresponding to volatile organic compounds. However, the Berkeley sample clearly shows the presence of other peaks which are absent in the thermogram of the New York sample. These peaks are not observed in samples collected in a highway tunnel or a parking garage and probably correspond to secondary species.

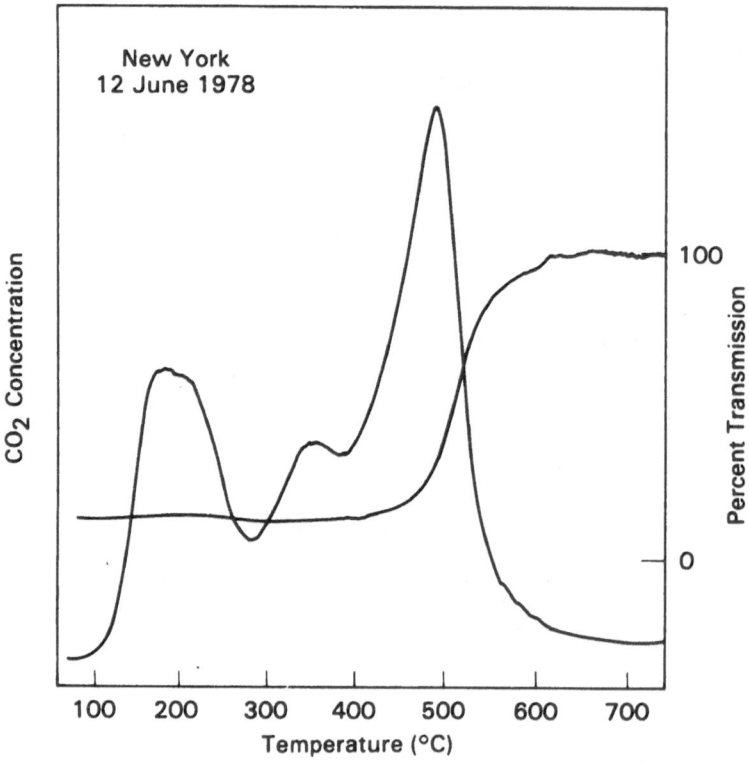

Fig. 12. Thermogram of a New York sample with high specific attenuation.

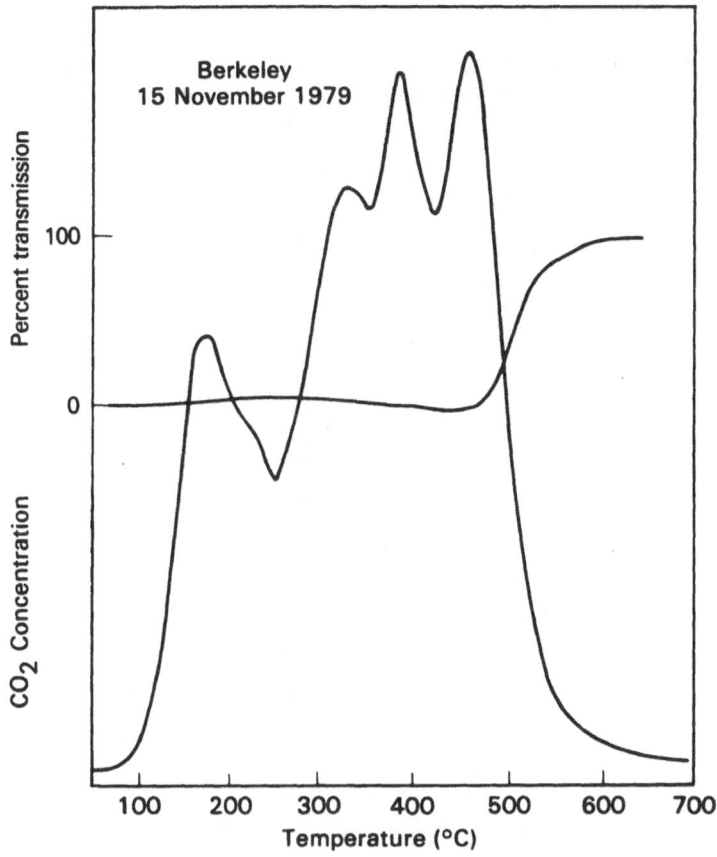

Fig. 13. Thermogram of a Berkeley, California, sample with low specific attenuation.

Secondary Organics and Ozone —It is clear from the results described so far that the ratio of black carbon to total carbon may vary on specific days. However, no large systematic differences are found as a function of the ozone concentration, which is viewed as an indicator of the photochemical activity [13]. This is graphically demonstrated in Fig. 14, which shows the distribution of the ratios of the optical attenuation to total carbon content for ambient samples from all the California sites taken together, subdivided according to peak hour ozone concentration. Clearly there is no trend for high-ozone days to be characterized by aerosols which have a significantly reduced black carbon fraction. This places a low limit on the importance of secondary organic particulates formed by the photochemical route. concentration.

References pp. 36-37.

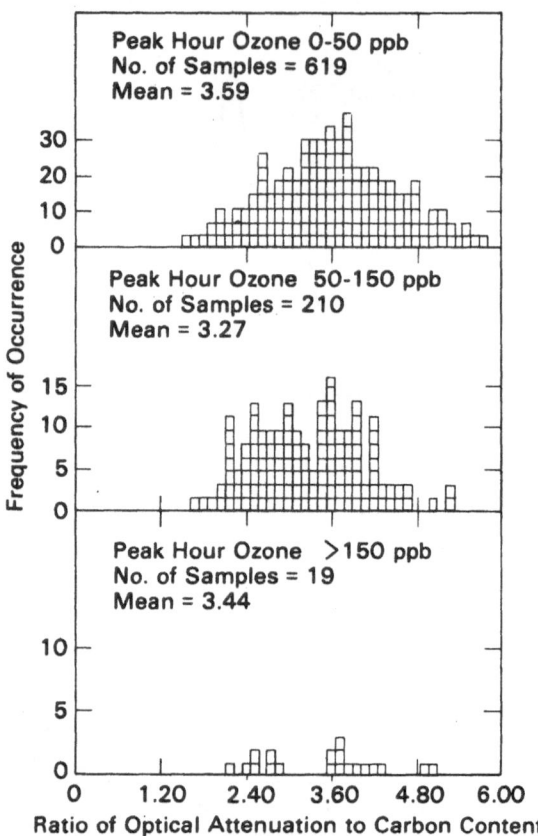

Fig. 14. Distribution of the ratios of specific attenuation subdivided according to the peak ozone concentration. Note that the means of the distributions are only marginally smaller at larger ozone concentrations, which puts a rather low limit on secondary organics produced in conjunction with ozone (from Ref. 13).

ACKNOWLEDGEMENTS

This work was supported by the Biomedical and Environmental Research Division of the U.S. Department of Energy under contract no. W-7405-ENG-48 and by the National Science Foundation under Contract No. ATM 80-13707.

REFERENCES

1. *For a review of the formation and characterization of secondary organics, see, for example, D. Grosjean, "Secondary organic aerosols: Identification and mechanisms of formation," in Proceedings, Conference on Carbonaceous Particles in the Atmosphere, Lawrence Berkeley Laboratory Report LBL-9037 (1979), p. 107 (available from NTIS).*
2. *S. G. Chang and T. Novakov, Atmos. Environ., Vol. 9 (1975), p. 495.*

3. *T. Novakov, A. B. Harker and W. Siekhaus, "Studies in aerosol chemistry by photoelectron spectroscopy – sulfur and nitrogen compounds," in Proceedings, First Annual NSF Trace Contaminants Conference, Oak Ridge National Laboratory Report CONF-730802 (1974), p. 379.*

4. *a) T. Novakov, "Chemical characterization of atmospheric pollution particulates by photoelectron spectroscopy," in Proceedings, Second Joint Conference on Sensing of Environmental Pollutants, Pittsburgh, Instrument Society of America (1973), p. 197.*
   b) T. Novakov, S. G. Chang and A. B. Harker, Science, Vol. 186 (1974), p. 259.

5. *H. Rosen and T. Novakov, Nature, Vol. 266 (1977), p. 708.*

6. *H. Rosen, A. D. A. Hansen, L. Gundel and T. Novakov, App. Opt., Vol. 17 (1978), p. 3859.*

7. *Z. Yasa, N. Amer, H. Rosen, A. D. A. Hansen and T. Novakov, Appl. Opt., Vol. 18 (1978), p. 2528.*

8. *H. Rosen, A. D. A. Hansen, L. Gundel and T. Novakov, "Identification of the graphitic carbon component of source and ambient particulates by Raman spectroscopy and an optical attenuation technique," in Proceedings, Conference on Carbonaceous Particles in the Atmosphere, Lawrence Berkeley Laboratory Report LBL-9037 (1979), p. 49.*

9. *A. D. A. Hansen et al., unpublished data.*

10. *H. Malissa, H. Puxbaum and E. Pell, Z. anal. Chem., Vol. 282 (1976), p. 109.*

11. *R. L. Dod, H. Rosen and T. Novakov, "Optico-thermal analysis of the carbonaceous fraction of aerosol particles," in Atmospheric Aerosol Research Annual Report 1977-78, Lawrence Berkeley Laboratory Report LBL-8696 (1979), p. 2.*

12. *Preliminary evidence for the correlation between optical attenuation and total particulate carbon was presented by H. Rosen, A. D. A. Hansen, R. L. Dod and T. Novakov, "Application of the optical absorption technique to the characterization of the carbonaceous component of ambient and source particulate samples," in Proceedings, Fourth Joint Conference on Sensing of Environmental Pollutants, Washington, American Chemical Society, 1978; and by A. D. A. Hansen, H. Rosen, R. L. Dod and T. Novakov, "Optical characterization of ambient and source particulates," in Proceedings, Conference on Carbonaceous Particles in the Atmosphere, Lawrence Berkeley Laboratory Report LBL-9037 (1979), p. 116 (available through NTIS).*

13. *H. Rosen, A. D. A. Hansen, R. L. Dod and T. Novakov, Science, Vol. 208 (1980), p. 741.*

DISCUSSION

J. Daisey *(New York University Medical Center)*

You implied that there are similarities between the tunnel and the New York City samples. I think that is wrong because of the contribution of space heating in New York City. You can actually see the black plumes coming out of the apartment houses. And, I think we may be able to demonstrate this source for some of that black carbon in the work we are doing with Jim Huntzicker and his group at the Oregon Graduate Center.

T. Novakov

I do not think I implied that the tunnel is the same as New York City. However, that particular location in New York is very enriched in primary emissions and there certainly is an automotive component.

L. Currie *(National Bureau of Standards)*

Have you any comments or experiments relating to various forms of inorganic carbon, such as carbonates?

R. Dod *(Lawrence Berkeley Laboratory)*

We believe that we see very little inorganic carbon. Dr. Novakov implied on one slide that we did see a carbonate peak. We have seen carbonates, bicarbonates, whatever they may be (but we assume they are carbonates) around 600°C on samples from the Argonne area. Also there were periods of a couple weeks in Berkeley when we detected carbonates. By and large, however, we do not see significant amounts of carbonates but we really have not investigated it extensively.

L. Currie

Were those carbonates in the fine fraction?

T. Novakov

That is total, no fractionation, so it is likely mineral.

R. Bradow *(U.S. Environmental Protection Agency)*

I think the literature contains a substantial number of references on the optical identification of the mineral calcite in ambient aerosol samples which are usually in the coarse fraction and are usually associated with roadways. For example, in many of the Eastern cities the primary mineral associated with roadway aggregate is calcite limestone. Consequently, it is quite possible that one will periodically find some carbonate in samples of coarse aerosol near roadways. Peter Mueller's work and all the other literature I am aware of indicates that far from roadways, fine aerosol carbonates and other similar forms of carbon are rare.

P. Mueller *(Electric Power Research Institute)*

Are you implying that there is a constant relationship between attenuation and soot concentration no matter where you sample?

T. Novakov

No. My definition of soot is black carbon plus whatever primary organic you have. The implication is that the attenuation is proportional to the black carbon, wherever it is.

P. Mueller

Is the black carbon elemental carbon?

T. Novakov

I hate to call it elemental carbon because that would mean that it contains no trace of hydrocarbon or oxygen. So, I think elemental carbon is not the best word to use because it implies the elemental form only with nothing else being present.

P. Mueller

In Denver, for instance, there is a highly variable relationship between the absorption and the elemental carbon concentration as measured by combustion techniques.

T. Novakov

If the black carbon is the principle absorbing material, then you would expect a correlation between attenuation and black carbon.

P. Mueller

Should the correlation be constant from site to site? We do not get that.

T. Novakov

That does not necessarily mean that the relationship does not exist.

P. McMurry (*University of Minnesota*)

Have you applied the thermal analysis technique to aerosols generated from organic aerosol precursers in smog chambers?

T.Novakov

We would really welcome anybody who can supply us some of that material.

R. Countess (*General Motors Research Laboratories*)

On your thermograms, you showed two peaks which identify the black carbon; one centered at 450 and one at 500 nanometers. Do you infer by this that there are two types of black carbon?

T. Novakov

That is possible based on the very limited knowledge that we have about thermal analysis. The morphological differences between carbons or soots would result in their combustion at different temperatures. So, if two sources produced black carbons that are different by morphology, or if you think in solid state physics terms, if the orientation of the crystallites on the surface of the particle is such that the cleavage plane is mostly exposed, these would be very difficult to oxidize. The residual hydrogen and particle size may also influence those differences. But we know very little about the basic processes which occur in this apparatus.

B. Appel *(California Department of Health)*

On one of your slides you showed a thermogram of a material previously extracted with methanol and chloroform-methanol, but you said you were not going to talk about that. Is somebody going to discuss that?

T. Novakov

That is Lara Gundel's work which I will discuss with you briefly. It shows that the least polar solvent removed almost the entire volatile fraction. The next solvent which was chloroform-methanol, removed what I suspect is secondary carbon around 350°C leaving black carbon plus some other non-extractable carbon which is not black.

E. Macias *(Washington University, St. Louis)*

I would like to comment on the relationship you reported between primary carbon and total or secondary carbon. From some work we did in St. Louis using similar kinds of relationships between black carbon and total carbon, we find that when you use 24 hour samples, which you have done, the correlation is improved over that found for shorter samples. Second, the daytime samples do not correlate as well as the nighttime samples. And, third, winter samples correlate better than summer. The conclusion that I would draw from this and from some other work involving the carbon to lead ratio is that there would be stronger influence of secondary carbon in the daytime and in the summer.

T. Novakov

I agree completely with what you have said. These were all 24 hour samples because it is of interest to know the average exposure level. Our point is that you can have similar enrichment in secondary material away from LA, for example, in Berkeley where we do not have much ozone.

D. Roessler *(General Motors Research Laboratories)*

If I take two particles, one of which has a certain amount of carbon, while the other particle has the same amount of carbon plus some secondary species like hydrocarbons, would I be able to distinguish optically between them?

T. Novakov

If one contains this absorbing material and the other one does not, then the attenuation would be different so you could distinguish them optically.

P. Mueller

Is there material in the particulate matter which absorbs light but which is not carbon.

T. Novakov

That is quite possible. In some cases you find that when you have burnt the carbon off the quartz filter, it has some reddish or brownish tint to it. However, what we are saying here applies to carbon and in most cases, other than those few cases when sampling was carried out during a dust storm, most of the material would leave the filter and it would be completely white as the filter was in the beginning.

J. Muhlbaier (*General Motors Research Laboratories*)

It seems that you are saying that all sources are the same or that all parts of the country have the same mixture of sources.

T. Novakov

This is a correct inference. However, in the paper we also take the New York City location as an example of primary enriched particles and you get basically the same results. It is not completely illogical to assume that sources are similar because people do drive in all parts of the country and similar fuels are used. Coal burning may or may not produce black carbon. I don't think that anybody knows if coal burning will produce soot.

R. Draftz (*I.I.T Research*)

In your optical attenuation measurements and then your subsequent analysis for total carbon, were the filters the same or did you use two different substrates?

T. Novakov

We used two different substrates. One for total carbon which is quartz and the other one for attenuation which was millipore.

R. Draftz

So, is it a possibility that the quartz filter might be sampled at a higher flow rate or the same flow rate?

T. Novakov

The flow rates are different but by less than a factor of two.

THE OPTICAL PROPERTIES OF
PARTICULATE ELEMENTAL CARBON

R. W. BERGSTROM

Systems Applications, Inc.
San Rafael, California

T. P. ACKERMAN

NASA Ames Research Center
Moffett Field, California

L. W. RICHARDS

Meteorology Research, Inc.
Santa Rosa, California

INTRODUCTION

Recent measurements of the absorption and scattering coefficients of tropo-spheric aerosols in both rural and urban areas have shown that these particles exhibit a significant (5 to 50% of the total extinction) amount of absorption in the visible spectrum. Elemental analyses have indicated that particulate elemental carbon is present and is the likely candidate to account for much of this observed absorption. Previous studies have demonstrated both experimentally and theoretically that a small amount of submicron carbon particles has a large effect on the absorption and scattering properties of tropospheric aerosols. This paper discusses the optical properties of particulate elemental carbon and illustrates the impact of these prop-erties on the horizon brightness.

References p. 48

CALCULATION OF THE ABSORPTION CHARACTERISTICS
OF CARBON MIXTURES

For the purpose of this work we have defined two different ways in which soot might be included in atmospheric aerosols:

1. External mixture. The soot and non-soot aerosols exist in the atmosphere as distinct particles which are mixed without interacting. In this case the properties of the mixture are sums of the properties of the individual distributions.
2. Internal mixture. The soot is deposited as a shell on the outside of a solid non-soot particle or is the core of a particle with a sulfate solution shell.

For the external mixture we have assumed that the non-soot aerosol may consist of either soil or sulfate particles. Log-normal distributions were chosen to represent the aerosol size distributions. The log-normal distributions is conventionally defined by

$$n(\ln r) = \exp\left[-(\ln r - \ln rgv)^2/(2 \ln^2 \sigma)\right] ,$$

where $n(\ln r)$ is the number per volume at radius r, rgv is the geometric mean radius, and σ is the geometric standard deviation. The soil particles were placed in the coarse mode with a rgv of 2.29 microns and σ of 2.11. These values were taken from the light loading measurements of Patterson and Gillette [1]. The sulfate particles were placed in the accumulation mode with $rgv = 0.2$ microns and $\sigma = 2.00$. These are the mean values of measurements made by five different experimenters as reported by Whitby [2]. The appropriate size distribution for the soot particles was not readily apparent. In order to cover the possible ranges of interest we mixed soot particles with the same two size distributions as those above but also included a nucleation mode distribution. Since the parameters for this small particle distribution are not well specified, we chose values of $rgv = 0.015$ microns and $\sigma = 1.70$, which are consistent with the values quoted by Whitby [2] for nucleation mode aerosols. Calculations were carried out at a wavelength of 0.5 microns. The appropriate values of the index of refraction at this wavelength are 1.94-0.66i for soot [3], $(1.53-10^{-7})$i for ammonium sulfate, and 1.50-0.0008i for soil [4]. Liquid sulfates have low imaginary indices similar to those of ammonium sulfate, but they have smaller real refractive indices approaching those of water as they become extremely dilute. Since we do not know the typical water concentration of soluble sulfates, we simply take the same refractive index as ammonium sulfate for the solutions to illustrate the effects of soot particles. No doubt objections can be raised to these specific refractive index values, but they are typical of the values in the literature, and small departures will not substantially alter our results.

In the case of internal mixtures we assumed that soot could be deposited on either soil particles or sulfate crystals as a shell and could act as a nucleus for sulfate solution droplets. Due to the limitations of Mie theory calculations, we were forced to assume that all particles were spherical and that the soot was deposited as a uniform, concentric shell or formed a spherical nucleus. From electron microscopy and from a knowledge of coagulation physics it is apparent that this is a poor assumption, but the error associated with the approximation is not known. Having assumed a concentric shell model for the internal mixture particles, it is then necessary to assume some functional relationship between the radius of the core and the radius of the particle, which will then be applied over the entire size distribution. For most of our calculations we have chosen to assume a constant ratio of core radius to particle radius, which has the practical advantage of allowing a simple calculation of the fraction of the particle volume contained in the core or shell. A few calculations were carried out assuming various other relationships between core radius and particle radius. Among these were constant thickness shells and distributions in which particles less than or greater than a particular radius were assumed to be homogeneous. The results showed that the most important factor in determining the optical behavior is the total amount of soot present. As long as the soot is distributed throughout the size distribution, the exact manner of the distribution is of secondary importance. However, if it is concentrated at either the small or large particle end of the distribution, then the effects can be notably different from the constant ratio results.

Values of the single-scattering albedo calculated for both internal and external mixtures of soot and sulfate are shown in Fig. 1 as a function of the ratio of soot volume to total particle volume. The solid curve represents an external mixture of accumulation mode sulfate and nucleation mode soot and the dashed curve represents an external mixture with both sulfate and soot assumed to have an accumulation mode distribution. The circles indicate internal mixtures with sulfate cores and soot shells, and the triangles represent soot cores and sulfate shells. The internal mixtures also were calculated assuming an accumulation mode distribution. The horizontal line at $\omega_0 = 0.85$ is included as a reference point representing the critical albedo calculated at Hansen et al [5]. Tropospheric aerosols with single-scattering albedos greater than this value act to cool the earth-atmosphere system, whereas values less than this act to warm the system.

It is obvious from Fig. 1 that for particle distributions of equal size and a given volume fraction of soot, the internal mixtures produce lower values of the single-scattering albedo than does the external mixture. For small soot particles in the external mixture the same is true for volume fractions less than about 30 percent. The fact that soot as a core is more effective than soot as a shell in absorbing radiation is due to two effects. A solution sheath around the core acts to focus photons on the core, thus increasing its effective cross-section, whereas a particle with a soot shell has a greater tendency to reflect or refract photons than does a particle with a solution shell due to the larger real part of the index of refraction. If we take a value of $\omega_0 = 0.6$, which represents an approximate lower limit of the urban albedo measurements, we see that the minimum amount of soot necessary to obtain this value is 20 percent as particle cores. Soot shells or small particles both require about 30 percent soot by volume and large particles require a mixture of over 50 percent by volume. To obtain a representative rural value of 0.85 requires considerably less soot, roughly 5 percent by volume for the internal mixtures and 10 to 20 percent for the external mixtures.

The results shown in Fig. 1 are in varying degrees dependent on the size distributions which have been used. Because the internal mixtures were assumed to have

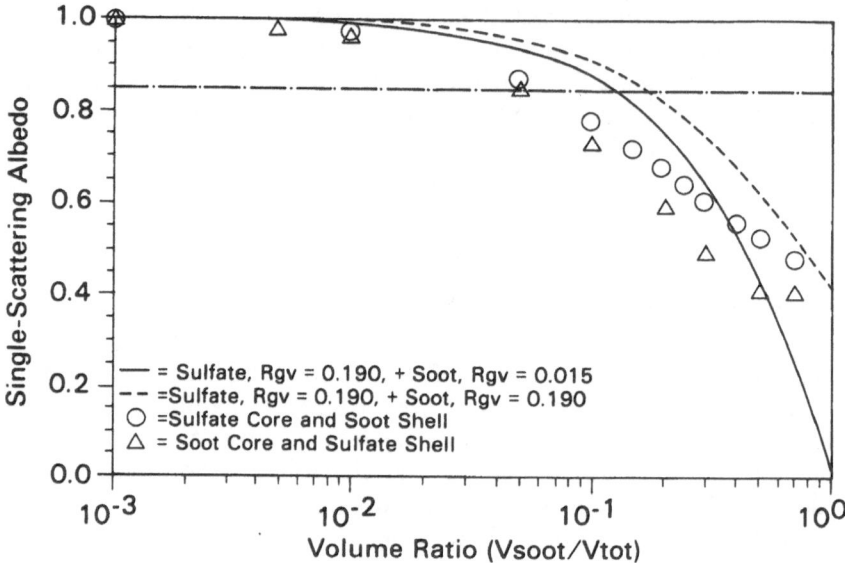

Fig. 1. Single scattering albedo as a function of the fractional volume of soot for a mixture of soot and sulfate.

constant core-to-shell radii ratios, their single-scattering albedo values are relatively insensitive to small changes (a factor of 2 or less) in the value of the geometric mean radius. The external mixtures are somewhat more sensitive, particularly to the choice of the mean radius of the soot distribution. For a given volume ratio of soot and a fixed standard deviation, there is a value of the mean radius intermediate between molecular-sized particles and accumulation-mode particles which will produce a minimum value of ω_0. In our calculations, this value of rgv turned out to be 0.05 microns. However, the difference between the value of ω_0 at rgv = 0.05 microns and the value at rgv = 0.015 is less than 3 percent. Thus, the volume fraction of soot needed to produce a given value of ω_0 is approximately the same in either case.

A similar plot of values for mixtures of soil and soot is shown in Fig. 2. In this case we have plotted values for three external mixtures corresponding to soil particles in the coarse mode and soot particles in each of the three modes. Only one internal mixture was assumed since we expect the soot to accumulate on soil particles and not the opposite. It is interesting to note that in this case the most effective mixture for producing low values of ω_0 is the external mixture of nucleation-mode soot. This is due to the mean size of the soil particles being much larger than the wavelength of the light, in which case they are for a given mass, inefficient at interacting with the photons. This in turn enhances the importance of the small soot particles as opposed to the same volume of soot spread over the surface of the soil particles. The amount of soot as small particles necessary to get $\omega_0 = 0.6$ is only 7 percent by volume, while about 15 percent is necessary if the soot is in the accumulation mode, and 20 percent if it is present as an internal mixture. The 0.85 value can be obtained

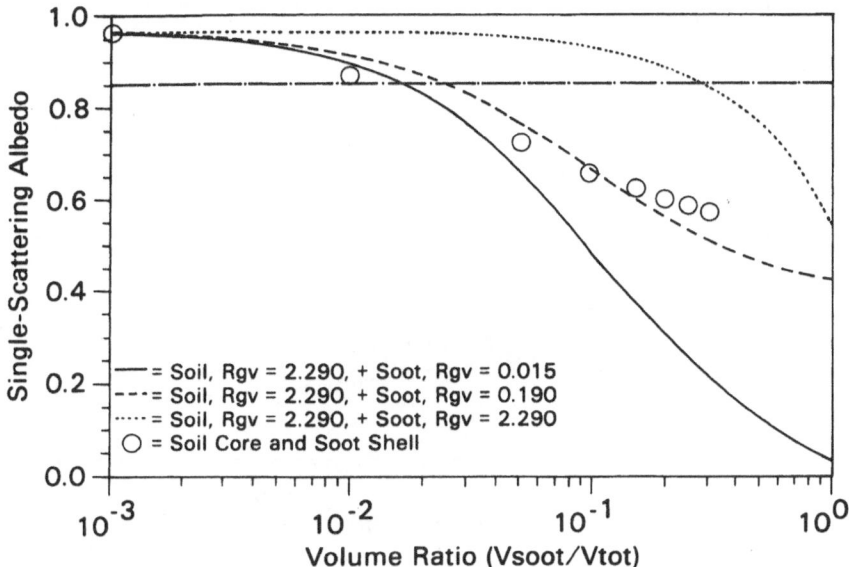

Fig. 2. Single scattering albedo as a function of the fractional volume of soot for a mixture of soot and soil material.

EFFECTS OF CARBON AEROSOLS ON HORIZON BRIGHTNESS

with about 1.5 percent soot present as either an internal mixture or an external mixture of small particles. As was the case with the sulfate mixtures, the value of ω_0 can be modified only slightly by small variations in the size distribution parameters.

EFFECTS OF CARBON AEROSOLS ON HORIZON BRIGHTNESS

The doubling and adding method was used to predict the horizon sky intensity of an urban area. The atmosphere above the surface layer was assumed to have global average optical properties [4]. The surface layer was assumed to be 1 km thick and have an accumulation mode volume of 32 $\mu m^3/cm^3$ (rgv = 0.20 μm, σ = 2.0) and a coarse mode volume of 31 $\mu m^3/cm^3$ (rgv = 3.0 μm, σ = 2.0). These aerosol volume values correspond to data from the recent Denver Winter Haze Study [6]. The index of refraction for the non-soot material was 1.5 - 0.0i. For simplicity, the surface reflection was neglected.

Fig. 3 shows the results of two calculations for the horizon sky intensity (zenith angle of 88.5°). The first calculation assumes that all of the aerosol is non-absorbing (called sulfate only) and the second assumes that 5 $\mu m^3/cm^3$ is carbon soot (rgv = 0.01 μm). The total amount of aerosol in the two cases is the same. The solar zenith angle is 30°, and the azimuthal angle is 120°. The intensity plotted is normalized by the solar intensity.

References p. 48

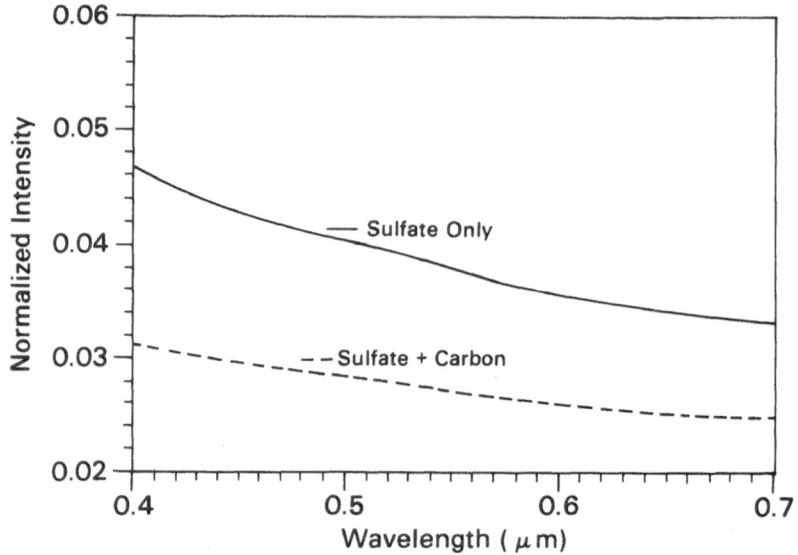

Fig. 3. Horizon sky intensity as a function of wavelength.

The results show that the effect of the soot is to reduce the horizon brightness substantially at all wavelengths. The blue wavelengths are reduced somewhat more than the red. The horizon intensity in both cases is relatively wavelength independent and would be perceived as white. If there is a delineation between the urban (soot-containing) haze and a non-soot-containing aerosol above, then the urban haze will appear as grey. This agrees with the measured horizon intensity values for the 1978 Denver Winter Haze Study [6]. Thus, these calculations tend to support the conclusion that the so-called Denver Brown Cloud is actually grey and is caused by carbon soot.

ACKNOWLEDGEMENTS

This research was supported in part by the CAPA-13 Committee of the Coordinating Research Council.

REFERENCES

1. *E. M. Patterson and D. A. Gillette, J. Geo. Res., Vol. 82 (1977), p. 2075.*
2. *K. T. Whitby, Atmos. Environ., Vol. 12 (1978), p. 135.*
3. *R. W. Bergstrom, Beitrage zur Physik der Atmosphare, Vol. 46 (1973), p. 223.*
4. *O. B. Toon and J. B. Pollack, J. Appl. Meteor., Vol. 15 (1976), p. 223.*
5. *J. E. Hansen, A. A. Lacis, P. Lee and W. C. Wang, Ann. N.Y. Acad. Sci., Vol. 338 (1980), p. 575.*
6. *S. L. Heisler, R. C. Henry, J. G. Watson and G. M. Hidy, "The 1978 Denver Winter Haze Study," ERT Report P-5417-1 prepared for the Motor Vehicles Manufacturers Association of the United States, Detroit, Michigan, 1980.*

DISCUSSION

J. Heintzenberg, *(Stockholm University)*

I suggest you use the refractive index of soot instead of graphite. Using 1.57-0.4i seems more realistic for atmospheric carbon attenuation.

T. Ackerman

There is considerable concern about what the index of refraction ought to be. We started using these numbers and thought rather than redoing the calculations in the middle, we would just keep on using the numbers. It turns out that whether the imaginary part is 0.6 or 0.4 does not make a great deal of difference in the results. The real part may affect it somewhat since it will make it a little less reflective. But, for the case of the carbon core and the sulfate around it, the real part has very little affect on it at all.

S. Dattner, *(Texas Air Control Board)*

Did you find that magnetite is brown or black?

T. Ackerman

It is almost as black as carbon. If you make a plot for carbon and magnetite, similar to the one that I have for sulfate, you cannot tell them apart.

P. Mueller, *(Electric Power Research Institute)*

You said something about magnetite being the only other possible absorbing species?

T. Ackerman

No, I said it was the only one which we are aware of that occurs naturally in atmospheric aerosols.

P. Mueller

I can imagine quite a few others . . . like, for instance, several metal sulfides, but I do not know if they would be present in significant quantities.

A. Waggoner, *(University of Washington)*

We have published data on attempts to extract blackness from filters with vigorous solvents such as aqua regia and HCL. These are chosen to extract most other materials including iron oxide, metallic sulfides, and absorbing organic materials. The result is that you can typically remove seventy to ninety percent of the fine particle mass from a filter but you leave 90 % of the blackness on the filter. This is secondary evidence that graphitic carbon is the absorber and the other possible absorbing species are not significant. The one exception was a sample from Barbados which included major amounts of dust from the Sahara which was yellowish. The absorption measured by the integrating plate method was much higher in the blue and linearly decreased with wavelength. That material was removed by the solvent washes.

J.Heicklen, *(Pennsylvania State University)*

Was your conclusion that the carbon would be increasing the average temperature of the earth due to absorption?

T. Ackerman

I did not say that. I would hesitate to say that because you are dealing with a local situation. To extrapolate from an urban area, to the global scale is unjustified even for a large urban area.

J. Heicklen

Where carbon exists, is the absorption more important that the scattering?

T. Ackerman

Climatologically, the absorption is more important than the scattering.

J. Heicklen

So that at least locally, in urban areas, it causes a temperature increase rather than a temperature decrease.

T. Ackerman

That is not clear. You have to be very careful about what happens here. If you decrease the amount of solar radiation reaching the surface, which is what the absorption will do, that will not necessarily cool the surface because the way the surface cools is by sensible heat flux into the atmosphere. Now if you absorb the heat in the atmosphere and do not let it reach the surface, then the surface is not going to cool as much because it's not going to evaporate as much water and it is not going to transfer as much sensible heat. So, it is not a simple relationship. The urban modeling which we have done suggests that the surface temperature changes relatively little due to the presence of aerosols*. Even the temperature in the near-surface layer of the atmosphere may not change much. What does change is the

*Ackerman, T.P., J. Atmos, Sci., Vol. 34, 1977, P 531.

height of the boundary layer and the rate of rise of the inversion. The rate of rise is slowed because the sensible heat flux from below is reduced. Heating occurs near the inversion level due to aerosol absorption rather than a sensible heat flux producing convective plumes within the boundary layer. It is an extremely complicated problem and our inversion rise models are not adequate to model the situation in full detail.

PERTURBATION TO THE ATMOSPHERIC RADIATION FIELD FROM CARBONACEOUS AEROSOLS.

G. E. SHAW

University of Alaska
Fairbanks, Alaska

ABSTRACT

Carbonaceous aerosols from combustion processes enter the atmosphere in the form of submicron particles. Subsequent processes involving coagulation and wet and dry removal give rise to a microparticle size spectrum whose shape continually evolves as the aerosol-laden air mass ages and disperses. After a transport time of $\sim$10-20 days, an asymptotic size distribution function seems to be approached, usually consisting of a single particle size mode with about 80% of the particle mass contained between particle radii limits of 0.06 to 0.3 μm; particles smaller than this range have been removed by coagulation or molecular diffusion processes while those larger have been removed by sedimentation, impaction and nucleation. It is found, empirically, that the carbonaceous aerosol at distant "background" locations, like the Arctic, are mixed with a sulfate aerosol which derives mainly from the nucleation of natural and anthropogenic trace sulfur-bearing gases. In the Arctic, the carbonaceous aerosols constitute 10-30% of the aerosol mass, while in the southern polar regions the percentage is much smaller, presumably due to the remoteness of anthropogenic and natural combustion sources.

The quantity of particles in the air column above the northern polar regions and their size is such that, apparently, significant interactions can occur between the particles and visible band radiative fluxes passing through the atmosphere. This raises the possibility that interactions with the radiation field may influence terrestrial climate by introducing heating into the earth-atmosphere system. It also introduces an opportunity to employ passive ground-based measurements of certain atmospheric optical parameters to deduce characteristics of the aerosols. The theory of determining aerosol parameters is described briefly with the aid of a two-stream approximation to the equation of radiative transfer and its use is illustrated for optical data taken near Fairbanks during an episode of Arctic-derived haze. It has been deduced that the albedo of single scattering for Arctic haze is lower than had been expected — about 0.6 to 0.8 — and that the optical thickness of the arctic aerosol during the spring months can be as large as 0.25 (at 500 nm wavelength). The cause of the arctic haze phenomenon seems to be associated with anthropogenic emissions at the mid-latitudes.

INTRODUCTION

Particles suspended in the free atmosphere over the northern polar regions contain carbonaceous compounds which absorb light and which are responsible for the dark grey coloration of air sampling filters exposed at Arctic latitudes [1, 2]. The particles are sometimes present in enough quantity to cause significant interference to visible band radiation fluxes passing through the atmosphere [3]. If such an aerosol cloud were placed over a snow-free continent, or an ocean, it would appear lighter or darker than the underlying surface, depending upon its scattering and absorption characteristics, and it would not be apparent a-priori whether it would introduce radiative heating or radiative cooling [4-9]. In the case of the arctic aerosols, however, there is no such ambiguity: the suspended particles introduce heating into the earth-atmosphere system. That heating, and not cooling will result is apparent by realizing that even a lightly colored aerosol layer will appear dark when placed over a reflecting ice or snow surface.

Obviously a problem of some practical, as well as scientific, importance is the possibility that carbon-containing haze particles in the Arctic influence the earth's climate. But quantifying this possibility involves evaluating numerous state variables, many of which are only poorly understood, and many of which interact with each other through feedback loops. So it would be premature to hazard a guess right now as to what the climatic alterations — if any — are from Arctic haze or what they could be if arctic haze becomes stronger in the future. In this paper certain radiative interactions caused by the carbon-containing aerosol spread out over the polar cap are considered which, because of the above-mentioned complications, stop short of predicting possible effects on the climate. Instead, at this preliminary stage in the research on Arctic haze, the optical interactions between Arctic aerosol particles and visible band radiation are discussed with a different idea in mind, namely to relate radiometric measurements of sky brightness, downwelling diffuse sky radiation, and atmospheric optical extinction, to the physical and micro-optical characteristics of the haze-laden air. This paper discusses the theory of such optical sensing methods and illustrates its use when applied to actual data taken during a spring haze episode in the Alaskan Arctic.

Some Pertinent Background Information on Arctic Haze — The earliest known report of enigmatic haze over the northern polar regions is a paper by Mitchell [10] describing visual observations made during the 1950's from Air Force weather reconnaissance flights in the Alaskan section of the Arctic. The haze then, as now, was pervasive and quite homogeneous over distances of $\sim$ 1000 km. From the sky color effects observed, Mitchell deduced that the responsible agent was light interacting with very small particles, on the order of size of the wavelength of visible light.

Fig. 1 illustrates a surprising and fascinating feature of Arctic haze. The haze undergoes a large and repeatable seasonal change in intensity, with maximum concentrations in spring and minimum concentrations in summer [11]. This empirical finding is surprising for the simple reason that it is opposite to that found at mid and low-latitude locations (see for example the seasonal variations in atmospheric

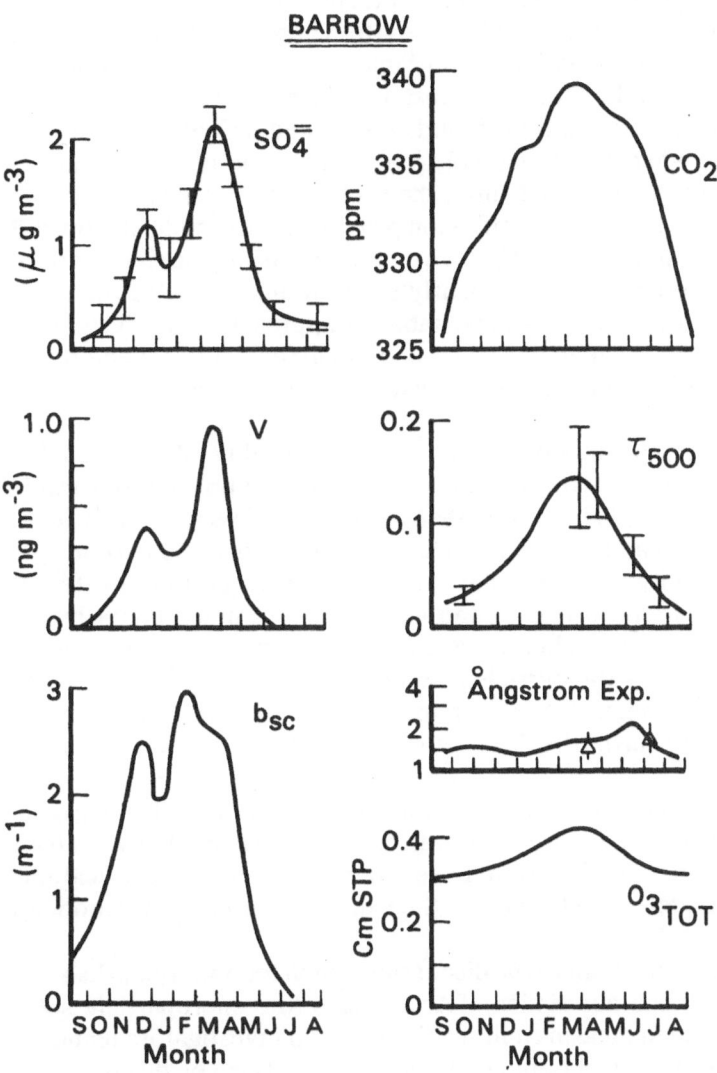

Fig. 1. Seasonal variations of sulfate aerosol, carbon dioxide, particulate vanadium, air column optical thickness, (500 nm wavelength), nephelometric scattering coefficient, Angstrom power law exponent and columnar ozone, measured at the Geophysical Monitoring for Climatic Change station at Barrow, Alaska.

turbidity in a paper by Flowers, *et al.* [12]). The reason behind the large seasonal variation of Arctic haze is probably connected with seasonal variation in particle removal mechanisms (atmospheric stability, cloudiness, changes in precipitation patterns, etc.) and, probably to a lesser extent, to seasonal variations in the transport pathways [13].

Another important discovery about Arctic haze is that its source is to at least a large extent *anthropogenic*. This has been deduced from the large relative proportions of pollution-associated tracers found in haze particles like vanadium. manganese and carbonaceous material [1,3,14-16]. There is also a tendency for Arctic haze to occur in association with poleward injections of air masses from mid-latitude locations in Europe, the USSR and eastern North America[17].

In comparison to conditions found at most mid-latitude, low elevation continental locations, where fresh aerosol from a multitude of sources are changing and mixing rapidly in space and time, the arctic haze is considerably more homogenized. This is suggestive of a diffusion process, for example like the diffusion of ink in water where initially one observes individual dispersing ribbons of color which in time blend together to cause a nearly uniform mix. Because the analogy suggests itself so strongly it seemed reasonable to approach arctic haze as a problem in diffusion theory. Such an approach is, of course, statistical and its use eradicates fine detail, but it also tends to clarify and emphasize certain major features which cause the phenomenon.

By using a diffusion theory approach (through eddy diffusion coefficients which have been estimated for the troposphere; (see for example Kao [18] and Czeplak and Junge [19]), Shaw [11] showed that anthropogenic-sulfur emissions at the mid-latitudes indeed can provide the necessary amounts of aerosol in spring at Arctic latitudes, but only provided that the sulfur residence time in the atmosphere is around 20 days in spring and around 1 or 2 days in summer. I have hypothesized that seasonal alterations in cloudiness over the polar regions may be one important variable that causes the particle residence time to vary by an order of magnitude from winter to summer. During summer the Arctic is covered by the most persistent cloud system on earth (the Arctic stratus) and during winter skies are clear about 80% of the time.

Other approaches have also been used to understand the seasonality of Arctic haze, chief among them are back trajectory analysis methods. In trying to understand a phenomenon like Arctic haze, where many of the classical meteorological theories or models are strained to their limit, a variety of methodologies may be needed.

Since they are relevant to the discussion which follows, remarks are now directed to the physical characteristics of Arctic haze: the microparticle size distribution spectrum has been measured in the field by two investigators using different techniques: this author used optical inversion methods (to be discussed in more detail later) and Bigg [20] counted and classified samples of haze particles with an electron microscope. Both investigators arrived at nearly identical results (Fig. 2). The outstanding feature of the particle size spectrum is that particles in the 0.06 to 0.3 μm radius range constitute around 80% of the total suspended material in the air. Bigg [20] also determined, by thin film chemical analysis methods, that the majority

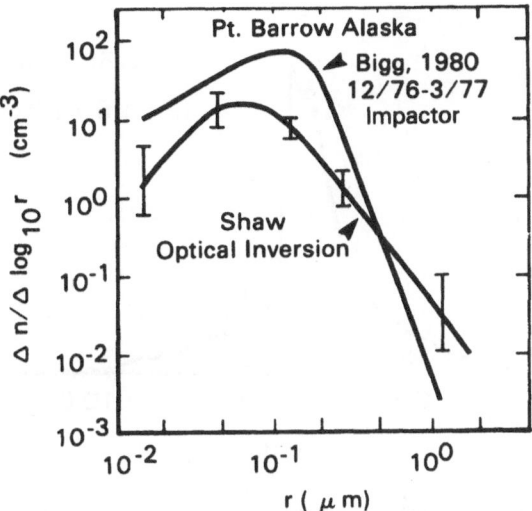

Fig. 2. Particle size spectra measured at high latitude. The solid curve was obtained by classifying particles collected by direct impaction at the surface [20]; the dotted curve was obtained from inverting aerosol optical extinction data subject to the constraint that particle number concentration agree with that measured simultaneously at the GMCC site. It is assumed that the aerosol was distributed homogeneously within a 2 km thick layer.

of particles in arctic haze are hydrated droplets of sulfuric acid. However, filters exposed at Pt. Barrow in the Alaskan Arctic when winds from the north become noticebly grey, sometimes when only 10 or 20 m³ of air passes through them. Patterson and Rahn [15], and Rosen [1] deduced that the coloration is due to *carbonaceous* material collected on the filter. It is estimated [2] that ∿ 10% of the total aerosol mass is carbon, which has profound ramifications when it is realized that,

1. particles around 0.1-0.3 μm in size interact *most efficiently* with optical band radiation,

2. that the black particles, even though they are a minority species, when placed over reflecting snow and ice reduce the planet's high latitude albedo and could therefore be of climatic significance.

A Coincidence of Nature Regarding Aerosols at "Background" Locations — Fig. 3 shows the optical extinction coefficient arising from 50 μg/m³ monodisperse clouds of particles of radius, r. The calculations (which are done using Maxwell's equations) show that particles in the size range ∿ 0.1-0.3 μm radius are most effective at removing optical energy, a fact easy to understand qualitatively. A cloud of particles which are large with respect to light wavelength (when r $\geqslant$ λ, geometric optics applies) have a decreasing cross section area to volume or weight ratio, whereas small particles (r $\ll$ λ) scatter light due to oscillating induced dipolar moments, as described by Rayleigh scattering theory. The efficiency of scattering decreases as particle size decreases and approaches an r⁶ dependency. In between, namely at

References pp. 67-68.

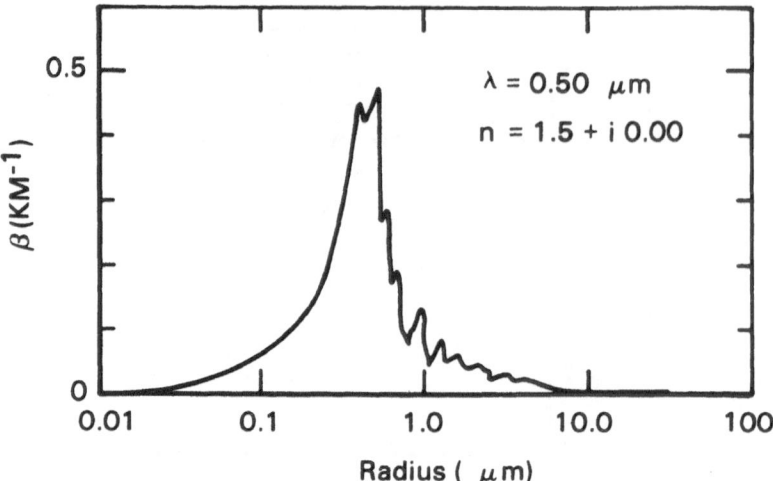

Fig. 3. Computed aerosol volume extinction coefficient, β, arising from 50 μg m^{-3} mass loading of monodisperse sulfate particles of radius r. For a given mass loading, particles of 0.3 μm radius are most effective at attenuating light.

particle radii around 0.1 μm, the particles resonate with visible-band radiation and interact with light most efficiently.

Now, coincidently, particles about 0.1μm in size are *also* those particles which remain airborne in the atmosphere for the longest times. Particles much smaller than 0.1 μm radius are constantly moving under Brownian influences and thus collide and attach to hydrometers or surface obstacles or to each other. Particles much larger than 0.1 μm radius, say 10 μm in radius, are massive enough to overcome constraining viscous forces and slip through the aerodynamical stream and impact on hydrometers or on the surface. But particles near 0.1 μm in radius do not diffuse effectively nor do inertial effects play much of a role, thus they remain in the atmosphere for the longest times, on the average, and tend to congregate at background locations.

Though Arctic haze seems relatively innocuous when its mass loading is compared to urban aerosol, its particle size, its composition, (containing absorbing particles), and the large area it affects turn it into something that must be given serious attention in the atmospheric sciences. It is interesting to point out here that, in order to have an equivalent optical thickness, blowing dust from sandstorms or giant sea salt particles from bubble rupturing by wave action in the seas (about 10μm in size) would have to be enhanced by about an order of magnitude in mass concentration! But even if this were the case, their optical effects still would not necessarily be competitive with arctic haze because the *residence* time for such giant particles is short.

Though Arctic haze is found in relatively small ($\sim$1 to 10 μg/m³) mass loads, it is important to realize that arctic haze remains in the atmosphere for long times ($\sim$20d) and, as anyone who has observed an Arctic haze episode can attest, causes readily apparent visual manifestations.

OPTICAL SENSING OF THE ABSORPTIVE PROPERTIES
OF SUSPENDED AEROSOL

What might be learned from measuring photon flux from the sun and from the sky in an atmosphere strongly affected by Arctic haze? Of capital importance, of course, is to have a measure of "haziness" which, for convenience and for eventual climatological considerations is best provided by the monochromatic *optical thickness* of the aerosol-laden atmosphere [21]. Concentrating on one narrow optical spectral region, the sky brightness, or even more relevant, the downwelling *flux* of radiant energy from the sky passing through a horizontal surface, will obviously depend on the "color" or absorption characteristics of the bulk material comprising the individual suspended particles. Strong light absorbing particles would darken the sky and lower the downwelling monochromatic flux; clear particles, like hydrated sulfuric acid droplets, would brighten the sky. Expressed per unit of haziness, or optical thickness, (remembering to define these quantities at individual wavelengths, since everything is strongly variable with wavelength) the brightness of the sky indicates, indirectly, the particle absorption or more precisely the albedo of single scattering, ω [22]. But unfortunately, a complicating factor arises: the *size* of the haze particles and, to a lesser extent, the reflectance of the underlying surface, which affect the downwelling diffuse radiation. These can be dealt with however.

The Schuster-Schwarzschild two-stream approximation is a useful technique to analyze diffuse sky intensity. It assumes that the sky intensity is isotropic in the downward-directed hemisphere and upward-directed hemisphere. Introducing this assumption into the equation of radiative transfer [23] and performing appropriate integrations [24], the equation of radiative transfer splits into the following two first order differential equations:

$$\frac{1}{2} \frac{d\bar{I}^-}{d\tau} = \bar{I}^- + \omega (1-\bar{\beta}) \bar{I}^- + \omega\bar{\beta} \bar{I}^+ + F_o \omega (1-\beta'(\mu_o)e^{-\tau/\mu}{}_o \quad (1a)$$

$$-\frac{1}{2} \frac{d\bar{I}^+}{d\tau} = - \bar{I}^+ + \omega(1-\bar{\beta})\bar{I}^+ + \omega\bar{\beta} \bar{I}^- + F_o \omega\beta'(\mu_o) e^{-\tau/\mu}{}_o \quad (1b)$$

where I^+ is the upward-directed, hemispherically-isotropic, diffuse radiant flux, I^- is the downward-directed, hemispherically-isotropic, diffuse flux, F_o is the solar irradiance entering the top of the atmosphere, ω is the albedo of single scattering, β is the backscattered fraction for incident *diffuse* radiation at optical depth τ, and $\beta'(\mu_o)$ is the backscattered fraction of diffuse radiation for *monodirectional* radiation incident at the sun's zenith angle $z = \cos^{-1} \mu_o$.

The boundary conditions for solving these equations are $I^-(o) = 0$ (sky at the top of the atmosphere is black and $I^+(\tau_t) = A I^-(\tau_t)$, where τ_t is the optical thickness of the entire atmosphere and A is the Lambertian reflectance, or albedo, of the surface. In equation 1, the hemispherical diffuse fluxes, I^- and I^+, and the incoming solar irradiance F_o are monochromatic quantities evaluated at spectral regions lying between optical absorption bands or lines. In regions of the optical spectrum affected by weak ozone absorption, it is satisfactory to replace F_O by F_O' $= F_O e - \tau_C/\mu_O$, where τ_C is the optical thickness of the stratospheric ozone layer due to optical absorption in the weakly-absorbing Chappuis bands.

Intercomparisons have been made between the two-stream solutions and exact solutions to the equation of radiative transfer carried out with the discrete-ordinant method [3] and these indicated that the two-stream approximation is accurate to about 2% for $\tau < 0.1$. In the work reported here, most measurements were conducted at red wavelengths where the condition $\tau < 0.1$ was met.

Note at this point that a measurable quantity I^-/F_O, at any particular absorption-free wavelength, is related by the solution of equation 1 to the variable ω, β, and β' (μ_O) (it is assumed that A and τ can be estimated from measurements).

THE BACKSCATTERED FRACTIONS β AND β' (μ_O)

β (backscattering of hemispherically-diffuse radiation) and $\beta'(\mu_O)$ (the backscattered fraction from incident unidirectional radiation at angle $z = \cos^{-1} \mu_O$) are both integrated moments of the scattering phase function, $P(\Theta)$, [25] where Θ is a primary scattering angle. The scattering phase function, in turn, is dependent on the size and composition of the suspended particles in the atmosphere. To be more specific and to remove ambiguity which sometimes arises in the definition of β and β' (μ_O), the definition of these quantities is explicity stated below:

$$\beta'(\mu_O) = \frac{1}{2} \int_0^1 \bar{p}(-\mu; \mu_O) \, d\mu \tag{2a}$$

$$\bar{\beta} = \int_0^1 \beta'(\mu) \, d\mu \tag{2b}$$

where P is the azimuthally integrated phase function,

$$\bar{P}(\mu, \mu') \equiv \frac{1}{\pi} \int_0^\pi P(\mu\mu' + (1-\mu^2)^{1/2} (1-\mu'^2)^{1/2} \cos \phi) \, d\phi \tag{2c}$$

Note that β and β' (μ_O) are triple and double integrals, respectively, over the scattering phase function P. Wiscombe and Grams [26] reduced the integrals defining β and β' (μ_O) to equivalent single integrals and evaluated them for Henyev-Greenstein [27] scattering phase functions, which have the functional form,

$$P_H(\mu) \equiv \frac{1 - g^2}{(1 + g^2 - 2\mu g)^{3/2}} \tag{3}$$

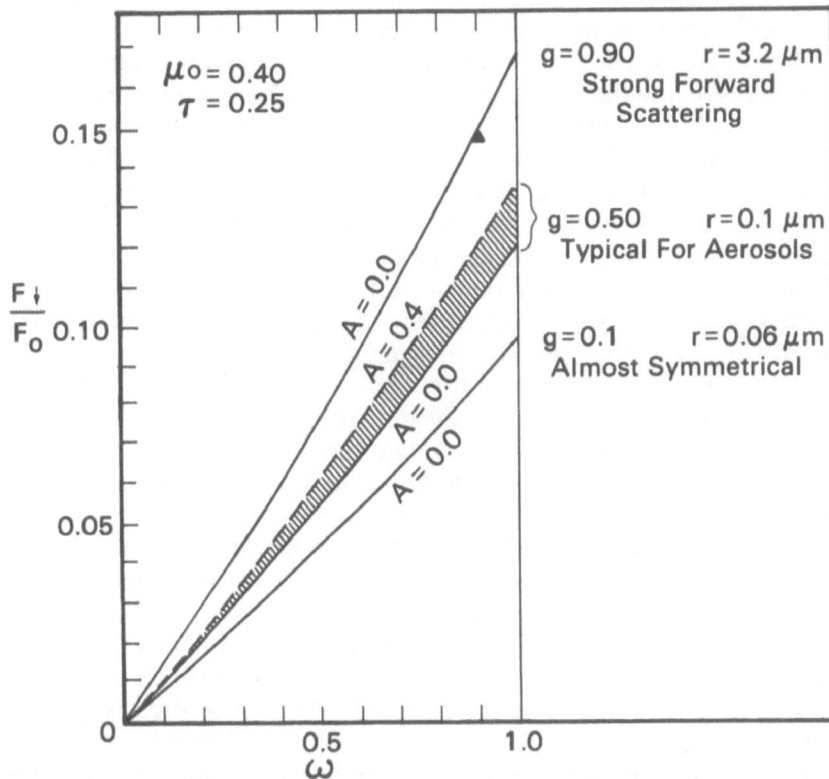

Fig. 4. Downwelling monochromatic diffuse radiant flux as a function of ω computed with the two-stream approximation for a plane parallel atmosphere with $\mu_0 = 0.4$, $\tau = 0.25$. A is the albedo of the underlying surface. Each value of g, the asymmetry factor, could arise from equivalent, monodisperse, aerosol clouds with particles of radius r [26].

where g, the asymmetry factor, lies in the range $0 < g < 1$. The Henyey-Greenstein phase function can be used to replace the Mie phase function in multiple scattering calculations with no more than a few percent error in computed fluxes [28]. Again, in principle, one can be more exact by using Mie theory if the extra complexity of calculations can be tolerated. Wiscombe and Gram's [26] calculations of β and β' (μ_0) transfer the dependency of I $(\beta, \beta' (\mu_0), \mu)$ to one less variable; I (g, ω).

DISCUSSION ON THE METHOD OF DECOUPLING I (g, ω) TO DERIVE g AND ω

Diffuse downwelling irradiance from the clear sky is the reappearance of radiation which has been removed from the direct solar beam by scattering. When particles are *large* with respect to their illuminating radiation, Fraunhofer diffraction theory applies and would indicate that the greatest portion of radiation removed by scattering is concentrated in a beam of angular radius $\Theta = \sin^{-1} (0.61 \, r/\lambda)$. For particles 1 μm in radius this means that about 80 % of the beam is scattered into a cone of half angle 20° from the sun. Thus, provided the sun is at an elevation angle larger than 20°, most of the scattered radiation would arrive at earth and contribute to the downward diffuse sky flux. If, on the other hand, the scattering particles are *small* with respect to the wavelength of incident light, the scattering phase function approaches symmetry about a plane perpendicular to the incident solar irradiance (the Rayleigh function for example) and half of the scattered radiation is scattered up and half is scattered down. Thus, in the case of pure scattering ($\omega = 1$) and in the optically thin regime where these arguments apply, the diffuse sky flux for a constant solar elevation angle will decrease as particle size decreases, or, equivalently, as g goes from 1 to zero. The conclusion is that when A, μ_0, τ and ω are held constant, I will be dependent on g, and thus on particle size, a fact not considered by Herman *et al.* [29] in their direct-diffuse method.

If particle size is held constant (i.e., g = constant) then it is also true that the downwelling flux of radiation from the diffuse sky will vary as ω varies. In particular, the ratio of scattered-radiation to that removed by scattering *plus* absorption, is, by definition, the single scattering albedo ω. Thus, as ω varies from one to zero, the intensity of the sky would change from some finite value to zero. And finally, when more detail is considered about what happens in the atmosphere one has to also consider the fact that the downwelling radiation will change as the underlying surface albedo, A, changes.

In Fig. 4, the downward-directed monochromatic radiative flux is shown for conditions $\tau = 0.25$, $\mu_0 = 0.4$ (solar elevation angle of 23.6°) and as a function of the particles' single scattering albedo, ω. The curves in Fig. 4 correspond to particle asymmetry factors for HG phase functions of g = 0.1 (almost isotropic scattering), g = 0.5 (typical for aerosols that one might find in the atmosphere at a background location) and g = 0.9 (strongly-peaked scattering phase functions corresponding to scattering for giant particles about 3 μm in radius). As expected from the previous discussion, the value of g is evidently found to seriously affect downward-directed diffuse radiation. The stippled region in Fig. 4 shows the variation that would occur if g were held constant (g = 0.5) and the underlying surface albedo, A, varied from A = 0 (black surface) to A = 0.4.

For the combination of variables used in Fig. 4, which were chosen to be representative for conditions that one might find during a spring haze episode at Barrow, the rate of change in relative sky brightness, dI/I, due to changes in A, g and ω are,

$$\frac{\partial (\ln I)}{\partial A} = 0.35 \tag{4a}$$

$$\frac{\partial (\ln I)}{\partial g} = 0.74 \tag{4b}$$

$$\frac{\partial (\ln I)}{\partial \omega} = 1.39 \tag{4c}$$

In a typical situation, one can experimentally judge (from satellite photos for example, or by actual photometry) A to about $A_0 \pm 0.15$, g to $g_0 \pm 0.1$, which would cause, effectively, an uncertainty in ω of $\omega_0 \pm 0.1$.

ESTIMATION OF THE ASYMMETRY FACTOR, g

The asymmetry factor, g, must be estimated, and in principle can be determined by measuring the size distribution of particles, dn/dr, say by collecting a sample of particles and examining them with an electron microscope, then computing g with Mie scattering theory. The value of g derived in this way for the representative Arctic haze distribution shown in Fig. 2 (for $\lambda = 0.7\,\mu m$) is $g = 0.32$ with an estimated uncertainty of $\Delta g = 0.05$. One can proceed next to evaluate ω by using the two stream method applied to actual measurements of monochromatic diffuse sky radiant flux, assuming that A can be reasonably well estimated, as it can be over a uniform surface like that around Barrow.

An alternative approach to deriving g would be to employ ground-based measurements of the optical transmissivity of the atmosphere at different wavelengths; this works because the wavelength-dependence of aerosol optical thickness (the aerosol optical extinction spectrum) is sensitive to the functional form of the size spectrum of the particles causing the extinction. One works the problem backwards, that is, combines the optical extinction spectrum with an appropriate theory to derive the particle size spectrum, dn/dr. Such formal inversions are not straightforward because of difficulties associated with poorly-conditioned kernels which appear in integral equations, but the difficulties have to some degree been overcome and are discussed in detail by Twomey [30]. An illustration of their use in recovering dn/dr from optical measurements is described by Shaw [31] and Walters [32].

It is not appropriate here to go into the mathematical details of inversion theory; they are adequately described in the references given. But in order to perform the optical inversion one must first specify the real and imaginary part of the particles' refractive index. By knowing that arctic haze is predominantly hydrated sulfuric acid droplets, there is no particular problem in specifying the real part of the refractive index. However, the imaginary part cannot be specified a-priori, since this would be equivalent to specifying the unknown variable ω! What allows an inversion to be carried through, even though ω is unknown, is that the particle optical extinction spectrum is only weakly dependent on the assumed value of the imaginary component of the refractive index, as illustrated by Table 1. One can obtain a reasonably good estimate of dn/dr (from the vector $\tau(\lambda)$), from which g can be computed even though ω is unknown. Once g is known, and assuming that one has a reasonable estimate of A, ω can be recovered by using the analysis described in the last section.

AN EXAMPLE USING EXPERIMENTAL DATA

The theory described in the last sections is now illustrated with an example using measurements made near Fairbanks, Alaska during a moderately intense haze episode in an air mass that had Arctic origin. Hemisperhical monochromatic ($\Delta\lambda = 10\,\mu m$ fluxes of radiant energy, the intensity of the direct solar beam and surface

TABLE 1
Sensitivity of Volume Optical Extinction Coefficient, β,
to Variations in the Imaginary Part, n_i, of the Particle Refractive Index.

n_i	$\beta_T(cm^{-1})$
0.000	91×10^{-8}
0.002	92×10^8
0.005	93×10^8
0.007	93×10^8
0.010	94×10^8
0.020	97×10^8

Data supplied by J. Reagan, 1975. Computations are for a Junge power law size distribution function, $dn/dr \propto r^{-4}$, for particles contained in the radii limits $0.02 < r < 10 \mu m$, with a real refractive index of 1.45 illuminated with light at wavelength 630 nm. The particle density is 2 g cm^{-3} and the particle mass loading is 100 μg m^{-3}.

albedo were measured at ten wavelengths. Since optical measurements like these are quite specialized, the instruments and the accuracy of which they are capable is briefly described.

Sun intensity was determined with a photometer which had been previously carefully calibrated at the Mauna Loa Observatory. The aerosol optical thickness spectrum, which was obtained by measuring the solar intensity at ten wavelengths, is shown in Fig. 5.

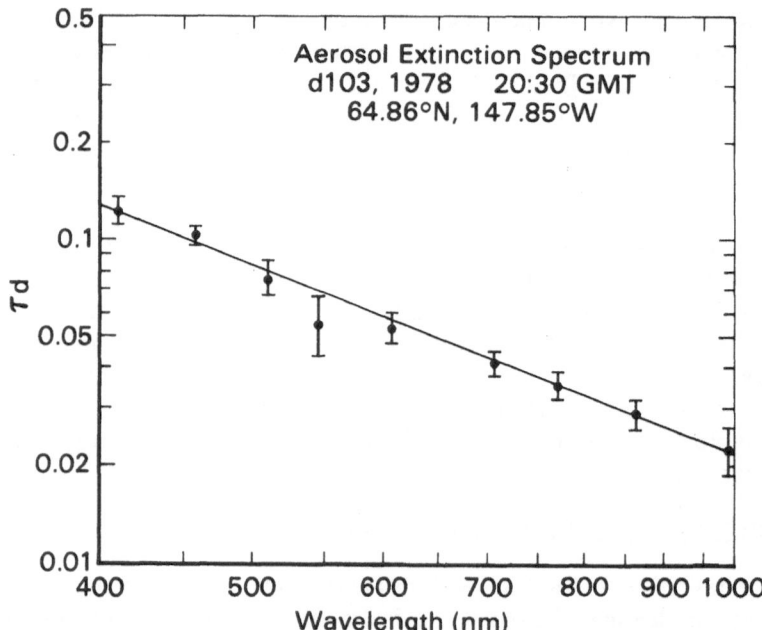

Fig. 5. Aerosol optical extinction spectrum $\tau(\lambda)$, measured near Fairbanks, Alaska during a haze episode.

Measuring the downwelling flux from the sky hemisphere at narrow, essentially monochromatic, wavelengths is a considerably more difficult task. It was done by arranging a stable solid state photodetector to view the radiance of an overlying circular diffusing plate which was accurately leveled, placed above all surrounding objects and illuminated by the sky; direct sun was kept off the diffuser by shading it with a circular disk which had a umbral shadow 1° in diameter and a penumbral shadow of 3°. The radiation which was missing in the near-sun aureole was corrected for (it was typically about 2%) by measuring the aureole's brightness with a photoelectric coronameter [33]. It has been impossible so far to obtain a diffusing material that obeys Lambert's law exactly, so a series of experiments were conducted on various combinations of cylindrical diffusers with circumferential shading rings. A combination was eventually found (Fig. 6) that obeyed Lambert's law to within one percent. Specific wavelengths were isolated with an interference filter placed in the optical path. The instrument was calibrated to 1% accuracy by unshading the diffuser and determining the direct solar irradiance from simultaneous measurements with a sun photometer.

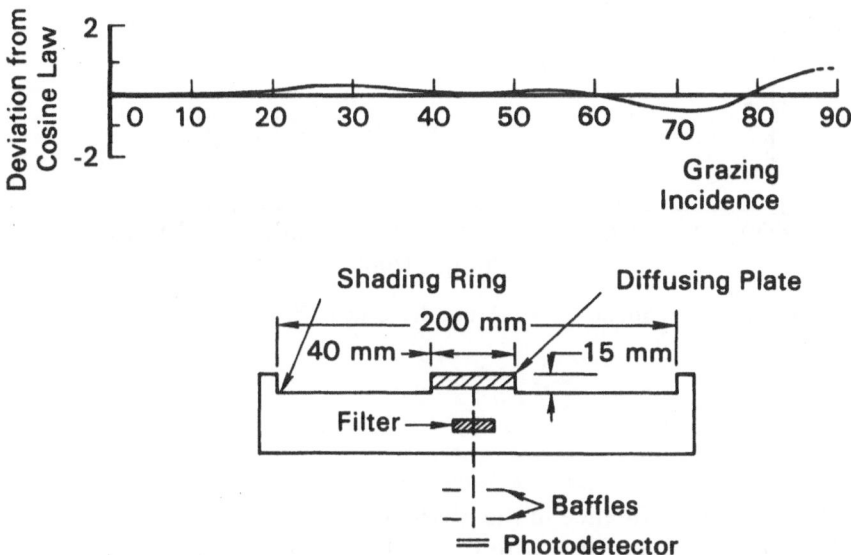

Fig. 6. Characteristics of the sensor used to determine monochromatic diffuse sky flux. Top: deviation of angular response from Lambert's cosine law. Bottom: cross-sectional diagram showing geometric relationships.

The albedo of the surrounding snow surface was estimated by photometry at individual wavelengths conducted from a building 4 m high. This is a weak point in the experiment since the accuracy is not very good (it is $\omega_0 \pm 10\%$) even though the surrounding environment is quite homogeneous.

Application of the two stream theory to the data gave values of ω as a function of

References pp. 67-68.

wavelength shown in Fig. 7. The numerical values are lower than one might expect and it is puzzling that the wavelength dependency is the reverse of what one might expect. Calculations have not yet been made to determine the relative amounts of carbon aerosol (assuming it is mixed in with a clear sulfate aerosol with the same size distribution). This is not thought to be profitable at this time because the increasing ω with increasing λ indicates that systematic errors may have affected the optical measurements. Future measurements will be made with new instrumentation designed so that it will be possible to assess the suspected systematic experimental error. Thus, until the problem is studied further experimentally, one should realize that the curve of $\omega(\lambda)$ in Fig. 7 is somewhat tentative. However, the methodology of inferring $\omega(\lambda)$ described in this paper remains valid.

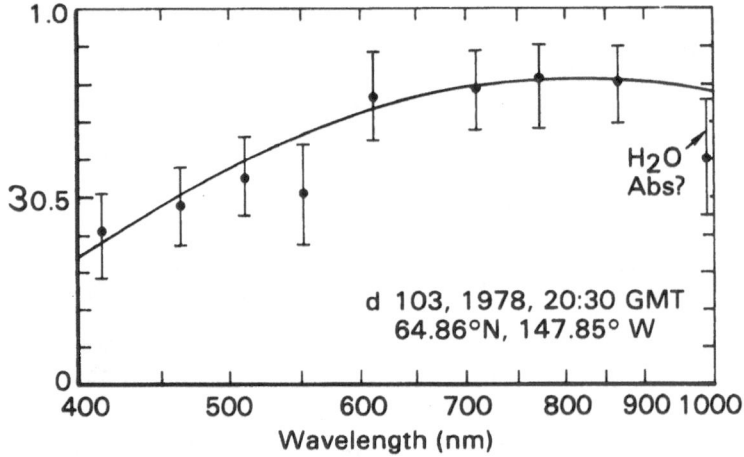

Fig. 7. Single scattering albedo, ω, of Arctic haze near Fairbanks deducted by optical passive remote sensing.

CONCLUSIONS

A certain degree of confidence in the method described comes from comparing the optically-inferred values for ω with those obtained by spectrometric measurements on filters exposed at Barrow during similar haze episodes [15]. Both gave low values of ω. A crustal aerosol would provide $\omega \sim 0.9$ to 0.95 and a pure sulfate aerosol would give $\omega \sim 1$. All this means, basically, that arctic haze appears dark when seen from above against a reflecting snow surface. This, in addition to the fact that the snow's albedo may also be lowered by particles that have fallen out and deposited in it [34], means that the earth-atmosphere system will absorb more sunshine and therefore introduce additional heating at the high latitudes of earth near the equinox times, especially during spring. Shaw and Stamnes [3] calculated that the heating may be as much as 2 kcal month^{-1} cm^{-2} which is significant for climatological consideration, although feedback mechanisms could add or detract from this effect.

The main thing learned is that Arctic haze is an absorbing haze and that passive optical measurements are realistic to carry out, provide useful climatological information, and can be done relatively inexpensively. The two stream method has the very great advantage of being simple. The two stream approximation is always good to revert back to when considering further refinements to the method, or when evaluating systematic errors.   •

Regardless of the degree of sophistication with which one pursues the analysis of radiation measurements, the point should not be missed that in the case of the relatively clear atmospheres, even approximate methods are sufficiently good to show that there is less diffuse sky radiation (about 20 %) than one would expect for a pure scattering aerosol. All the evidence indicates that Arctic haze introduces heating into the high latitudes. With the heating predicted by anthropogenic emissions of CO_2 combined with the heating introduced by Arctic haze, one can only wonder why the planet has been cooling since about 1940. Increasing emissions of carbon aerosol might reverse the situation.

ACKNOWLEDGEMENTS

This research has been sponsored by a grant from the Office of Naval Research (Grant No. N-00014-76C-0435) and by the National Science Foundation (Grant No. DPP77-27242). The author expresses thanks to K. Rahn for helpful remarks during the preparation of this paper.

REFERENCES

1. H. Rosen, "Determination of soot in the arctic haze near Barrow, Alaska," Presented at 2nd symposium on Arctic Air Chemistry, University of Rhode Island, 1980.
2. K. A. Rahn and G. E. Shaw, (submitted for publication).
3. G. Shaw and K. Stamnes, Ann. NY Acad. Sci., Vol. 338 (1980), p. 533.
4. M. A. Atwater, Science, Vol. 170 (1970), p. 64.
5. R. Charlson and M. Pilat, J. Appl. Meteor., Vol. 8 (1960), p. 1001.
6. P. Chylek and J. A. Coakley, Jr., Science, Vol. 183 (1974), p. 75.
7. P. Halpern and K. L. Coulson, J. Appl. Meteor., Vol. 15 (1976), p. 464.
8. J. M. Mitchell, Jr., J. Appl. Meteor., Vol. 16 (1971), p. 703.
9. S. I. Rasool and S. H. Schneider, Science, Vol. 173 (1971), p. 138.
10. J. M. Mitchell, Jr., J. Atmos. Terr. Phys., (1956), p. 195.
11. G. Shaw, "An eddy diffusion model of Arctic Haze," Presented at the 2nd symposium on Arctic Air Chemistry, University of Rhode Island, 1980.
12. E. G. Flowers, R. A. McCormick, K. R. Kurfis, J. Appl. Meteor., Vol. 8 (1969), p. 955.
13. K. A. Rahn and R. J. McCaffrey, Ann. NY Acad. Sci., Vol. 338 (1980), p. 486.
14. J. M. Daisey, R. J. McCaffrey and R. A. Gallagher, "Particulate organic matter in the Arctic aerosol," Presented at the 2nd symposium on Arctic Air Chemistry, University of Rhode Island, 1980.
15. E. Patterson and K. A. Rahn, "Coefficients of absorption of visible radiation by the Barrow, aerosol and their seasonal variations," Presented at 2nd symposium on Arctic Air Chemistry, University of Rhode Island, 1980.
16. C. J. Weschler, "Identification of selected organics inArctic aerosols," Presented at the 2nd symposium on Arctic Air Chemistry, University of Rhode Island, 1980.

17. K. A. Rahn, "The Eurasian sources of arctic aerosol," Norwegian Institute for Air Research Report, Lillestrom, Norway, 1979.
18. S. K. Kao, "Basic Characteristics of global scale diffusion in the troposphere," In: Turbulent Diffusion in Environmental Pollution, F. N. Frenkiel, E.D., eds., Academic Press, New York, (1974), pp. 15-32.
19. G. Czeplak and C. Junge, "Studies of inter-hemispheric exchange in the troposphere by a diffusion model," in: Turbulent Diffusion in Environmental Pollution, H.E. Landsberg and J. Van Mieghem, E.D. eds., Advances in Geophysics Series, Academic Press, New York, (1974), pp. 57-72.
20. E. K. Bigg, J. Appl. Meteor, Vol. 19 (1980), p. 521.
21. G. E. Shaw, J. A. Reagan and B. M. Herman, J. Appl. Meteor., Vol. 12 (1973), p. 374.
22. S. G. Jennings, R. G. Pinnick, J. B. Gillespie, Appl. Opt., Vol. 18 (1979), p. 1368.
23. S. Chandrasekhar, Radiative Transfer, Dover Publications, New York, (1960), p. 393.
24. J. Coakley and P. Chylek, J. Atmos. Sci., Vol. 32 (1975), p. 409.
25. D. Deirmendjian, Rev. Geophys. & Spa. Phys., Vol. 18 (1980), p. 541.
26. W. Wiscombe and G. Grams, J. Atmos. Sci., Vol. 33 (1976), p. 2440.
27. L. G. Henyev and J. L. Greenstein, Astrophys. J., Vol. 93 (1941), p. 70.
28. J. Hansen, J. Atmos. Sci., Vol. 26 (1969), p. 478.
29. B. Herman, R. S. Browning and J. Delusi, J. Atmos. Sci., Vol. 32 (1975), p. 918.
30. S. Twomey, Introduction to the Mathematics of Inversion in Remote Sensing and Indirect Measurements, Elsevier Scientific Publishing Co., New York, (1977), p. 237.
31. G. E. Shaw, Applied Optics, Vol. 18 (1979), p. 988.
32. P. T. Walters, Appl. Optics, Vol. 19 (1980), p. 2353.
33. G. E. Shaw and C. S. Deehr, J. Appl. Meteor., Vol. 14 (1975), p. 1203.
34. S. G. Warren and W. J. Wiscombe, J. Atmos. Sci. (Submitted for publication).

DISCUSSION

J. Ogren, *(University of Washington)*

In addition to the radiative heating terms, you need to look at terms of the eddy heat fluxes into the polar regions in order to really assess the significance of just the solar radiated transfer heating rates.

K. Rahn *(University of Rhode Island)* *

We are aware that eddy heat fluxes must eventually be considered in evaluating climatic changes. Our direction has been so far limited to assessing changes in the solar heat flux terms.

J. Heintzenberg, *(Stockholm University)*

In summer the Arctic is mostly covered by stratus clouds. Does this situation apply to that time of year?

K. Rahn

These calculations are strictly for springtime.

Editor's Note: Since Prof. Shaw was unable to attend the conference, Prof. Rahn, a co-investigator of the Arctic Haze phenomenon, gave this paper and answered the questions after the presentation.

J. Heintzenberg

The carbon, according to my limited number of samples and Hal Rosen's results, disappears in summer. There is essentially no absorbing aerosol during the summer.

K. Rahn

Correct. I agree with you. A point not mentioned is that we are limiting our discussion to the spring season when you have both sun and aerosol. In the winter there is aerosol but no sun. In the summer there is sun but no aerosol.

R. Charlson, *(University of Washington)*

I think it is interesting to speculate as to reasons why the aerosol concentrations are this large in the Arctic. A possibly important explanation is that the Arctic is a desert. There is essentially no sink term functioning. There is a lack of precipitation and cloud activity in the winter. You notice the aerosol concentrations are much lower as soon as those Arctic air masses collide with the maritime air just outside the Arctic region. The extensive distribution in the Arctic may very well be due to a lack of a sink term and results in much longer transport distances.

K. Rahn

I agree, but there is more to it. The time of the maximum aerosol concentration is spring, which is the time of minimum precipitation in the Arctic. But, one must transport first the aerosol to the Arctic from the sources, so the conditions along the path at different times of the year must be considered. That is a much more complicated subject but the answer seems to be related to a coincidence of nature. Out of all the possible pathways by which air reaches the Arctic, there is only one which is really suitable for transporting aerosols, and that is the one from Eurasia, one of the world's most polluted regions. Along almost the entire length of this pathway there is very little precipitation, particularly during the winter.

M. Shelef, *(Ford Motor Company)*

In your presentation you said that the aerosol is mostly sulfuric acid. Are there any cations associated with the sulfate?

K. Rahn

The work on Arctic aerosols has been rather limited in that regard. We have not measured cations directly. I do not know of any thorough, on-going measurements of cations. I have seen limited work undertaken by the Canadian Atmospheric Environment Service. Normally, in mid-latitudes one thinks of a mixture of ammonium and hydrogen ions. I cannot say for certain what it is in the Arctic aerosol. It is probably weighted toward the hydrogen ion, but we do not know.

G. Wolff, *(General Motors Research Laboratories)*

What concentration of elemental carbon and what values of b absorption are we talking about?

K. Rahn

It is on the order of a few tenths of a microgram per cubic meter during much of the winter. This is derived from converting measurements of absorption on filters to carbon concentrations using a conversion factor. There is a lot of discussion on what the conversion factor is.

H. Rosen, *(Lawrence Berkeley Laboratory)*

It ranges between about 0.2 to 0.5 microgram per cubic meter in the springtime.

G. Wolff

Note added in printing concerning the concentrations of elemental carbon at Pt. Barrow. Since the symposium was held, 10 high volume filter samples collected in Pt. Barrow at the NOAA-GMCC site were analyzed for elemental carbon by GMR. The filter media was spectrograde and the technique was the one used by Wolff et al.* on the Luray, Va. samples. The results are summarized below.

SUMMARY OF CARBON MEASUREMENTS AT PT. BARROW, ALASKA

Starting Date	Starting Time	Ending Time	Mass $\mu g/m^3$	Elemental C $\mu g/m^3$	Elemental C ÷ Mass
3-24-80	1150	1025	5.47	0.30	0.05
3-25-80	1030	1400	5.36	0.26	0.05
3-26-80	1405	0810	7.31	0.33	0.05
3-27-80	0820	0910	5.36	0.25	0.05
3-28-80	0915	0910	6.02	0.27	0.04
3-29-80	915	1120	—	0.25	—
3-30-80	1110	0925	3.00	0.27	0.09
3-31-80	0930	0925	6.97	0.29	0.04
4-1-80	0930	1045	—	0.29	—
4-2-80	1050	0735	6.96	0.31	0.04
MEAN			5.81	0.28	0.05

K. Rahn

That makes carbon slightly less than 10% of the mass of the total sub-micron fraction.

---

*Wolff et al., These Proceedings.

N. Kelly, *(General Motors Research Laboratories)*

You said that the filter paper was degraded by sulfuric acid. In aged aerosol like that, why would it not be neutralized by background ammonia?

K. Rahn

Professor Brosset's work in Sweden has shown that the ammonia concentration over continents is dramatically less in winter than it is in summer. We are talking about winter transport or later spring transport, well before the ammonia has had a chance to rise. There just is not enough ammonia around to neutralize the acid.

P. Coffey, *(New York State Department of Environmental Conservation)*

These were high volume samples. How do you separate the local combustion sources from long range transport since you have some intense inversions around Point Barrow in winter?

K. Rahn

For many constituents you do not have to worry about local pollution. Sampling at Barrow is sector-controlled. There is a clean sector of approximately one-third of a total circle where the air has not passed over any substantial human activity. When the flow controllers function properly, our sample comes strictly from that sector. During the first year the controller worked but during the second year it did not. However, we got virtually the same answers for both sulfate and for vanadium. Subsequently, work by the NOAA and GMCC groups have shown that there is little or no directional dependence of light scattering which is another measurement of fine aerosols. The village of Barrow has a population of about 3,000 people. The life style is rather simple. There is no industry and there is a limited number of automobiles. Consequently, there is not a substantial local affect.

V. Mohnen, *(State University, Albany, New York)*

I agree with the ammonia assessment you make. The data that we saw in the northeast United States throughout a three year period show low winter concentrations and low winter deposition of NH_3. However, I think that your conclusion about cations premature unless you have evidence on the nature of the acid, which, you say, is sulfuric acid. I would say from the experience of the MAP3S network that nitric acid is the predominant acid in the wintertime, sulfuric acid in the summertime. If this is reversed in Alaska this would be quite interesting.

K. Rahn

It is reversed in the winter. We have not measured it but it has been measured by Len Barrie of Atmospheric Environment, Canada. There is ten times as much sulfate as nitrate in the wintertime at Mould Bay which is roughly similar in location to Barrow. In the summertime there is more nitrate but it seldom reaches the concentration of sulfate. So this is really a sulfate acidity question and not a nitrate acidity question. We may be dealing with loss of nitrate and loss of the more volatile nitric acid from the particles because of its huge acidity which is presumed, and I stress presumed, to build up in these particles during transport.

L. Atkinson, *(U.S. Environmental Protection Agency)*

Could you elaborate on the trajectory analysis method?

K. Rahn

That is very hard to do in a sentence or two. At this point, it is almost impossible to construct meaningful trajectories for Barrow, Alaska backwards to the source regions. The distances are too long, between 5,000 kilometers minimum, 10,000 kilometers maximum. The travel time is between five and twenty days — some people feel longer. During the lst few thousand kilometers when they pass over the Arctic ice pack, there are no stations of any kind. Everything has to be done by extrapolation to generate the basic maps from which the trajectories were calculated. It is not possible to connect Barrow definitively to any sources using the meteorological data we have now. We have taken the alternate approach of studying the overall air circulation in that region. We know how the air flow comes into Barrow and we know how it leaves various other areas. There is really not much question that during most of the winter, the basic air flow over the polar ice pack is from the Siberian side across to the North American side and the flow is very strong. Beyond that, we rely on chemical data to link Barrow to the presumed source. There is very strong chemical evidence that the Barrow aerosol is compatible with Eurasian sources and incompatible with sources in the eastern part of North America.

C. Brosset, *(Swedish Water & Air Pollution Research Institute)*

I would just like to comment on Ken's discussion of my work. I think there is some misunderstanding. If you have sulfate in your sample, you can think of it as sulfuric acid which has equilibrated with water pressure, ambient pressure and nitric acid pressure in the air. These equilibria determine the composition of the droplets and it is always a droplet even if you have crystalization conditions because it usually is supercooled or supersaturated.

J. Hummel *(General Motors Research Laboratories)*

In Tom Ackerman's talk*, he had essentially a wavelength independent single scattering albedo; but in your results there is clearly a strong wavelength dependence in the single scattering albedo. Do you know why your results differ?

*Bergstrom, R.W., Ackerman, T.P. and Richards, L. W., These Proceedings.

K. Rahn

No.

T. Ackerman, *(NASA Ames)*

I would like to comment on that point. I think there is a problem with the inversion. If you look at the size distribution that Shaw got from his inversion, there seems, in my opinion, to be too many small particles. There may be a problem in Shaw's assumptions. I do not think that his albedo of single scattering at wavelength 0.4 microns is valid.

A. Waggoner, *(University of Washington)*

Shaw's results seem to me to be inconsistent. The aerosol optical depth measured from the sun photometer is higher in the blue than it is in the red. Therefore, there is more scattering in the blue which is consistent with the size distribution that Shaw is suggesting is present. If you have an aerosol that has higher scattering in the blue than the red and it has the same absorption in the blue and red, then the albedo would be higher in the blue than in the red. The results presented are exactly opposite.

G. Shaw

A complete error analysis has been conducted including the evaluation of systematic errors; I stand behind the results. There is more scattering in the blue spectral regions.

W. Richards, *(Meteorology Research, Inc.)*

Fine particle carbon blacks are typically brown and experimentally, for very small carbon particles, the absorption is stronger in the blue than in the red.

75

THE IMPORTANCE OF ELEMENTAL CARBON
SESSION SUMMARY

George M. Hidy

Environmental Research & Technology, Inc.
Westlake Village, California

I would like to take this opportunity to summarize the first session. The title of this session was the importance of elemental carbon, or other particulate carbonaceous material, in the atmosphere. I think from the discussion this morning by Professor Charlson we can view the carbon component of the aerosol as a universal part of the general phenomenon in the atmosphere, not just an urban pollution question. As part of the general aerosol, particulate carbon is exposed naturally to all of the physical processes that influence aerosol behavior in the atmosphere. We also have had some allusion to the fact that the carbon in the atmosphere is chemically reactive in some way, either as a catalyst for other reactions or as a reactant itself, whereby it is oxidized to a chain of products.

Dr. Novakov has indicated that there are many ways to define particulate carbon, and the terminology, as others have pointed out, is fairly loose at this point. We have a nomenclature that ranges from elemental or graphitic carbon to soot, to total non-carbonate carbon and various solvent soluble fractions. Soot could include only black carbon or it could include the primary organic fraction. This component also encompasses a variety of organic materials, which have been characterized only in a limited way. Finally, we have the secondary organic carbon fraction which is produced by chemical reactions; these are probably relatively highly oxygenated materials of certain specific form, which have not been studied extensively.

Dr. Novakov unfortunately did not spend very much time on the considerable amount of work he and his colleagues have done looking at carbon or soot or what have you as catalytic factors in acid gas reactions that take place in the

Reference p. 76

atmosphere. These reactions include the oxidation of sulfur dioxide in the presence of carbon in liquid droplets or conceivably on carbon that is somewhat dry. I would like to make a comment on our experience in applying the reaction rates derived from these laboratory experiments to Los Angeles. The rate expression from the Livermore work shows that the SO_2 oxidation reaction is nearly zero order in SO_2. This implies that if you have any reactive carbon or catalytic carbon present in an urban atmosphere, you have, in effect, an area source for sulfate which is basically independent of the SO_2 present. We have added to our calculations a term which includes the oxidation rate of SO_2 by this mechanism for a test case in Los Angeles (1). Using this addition, we have attempted to explain anomalies which are well known in numerical modeling of Los Angeles sulfate behavior. The anomaly basically relates to the fact that there is more sulfate present on the west side of the Los Angeles basin than can be accounted for by daily carryover, or reaction mechanism proportional to SO_2 concentration. We find that although the carbon mechanism for sulfur dioxide oxidation does not substantially increase the amount of sulfate found in the total volume of air over the basin, it does provide a possible explanation for a significant part of the western Los Angeles basin sulfate anomaly. Thus, there is an interesting potential link between carbon and sulfate behavior in an urban atmosphere, which has at least one practical application.

Another area, which was discussed this morning, is the issue of carbon's importance to the optical properties of the atmosphere. It is quite clear that black carbon potentially has a substantial influence on the overall light extinction of the atmosphere, because of its absorptive properties. I think you are all aware that carbon has a "double bang" for optical properties in the sense that it both scatters and absorbs light. In effect, carbon creates an optical effect which is roughly twice that of other nonabsorbing aerosol material per unit mass.

Finally, we have seen that at least in certain urban conditions and potentially in the Arctic region, there is an incremental heating rate associated with an absorbing aerosol. The carbon aerosol potentially can be a significant factor in trapping pollutants in the urban situation by intensifying and maintaining an inversion aloft through differential heating of the air layers. If we can accept the numbers of Dr. Shaw, a potentially large radiation perturbation can be expected in the Arctic which may be the same order of magnitude as the CO_2 absorption effect.

In conclusion, we can see that from this initial portion of the symposium, we have very good reason for being here to talk about the environmental consequences of particulate carbon in the atmosphere.

REFERENCE

1. *Henry, R. C., D. G. Godden, G. M. Hidy and N. J. Lordi, 1981. "Simulation of Sulfur Oxide Behavior in Urban Areas." 3rd Year Report, Sulfate Task Force, American Petroleum Institude, Washington, D.C.*

SESSION II
ANALYTICAL MEASUREMENT TECHNIQUES
AND CHEMISTRY OF ELEMENTAL CARBON

Session Chairman

C. BROSSET

Swedish Water and Air Pollution Research Laboratory
Gothenburg, Sweden

ANALYSIS OF ORGANIC AND ELEMENTAL CARBON IN AMBIENT AEROSOLS BY A THERMAL-OPTICAL METHOD

J. J. HUNTZICKER, R. L. JOHNSON, J. J. SHAH and R. A. CARY

Oregon Graduate Center
Beaverton, Oregon

ABSTRACT

A thermal-optical method has been developed for the analysis of organic and elemental carbonaceous aerosol on glass or quartz fiber filters. Organic carbon is volatilized in two steps: at 350°C in an O_2 (2%)-He mixture and at 600°C in He. The volatilized organic carbon is oxidized to CO_2, reduced to CH_4, and measured by a flame ionization detector. Elemental carbon is combusted to CO_2 in O_2 (2%)-He at 400, 500, and 600°C, and the CO_2 is measured as above. The reflectance of the filter segment, which is continuously monitored with a He-Ne laser system, decreases during the organic analysis because of pyrolytic conversion of organic to elemental carbon and increases during the combustion of elemental carbon. Correction for pyrolytic production of elemental carbon is accomplished by measuring the amount of elemental carbon oxidation necessary to return the filter reflectance to the value it had before pyrolysis occurred. This is facilitated by the slow, three step elemental carbon combustion process. All switching of gas flows, timing, temperature control, pyrolysis correction, analog to digital conversion electronics, electrometer functions, signal integration, data storage, and data outputs are controlled by a microcomputer system built around a Motorola 6802 microprocessor. The instrument has been used to measure organic and elemental carbon concentrations at 42 urban sites in the United States.

INTRODUCTION

The recognition of carbon as a significant component of atmospheric aerosols has stimulated research into new analytical methods for aerosol carbon. Methods currently in use range from detailed gas chromatographic-mass spectrometric measurements to determination of total aerosol carbon by combustion or other

References pp. 84-85.

means. Because of the complexity of the carbonaceous aerosol, analytical methods which separate the aerosol into organic, elemental (i.e., soot), and carbonate classes have the potential for becoming valuable tools in aerosol characterization studies.

Of the three classes, analysis for carbonate carbon is the easiest and is accomplished by acidification of a filter segment and measurement of the evolved CO_2 [1]. Speciation between organic and elemental carbon generally relies on either solvent extraction [2-9], selective volatilization [2, 10-13], or optical methods [14]. Grosjean [15] has emphasized that current analytical procedures for elemental carbon rely on "operational" or method-dependent definitions of elemental carbon. For example, solvent extraction methods generally assume that solvent-insoluble carbon corresponds to an upper limit to elemental carbon. Methods involving selective volatilization assume that all carbon not volatilized from a filter at some arbitrary temperature (e.g., 600°C) in an inert atmosphere is elemental carbon. Finally, optical methods assume that the only (or predominant) absorber of visible light in the aerosol is elemental carbon. Ideally, all methods should converge on the same value for elemental carbon, but insufficient comparison studies have been performed to verify this.

THE OREGON GRADUATE CENTER CARBON ANALYZER

Developmental work on an organic and elemental carbon analyzer has been in progress at the Oregon Graduate Center since 1976. The initial configuration of the carbon analyzer was conceptually simple [10]. An aerosol sample on a glass or quartz fiber filter segment was heated to 600°C in a He atmosphere for the purpose of volatilizing organic carbon. The volatilized organic carbon was oxidized to CO_2 in a MnO_2 bed at 950°C, reduced to CH_4, and measured by a flame ionization detector (FID). Elemental carbon was measured by introducing O_2 into the sample oven and combusting the carbon to CO_2 which was measured as above. A difficulty in this method of speciation was discovered, however; namely, during the organic analysis a fraction of the organic carbon was pyrolytically converted to elemental carbon. This was manifested by an increase in the "blackness" of the filter at the end of the organic analysis.

To overcome this difficulty, the system which is now in use was developed. It involves modifications in the combustion method itself, the addition of an optical system for the continuous monitoring of filter reflectance, and the automation of the analytical sequence by microprocessor technology. The system is shown schematically in Fig. 1. As described below, the optical system is used to correct for the pyrolytic conversion of organic to elemental carbon. Dod *et al.* [16] have also developed a system to monitor optical attenuation of the filter during combustion but have used optical transmission rather than reflectance.

In our current method, four filter disks (0.25 cm² each) are placed in the quartz sample boat, the combustion zone temperature set to 350°C, and the oven purged with an O_2 (2%)-He (98%) mixture. The boat is then inserted into the combustion zone in which oxidation and volatilization of organic carbon into the flowing O_2-He

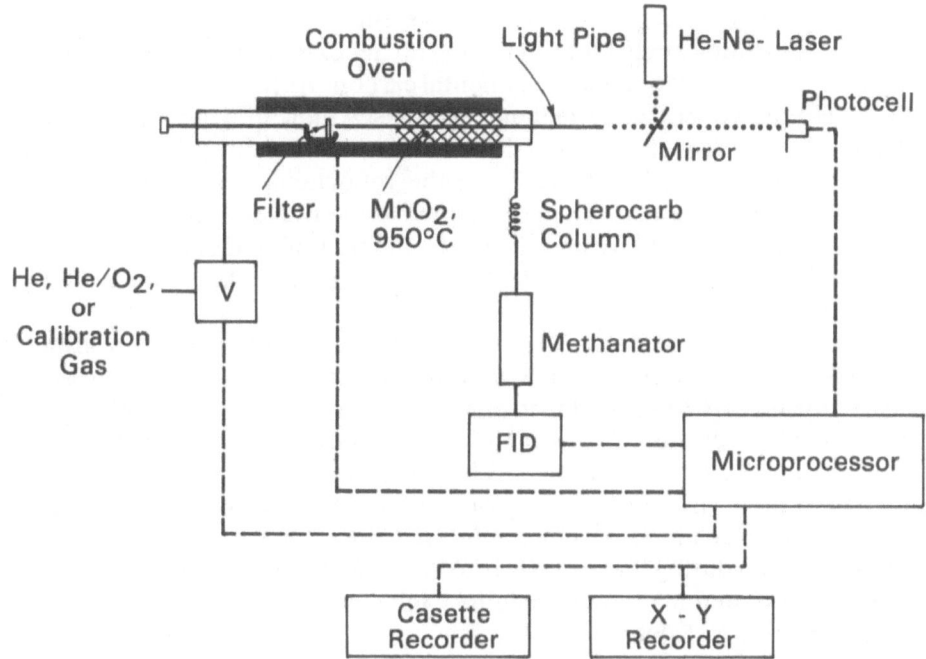

Fig. 1. Carbon analyzer. The dotted lines correspond to the laser light path, and the dashed lines are electrical connections. The methanator is a tube packed with Ni on firebrick at a temperature of 450°C. FID: flame ionization detector. V: valve. Details of the optical system and the gas flow system are discussed by Johnson [17].

stream occur. The volatilized carbon is oxidized to CO_2 in the MnO_2 bed, reduced to CH_4, and measured as described above. Approximately 2/3 of the organic carbon is removed in this step. Reflectance measurements indicate no net oxidation of elemental carbon and minimal conversion of organic to elemental carbon during this step. The sample oven is then purged with helium to remove all oxygen from the oven. During this process the sample remains in the oven, and the oven temperature is maintained at 350°C. After 280 seconds, the combustion zone temperature is raised to 600°C, and the remaining organic carbon is volatilized into the helium carrier gas. This carbon is oxidized to CO_2 in the MnO_2 bed and measured as described above.

For the measurement of elemental carbon the combustion zone temperature is dropped to 400°C, following which the carrier gas is changed to O_2 (2%)-He (98%). The sample remains in the oven during this process. After 100 seconds under these conditions, the temperature is raised to 500°C where it is held for 120 seconds and finally to 600°C where it is held for 200 seconds. During this step-wise combustion the evolved CO_2 is measured as usual. The purpose of the step-wise combustion is discussed below. After the combustion of elemental carbon is complete, a calibration is performed by injecting a known amount of CH_4 into the oven and measuring the FID response.

Throughout the combustion process the reflectance of the filter sample is continuously monitored by a 633 nm He-Ne laser system. At 350°C in O_2/He the reflectance remains essentially constant, indicating no net oxidation of elemental carbon or pyrolytic conversion of organic to elemental carbon. As the temperature is raised to 600°C in He, however, the reflectance decreases indicating an increase in elemental carbon. The step-wise combustion of elemental carbon in the third phase of the analysis results in a slow increase in the reflectance corresponding to the oxidation of elemental carbon. The correction for the pyrolytic production of elemental carbon is determined by measuring the amount of elemental carbon oxidation necessary to return the filter reflectance to the value it had before pyrolytic conversion occurred.

The analytical system is under the control of a microcomputer built around a Motorola 6802 microprocessor. All switching of gas flows, timing, temperature control, pyrolysis correction, analog to digital conversion electronics, electrometer functions, signal integration, data storage, and data outputs are controlled by the computer system. All data (FID output, reflectance, combustion zone temperature, integrated peak areas, pyrolysis correction, and time) are stored on a cassette tape which can be analyzed at a later date on the OGC PRIME 350 computer. The only operator interaction during the analysis is to enter the filter identification code into the microcomputer and load the sample into the system. The complete analytical cycle is summarized in Table 1.

TABLE 1
Analytical Cycle for Organ and Elemental Carbon

1) Load sample at room temperature and purge with He.
2) Volatilize organic carbon at 350°C in He/O_2.
3) Purge O_2 from system with He.
4) Volatilize residual organic carbon at 600°C in He.
5) Reduce oven temperature to 400°C.
6) Reintroduce He/O_2.
7) Oxidize elemental carbon at 400, 500, 600°C.
8) Return oven temperature to 350°C.
9) Inject calibration gas (CH_4).
10) Correct for pyrolytic production of elemental carbon, integrate peaks, and transfer data to cassette and x-y recorders.

Carbonate analysis is accomplished in a separate system in which a filter segment is acidified with 1% aqueous H_3PO_4. The evolved CO_2 is reduced to CH_4 in the methanator and measured in the FID. The output of the FID is measured by a microcomputer system and stored on a cassette magnetic tape.

INSTRUMENT CHARACTERIZATION

The instrument has been characterized with respect to its response to pure compounds and to mixtures. The former included tetracosane, tetradecane, tetratriacontane, coronene, perylene, glutaric acid, oleic acid, stearic acid, dioctylphthalate, and mannitol. Solutions of these were doped onto filters and analyzed in the normal manner. The average percent recovery was $99 \pm 6\%$, indicating quantitative measurement. Significant pyrolytic production of elemental carbon was observed only in the case of mannitol for which 6% of the organic carbon was converted to elemental carbon. Lampblack was also analyzed in this manner and gave 1% as organic carbon and 98% as elemental carbon for a total recovery of 99%.

For more complex substances the degree of pyrolytic conversion was substantially higher. For aerosol collected from the combustion of distillate oil, 31% of the organic carbon was converted to elemental carbon. Of the materials studied, the largest amounts of pyrolytic conversion were observed for biological samples. For wood fiber, plant leaf material, and ragweed pollen 45, 64 and 18% respectively of the carbon was pyrolytically converted to elemental carbon. Further indication of the importance of pyrolytic conversion is given in Table 2 where the ratios of pyrolytically produced elemental carbon (PEC) to corrected organic carbon (OC) and PEC to corrected elemental carbon (EC) are listed for samples collected in nine different cities. The samples were high volume, glass fiber filters collected at National Air Surveillance Network stations during 1975. The listing for each city corresponds to about 25 different samples. In all cases the amount of pyrolytically produced carbon is significant and highly variable. For example, in the most extreme case, Denver, 42% of the organic carbon was pyrolytically converted to elemental carbon. For all sites and filters the average ratio of pyrolytically produced elemental carbon to total carbon was 0.15.

TABLE 2
Ratios of Pyrolytically Produced Elemental Carbon (PEC)
to Corrected Organic Carbon (OC) and to
Corrected Elemental Carbon (EC)

Site	PEC/OC	PEC/EC
Elizabeth (N.J.)	0.26 ± 0.15	0.21 ± 0.17
Berkeley (Calif.)	0.21 ± 0.15	0.36 ± 0.30
Los Angeles	0.21 ± 0.20	0.28 ± 0.29
Chicago	0.30 ± 0.14	0.35 ± 0.24
Pasadena	0.19 ± 0.15	0.33 ± 0.26
Philadelphia	0.28 ± 0.13	0.28 ± 0.17
East Chicago	0.34 ± 0.19	0.28 ± 0.20
Denver	0.42 ± 0.18	0.58 ± 0.41
Portland (Ore.)	0.29 ± 0.12	0.34 ± 0.19

To investigate the problem of pyrolysis further, ambient aerosol filters were analyzed before and after solvent extraction. (Details of the solvent extraction procedure are discussed by Johnson [17].) After extraction by a solvent mixture consisting of 2:1 v/v chloroform-methanol, the amount of pyrolytically produced elemental carbon on the filter was reduced by more than 80%. This is consistent with the results of Colodny *et al.* [18] who found that carbon analysis of chloroform-methanol *extracts* resulted in a large (up to 40%) degree of conversion of organic to elemental carbon during the analysis. These results suggest that biological material, which is likely to be relatively insoluble, is not primarily responsible for the pyrolytic production of elemental carbon observed in typical ambient samples. This conclusion is supported by the observation that the degree of pyrolytic conversion is independent of whether the filters correspond to fine aerosol (particle size $\lesssim 2\mu$m) or total aerosol. (Biological materials would be largely absent in the fine aerosol.) Because very little pyrolytic conversion is seen for pure compounds, it is likely that the effects observed on actual ambient aerosol filters result from the highly complex mixtures present in the aerosol.

The fundamental limitation on the analytical sensitivity is the uncertainty in the response to blank filters. For glass fiber filters (Gelman A/E) the blank values are equivalent to 2.8 ± 1.4 μgC/cm^2 for organic carbon and 0.2 ± 0.1 μgC/cm^2 for elemental carbon. For Pallflex QAST quartz fiber filters the values are 1.0 ± 0.5 and $0.3 \pm 0.2 \mu$gC/cm^2, respectively. An additional source of uncertainty is the pyrolysis correction. If the most conservative estimate is assumed, the uncertainty in the pyrolysis correction can be taken as $\pm 100\%$. The resultant uncertainties in organic and elemental carbon are, on the average, about $\pm 20\%$. Work is continuing to refine this limit.

SUMMARY

An instrument has been developed for the analysis of organic, elemental, and carbonate carbon on aerosol filter samples. Speciation between organic and elemental carbon is achieved by a combined thermal-optical process. An important and essential feature of this instrument is its ability to correct for the pyrolytic conversion of organic to elemental carbon during the organic analysis. The instrument has been used to analyze samples from the National Air Surveillance Network (NASN) filter bank for 1975, from the Ohio River Valley, and from the vicinity of a coal gasification plant in Yugoslavia [19]. For the urban NASN samples approximately 60% of the carbon is organic and 40% is elemental. These results will be reported in detail in subsequent publications.

ACKNOWLEDGEMENT

This work was sponsored in part by National Science Foundation Grant No. PFR-7824554 and U.S. Environmental Protection Agency Grant No. R806274.

REFERENCES

1. *P. K. Mueller, R. W. Mosley and L. B. Pierce, in "Proceedings of the Second International Clean Air Congress" (H. M. Englund and W. T. Berry, editors, Academic Press, 1971), pp. 532-539.*

2. D. Grosjean, Anal. Chem., Vol. 47 (1975), pp. 797-805.

3. B. R. Appel, P. Colodny, J. J. Wesolowski, Environ. Sci. Technol., Vol. 10 (1976), pp. 359-363.

4. B. R. Appel, E. M. Hoffer, E. L. Kothny, S. M. Wall, M. Haik and R. L. Knights, Environ. Sci. Technol., Vol. 13 (1979), pp. 98-104.

5. W. R. Pierson and P. A. Russell, Atmos. Environ. Vol. 13 (1979), pp. 1623-1628.

6. J. M. Daisey, M. A. Leyko, M. T. Kleinman and E. Hoffman, Ann. New York Acad. Sci., Vol. 322 (1979), pp. 125-141.

7. R. McCarthy and C. E. Moore, Anal. Chem., Vol. 24 (1952), pp. 411-412.

8. V. P. Kukreja and J. L. Bove, Environ. Sci. Technol., Vol. 10 (1976), pp. 187-189.

9. J.A. Pimenta and G. R. Wood, Environ. Sci. Technol., Vol. 14 (1980), pp. 556-561.

10. R. L. Johnson and J. J. Huntzicker, in "Proceedings: Carbonaceous Particles in the Atmosphere," T. Novakov, Ed., Lawrence Berkeley Laboratory, Berkeley, California, June 1979, pp. 10-13.

11. R. L. Johnson, J. J. Shah and J. J. Huntzicker, in "Conference on Sampling and Analysis of Toxic Organics in the Atmosphere," American Society for Testing & Materials, STP721, Philadelphia, Pa, (1980), pp. 111-119.

12. S. H. Cadle, P. J. Groblicki and D. P. Stroup, Anal. Chem., Vol. 52 (1980), p. 2201.

13. D. Grosjean, S. Heisler, K. Fung, P. Mueller and G. Hidy, "Particulate organic carbon in urban air: concentrations, size distribution and temporal variations," presented at the American Institute of Chemical Engineers 72nd Annual Meeting, San Francisco, California, November 1979.

14. R. G. Delumyea, L. C. Chu and E. S. Macias, Atmos. Environ. Vol. 14 (1980), pp. 647-652.

15. D. Grosjean, comments at the Second Chemical Congress of the North American Continent, Las Vegas, Nevada, August 1980.

16. R. L. Dod, H. Rosen and T. Novakov, in "Atmospheric Aerosol Research: Annual Report 1977-78," Lawrence Berkeley Laboratory, Berkeley, California LBL-8696, pp. 2-10.

17. R. L. Johnson, M. S. Thesis, Oregon Graduate Center 1981.

18. P. Colodny and B. R. Appel, Report No. 167, Air and Industrial Hygiene Laboratory, California State Department of Health, Berkeley, California, June 1974.

19. J.J. Huntzicker, R. L. Johnson and J. J. Shah, "Carbonaceous aerosol in the vicinity of a Lurgi gasifier," presented at the Second Chemical Congress of the North American Continent, Las Vegas, Nevada, August 1980.

DISCUSSION

P. Mueller, *(Electric Power Research Institute)*

At ERT speciation between organic and elemental carbon is achieved by placing the filter disk on a MnO_2 bed and heating the sample to 550°C to remove organic carbon. Elemental carbon is then measured by heating the sample to 850°C. Do you expect any "charring" under these circumstances?

J. Huntzicker

We have tested this approach by running a set of filters both with our standard method and with the filters lying on a MnO_2 bed. Comparison of the two sets of results indicated that pyrolytic conversion of organic to elemental carbon was not significantly reduced by addition of the MnO_2. We would want to be cautious, however, in extrapolating these results to someone else's analytical system. In this regard George Wolff's comments* on the need for an intercomparision study are quite appropriate.

R. Dod, *(University of California, Berkeley)*

Looking at the optical properties during the pyrolysis with quartz filters, I find that the coloration due to carbonization seems to extend completely through the filter. In other words, if you look at the backside of the filter, it also is black after this pyrolysis step. What is this going to do to the reflectance?

J. Huntzicker

We have observed this also. It appears that the pyrolytic conversion or carbonization occurs — at least partially — as a result of the interaction between organic vapors moving through the filter and the filter fibers. This does not seem to pose a serious problem, however. Using a common set of filters, we have compared our method with both solvent extraction and optical absorption procedures. Agreement between the three methods was within about 10% for elemental carbon.

R. Dod

From the output you showed, it looked like the change in reflectance did not correlate with the area under your carbon peaks.

J. Huntzicker

The reflectance (R) is a non-linear function $\left(\dfrac{1-R^2}{2R}\right)$ of the elemental carbon concentration ($\mu gC/cm^2$).

R. Stevens, *(U.S. Environmental Protection Agency)*

Have you considered in the NASN analysis the possibility of acidic organics interacting with the alkaline glass fiber surface to produce an over-estimate of the volatile carbon? It would be similar to the SO_2-sulfate artifact problem. Perhaps acetic acid would do the same.

These comments appear in the preface.

J. Huntzicker

Any carbonaceous artifacts on the filter would certainly be included in the carbon analysis. We have not yet made any measurements of artifact formation on glass fiber filters.

R. Bradow, *(U.S. Environmental Protection Agency)*

In the case of source sampling with glass fiber filters, the feature that Bob was talking about is fairly well documented. At reasonably high concentrations of say, ten parts per million of organics from number two diesel fuel or similar fuel, glass fiber filters have been shown to retain hydrocarbon material. I believe this was published about a year and a half ago by my group in the Society of Automotive Engineering literature. I would suggest a strong possibility that certain types of filtering media adsorb gaseous organic material. Clearly, your technique sees what is on the filter but not necessarily only aerosol material.

J. Huntzicker

That is right. We would prefer to do this on quartz as opposed to glass for a variety of reasons. For one thing, we would take quartz much hotter and quartz is a much cleaner filter. The blanks are lower, and, perhaps, the artifact problem might be smaller.

R. Draftz, *(I.I.T. Reserach)*

One of the problems we have been running into is that in trying to determine the inorganic carbonate by acid dissolution, we find about the only way to get the carbonate, limestone, dolomite, or calcite dissolved is to heat it in perchloric acid. This releases it because it does not dissolve very rapidly at all. When most people do their initial baselines for recovery and standardization, they use things like sodium carbonate, phosphoric acid or HCl. With these you get beautiful recoveries. Working with real aerosols is different. For example, we looked at some episodes in Miami and in Denver and a few other places and the calcite or dolomite content was about 40 to 60 micrograms per cubic meter.

J. Huntzicker

When the sample is acidified, it is heated to about 50°C in phosphoric acid. We have run calcium carbonate, sodium carbonte, potassium carbonate, and magnesium carbonate but not dolomite. I do not think it is going to be a big problem, but that is certainly something to look into.

T. Hansen, *(Lawrence Berkeley Laboratory)*

I would like to ask you about the choice of temperatures for the organic-elemental analysis. It would seem that with microprocessors you could do a continuous temperature run similar to the method at Berkeley.

J. Huntzicker

We have done that in the hope that the thermogram would be indicative of different source types. We found that specific source types gave characteristic thermograms on the basis of source sample analysis. In ambient samples the thermograms were sufficiently complex that individual source type contributions could not be resolved.

AN EVALUATION OF METHODS
FOR THE DETERMINATION OF
ORGANIC AND ELEMENTAL CARBON
IN PARTICULATE SAMPLES

S. H. CADLE and P. J. GROBLICKI

General Motors Research Laboratories
Warren, Michigan

ABSTRACT

Methods for separating and measuring organic and elemental carbon in particulate samples collected on filters are evaluated. Included in this evaluation are solvent extraction, nitric acid digestion, optical absorption, and various thermal techniques. Emphasis is placed on the measurement of ambient samples, but samples from diesel- and gasoline-fueled vehicles, as well as wood burning particulate, are also analyzed. It was found that all of the methods have limitations and must be applied carefully. The optical absorption techniques rely on a poorly determined absorption coefficient and are subject to interference by other species. Thermal methods are plagued by carbonization of organic material resulting in overestimation of elemental carbon and underestimation of organic carbon. Despite these problems, an extensive comparison of ambient samples from Denver showed excellent correlation between the various methods. Recommendations are presented for analyzing various sample types.

INTRODUCTION

A variety of methods has been used to determine the organic and elemental carbon content of ambient and source particulate samples. These have included solvent extraction of the organics followed by total carbon analysis (1, 2), nitric acid digestion of the organics followed by total carbon analysis (3-5), and various thermal methods including both temperature programmed (6, 7) and step-wise pyrolysis and oxidation procedures (8-11) utilizing either CO_2 or CH_4 detection techniques. Optical methods which have been used to determine the elemental carbon content include infrared absorbance (12), Raman spectroscopy (13), visible absorbance by the integrating plate method (14, 15), as well as reflectance (16) and optoacoustic methods (17, 18). In one technique, optical and thermal methods have been combined (19).

References pp. 100-101.

Since the separation of organic and elemental carbon is extremely difficult, it is expected that the different analytical techniques will give somewhat different results. Therefore, we felt it important that the differences between the methods be investigated. In this paper, we compare results between extraction, nitric acid digestion, various thermal methods, and the integrating plate method. Advantages and limitations of the different methods are discussed. Emphasis is placed on thermal methods as applied to ambient particulate.

EXPERIMENTAL

Samples — Samples of atmospheric suspended particulate were collected on Gelman Spectro Grade Type A glass fiber filters using high-volume samplers and on Gelman Micro-Quartz filters using low-volume, size-segregating samplers during field studies conducted in Denver, Colorado (20), and in rural areas near Pierre, South Dakota (21), and Abbeville, Louisiana (22). Samples of particulate from gasoline and diesel cars were obtained from passenger cars run on a chassis dynamometer. The test stand, which includes a dilution tube, has been described previously (23). Wood burning particulate was collected from the outlet of a residential fireplace.

Extractions — Extractions were carried out for 8 hours using reagent grade or spectrograde solvents in a soxhlet extraction apparatus. In addition to using standard organic solvents, we also used Fluorinert FC-78, a nonpolar, low-boiling solvent marketed by 3M Company. The filters to be extracted were folded with the deposit toward the inside, wrapped in an additional piece of filter material, and placed in a glass sleeve which was inserted into the extractor. After extraction, the filters were air dried for at least two days. The amount of carbon extracted was determined by total carbon analysis of a small piece of the filter before and after extraction.

Nitric Acid Digestion — Filter samples for nitric acid digestion were placed in a glass sleeve in the same manner as for extractions. The sleeve was then placed in a beaker of concentrated nitric acid. Two digestion procedures were used. The first was similar to the procedure used by McCarthy and Moore (3). The sample was heated to boiling on a hot plate, allowed to boil for 30 minutes, then was diluted to approximately 6 N HNO_3 and allowed to stand overnight. We will refer to this procedure as the 24-hour nitric acid digestion method. The second procedure was that used by Pimenta and Wood (5) who heated the nitric acid on a steam bath for 2 hours. In both cases, the filter sample was washed with water after the digestion, then dried in an oven at 110°C. The amount of carbon digested was determined by total carbon analysis of a small filter section before and after digestion.

Vacuum Stripping — Organic carbon was removed by placing a filter in a National Appliance Model 5831 vacuum oven at 180°C or in a quartz tube in a Lindberg tube furnace held at 350°C. A Vac Torr S35 vacuum pump was used in both cases to reduce the pressure to less than 1 torr.

Thermal Methods — An automated carbon analyzer which we previously described (10) was used to determine total organic and elemental carbon. In reference 10 and this paper, we use the terms ''apparent organic'' and ''apparent elemental'' carbon to emphasize the uncertainties in the separation performed by this analyzer. Similar uncertainties apply to all the methods compared in this paper. The analyzer operates by dropping a sample into a heated zone at 650°C under helium. The volatilized and pyrolyzed organics are catalytically oxidized at 650°C to CO_2 and measured with a nondispersive infrared (NDIR) analyzer. Air is then swept over the sample to remove the remaining carbon which is also detected as CO_2. The amount of carbon removed at 350°C in air or helium was determined in a number of experiments by measuring the total carbon before and after the sample was heated in a separate oven.

Temperature-programmed carbon analysis was performed in a separate apparatus consisting of a segmented quartz tube heated by two Model 123-8 Lindberg tube furnaces. One furnace was used to temperature program the sample while the other furnace, which was immediately downstream, held a catalyst at 700°C. Oxygen was introduced before the catalyst which oxidized organics and CO to CO_2 for detection with an NDIR analyzer. Temperature programming and data acquisition were handled by a Hewlett-Packard 9845A system. Programming was performed either in a He-air mixture or in pure helium at 25°C/min. The shape of the thermograms did not change significantly when programmed at 10°C/min. Oxygen and water were removed from the helium with a Supelco Model 2-3802 gas purifier followed by a column packed with activated charcoal held at 700°C. A separating orifice prevented the oxygen, which was introduced in the catalyst section, from reaching the sample. Operation of the instrument produces thermograms of CO_2 concentration vs. temperature. Since the gas flow in the instrument is constant, the CO_2 concentration is directly proportional to the rate of carbon removal.

Integrating Plate Method — This method is similar to that described by Lin *et al*. (14) and has been described in detail elsewhere (24). Basically, it consists of determining the amount of light absorbed by the particulate on the filter. This absorbance is a measure of the elemental (i.e.,black) carbon. In this work, filter blanks were taken either from the unexposed edge of the filter or from the same filter after the carbon had been removed by combustion. Absorptivity of the carbon was taken as 10 m²/g, which is intermediate in the 2-17 m²/g range reported in the literature (17, 24, 25, 26). This method was developed for particulate collected on Nuclepore filters. When fibrous filters are used, such as in this study, a correction must be made to account for the fact that many of the particles are not located on the filter surface. We used a factor of three as described by Sadler (27) to correct for this.

RESULTS

Removal of Organic Carbon — All but the optical methods of carbon analysis require the separation of the organic carbon from the elemental carbon. This is not a trivial task since the distinction between organic carbon, which includes high molecular weight, oxygenated compounds, and elemental carbon, which contains significant quantities of hydrogen and oxygen in surface functional groups, is not clearcut. Current methods for removing the organics include solvent extraction, nitric acid digestion, and thermal and/or vacuum stripping. We determined the amount of carbon removed by each of these methods from six high-volume samples of Denver's ambient particulate.

Typically, extraction efficiencies have been determined by mass measurements. In this work, carbon measurements were used to directly determine the amount of organic carbon extracted. This avoids problems encountered with some solvents which can also remove inorganic species such as nitrates and sulfates. Table 1 shows extraction results for Fluorinert FC-78, hexane, and water. These three solvents are not expected to remove all the organic carbon but are included to give an idea of the distribution of nonpolar and polar compounds. The other solvents, however, have been used in carbon analysis. O-dichlorobenzene extractions performed under an oxygen atmosphere have been used by Pierson and Russell in their measurements of carbon in the Denver area (28). Extraction by benzene followed by 1:2 methanol-chloroform has been recommended by Appel *et al*. (2) for the removal of the primary and secondary organics from the ambient samples. Dichloromethane is commonly used to determine the organic fraction of diesel particulate (29). In our laboratory, 4:1 benzene-ethanol is used for diesel particulate extractions because of its higher efficiency (29). Similar mixed solvents have been recommended by Grosjean (1) for ambient particulate. Results in Table 1 show that these solvents remove from 40-50% of the total carbon with benzene-ethanol removing the most. Since the 4:1 benzene-ethanol had the highest extraction efficiency, it will be used in the remainder of this work.

The 24-hour nitric acid digestion consistently removed more carbon than the 2-hour digestion on a steam bath. Comparison between the two batches of 24-hour digestions shown in Table 1 and additional results not included in this table showed that the method produced erratic results with ambient particulate samples. Similar erratic results were found by Pimenta and Wood (5). We determined that much of the carbon remaining after both the 2-hour and 24-hour digestions could be removed at 650°C under helium. In fact, all of the carbon remaining after a 24-hour digestion could be pyrolytically removed from four of the six samples in batch 2, which clearly shows that the nitric acid is attacking the elemental carbon. These results contrast with those of Pimenta and Wood (5) who reported that nitric acid did not attack graphite or coke dust and of McCarthy *et al*. (3) who showed it did not attack carbon black used in tire tread formulations. Apparently, the elemental carbon in these ambient samples is not graphitic and does not resemble the carbon black used in tires. If nothing else, these results point out the difficulty in defining elemental carbon in particulate samples. In the remainder of this work, the 2-hour nitric acid digestion procedure will be used since it has been reported to be reproducible (5) and appears to minimize attack on the elemental carbon.

The data in Table 1 show that heating in a vacuum at 180°C removed only 31% of the carbon compared to 62% at 350°C in a vacuum. Heating in air at 350°C, which for the sake of brevity is referred to as oxidation, removed even more carbon, i.e., 67%. Therefore, no advantage to vacuum removal was found and it was dropped from further consideration. It is also seen that oxidation at 350°C removed more

TABLE 1

Removal of Organic Carbon from Six Ambient Particulate Samples

Method	Average Percent Carbon Removed	Standard Deviation
Extraction		
Fluorinert FC-78	20	5.0
water	27	11.6
hexane	32	12.4
o-dichlorobenzene (O$_2$ atmosphere)	40	15.3
dichloromethane	40	12.0
toluene	47	9.8
benzene followed by 1:2 methanol-chloroform	51	13.1
4:1 benzene-ethanol	54	15.0
Digestion		
nitric acid — 2 hours on steam bath	68	6.0
nitric acid[a] — 24 hours — batch 1	73	8.5
nitric acid[a] — 24 hours — batch 2	85	1.4
Thermal Analysis		
180°C vacuum	31	3.7
650°C pyrolysis	58	7.2
350°C vacuum	62	7.5
350°C oxidation	67	5.0

[a] 0.5 hours on hot plate, dilute, and let stand overnight. Batch 1 and 2 are repeats from the same filters.

material than pyrolysis at 650°C. This is likely due to carbonization of organics during the pyrolysis, a phenomenon which will be discussed in detail later in this report.

Both 350°C oxidation and 2-hour nitric acid extraction removed 67-68% of the carbon compared to a maximum of 54% removal by the most effective solvent. This suggests that 14% of the organic matter is not extractable and would remain on the filter to complicate subsequent analysis. We decided, therefore, to concentrate on pyrolysis and oxidation methods rather than the more time-consuming extraction or digestion methods.

Thermal Methods — In order to choose conditions for thermal methods, data about the nature of the carbon present was obtained from carbon thermograms of ambient and source particulate. Fig. 1 shows the rate of carbon loss from particulate heated in air and He. The top curve is typical of results obtained with automotive diesel particulate. Most of the carbon in this sample is contained in the large oxidation peak which occurs between 500 and 700°C, indicating that this is elemental carbon. The ambient sample was collected in Warren, Michigan. It is fairly representative of ambient particulate thermograms, although considerable variation is observed. Thermograms of wood-burning particulate samples have shown a wide variation in shape and are strikingly different from most ambient samples. The difference between the thermograms run in air and in He is normally attributed to the presence of elemental carbon, but it can also be due to the carbonization of organic material (which occurs to a great extent in helium). This carbonization

References pp. 100-101.

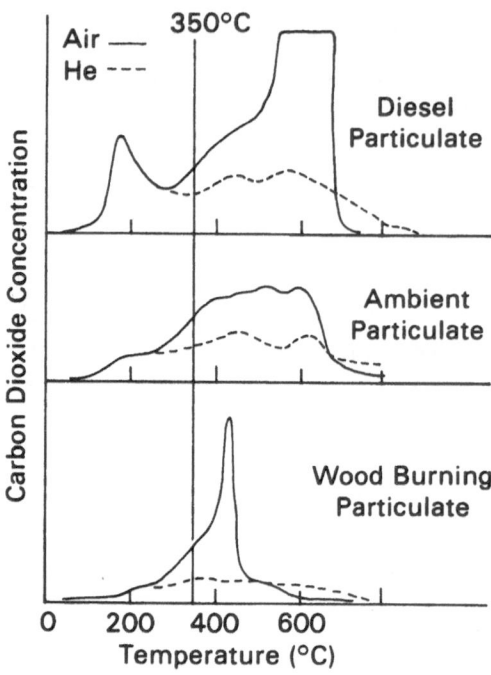

Fig. 1. Carbon thermograms in air and helium.

problem was investigated most intensively because it causes errors in the analysis of both organic and elemental carbon.

Johnson *et al.* (9) have suggested that oxidative removal of organics at 340°C minimizes the carbonization process and does not remove elemental carbon. Dod *et al.* (19) have also found that little elemental carbon is removed by oxidation at 350°C in air. We obtained similar results. The effect of this oxidation step on the thermogram is shown in Fig. 2 for a sample of wood-burning particulate. The top portion of the figure includes the thermograms obtained in air, in helium, and in air after pyrolysis in helium. By our definition, apparent organic carbon is the carbon removed in helium (curve Ib), while apparent elemental carbon is that carbon oxidized after the pyrolysis step (curve Ic). The sum of curves Ib and Ic equalled the total carbon removed in air (curve Ia). Integration of the curves suggested that 60% of the carbon was apparent organic carbon and 40% was apparent elemental carbon. The onset of carbon removal at 400°C in curve Ic confirms the observation of Johnson *et al.* (9) that samples can be heated in air at 340°C without loss of elemental carbon.

The lower half (II) of Fig. 2 shows the analysis after the sample had been oxidized at 350°C. Curve IIa is the total carbon as in Ia. Curve IIb represents the volatile and pyrolyzable material and curve IIc represents the elemental carbon remaining after 350°C oxidation. With the 350°C oxidation pretreatment, the apparent elemental carbon is now 26% of the total carbon present. This decrease in elemental carbon from 40 to 26% provides evidence that the carbonization of organics is indeed occurring during the pyrolysis step.

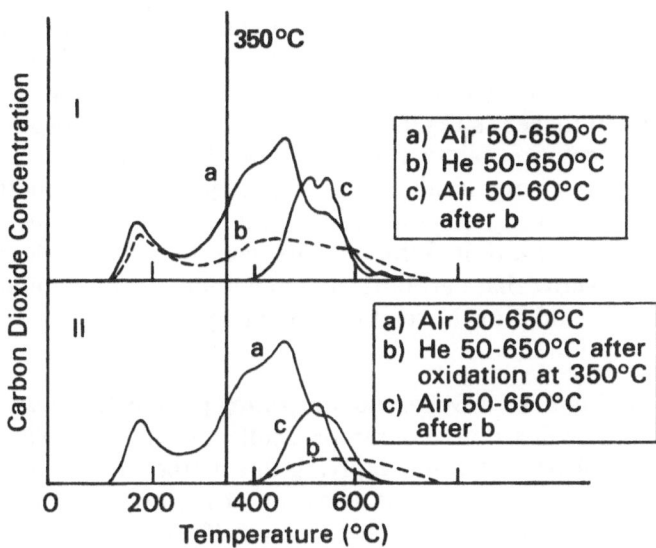

Fig. 2. Carbon thermograms of wood fire particulate samples.

Additional studies on the extent of carbonization were conducted using three techniques. The first technique combined the step-wise pyrolysis-oxidation procedure using the instrument described above with an identical analysis of another portion of the sample which had been previously extracted with benzene-ethanol. The decrease in elemental carbon after solvent extraction was taken as a measure of the amount of carbonization occurring during pyrolysis. The second technique was similar, except that organics were removed by oxidation at 350°C instead of solvent extraction. The third method was to measure the optical density of filters by the integrating plate method before and after various thermal treatments. Increases in the optical density are taken as a measure of elemental carbon formed by carbonization. Results of these carbonization studies are summarized below. Details of the extraction and integrating plate studies are given in Appendix A.

Benzene-ethanol extraction of thirty-two Denver high-volume filters showed an average decrease of 30% in the apparent elemental carbon. Oxidation of these same samples at 350°C also removed 30% of the apparent elemental carbon. In addition, oxidative removal of the organic carbon was performed on 28 size-segregated samples from Denver, 24 gasoline car particulate samples, 3 diesel car particulate samples, and 3 fireplace particulate samples. The reductions in apparent elemental carbon, i.e., the amount of carbonization, were 18, 14, 5 and 18%, respectively. The amounts of the total organics which carbonized during the pyrolysis were 23, 12, 14, and 44%, respectively, for these samples. One of the reasons for the large amount of carbonization in the Denver samples is that one third of the carbon comes from wood burning (30). Additional measurements reported elsewhere (31) have shown the average decrease in apparent elemental carbon in ambient samples caused by oxidative removal of the organics varies from 3 to 40% depending on the collection site.

The integrating plate method was used to estimate the percent carbonization during 650°C pyrolysis of the organics in ambient particulate samples from Denver, South Dakota, and Louisiana, as well as wood fire and gasoline car samples. Results were 15, 17, 14, 8, and 34% carbonization, respectively. Although large uncertainties exist in these results due to both an uncertain absorption coefficient and an uncertain correction factor for the use of fibrous filters, it was clear that some carbonization occurred in all cases. The integrating plate method was also used to investigate a variety of thermal methods in order to select a procedure which minimized carbonization. While results were not always clearcut (see Appendix A), they did show that carbonization in helium can be minimized by rapid heating but that air oxidation instead of helium pyrolysis prevents most of the carbonization.

Integrating Plate Method — Since the integrating plate method was designed to measure the absorption coefficient of samples collected on filters, several potential problems must be considered when applying this method to the measurement of elemental carbon. One problem is the selection of the proper absorptivity of the elemental carbon. Literature values range from 2-17 m^2/g (17, 24, 25, 26, 31) but center around 10 m^2/g for combustion-produced particulate. Another problem is introduced when samples are collected on fibrous filters, as they were in this study, because an empirically-derived correction factor must be applied to account for the penetration of some particles into the filter (27). Errors are also introduced when light-absorbing organics are present. As shown in Appendix A, this appears to be a major problem with some samples from wood-burning and may affect diesel samples as well. The extent of this error for six Denver high-volume filter samples was determined by extracting the filters with benzene-ethanol and measuring the absorbance of the resulting solution. This experiment showed that 6% of the absorbance at 550 nm on these high volume samples was caused by the extractable organics. It should be possible to minimize this error by analysis at different wavelengths. Care must be taken that significant amounts of other absorbing species are not present. For example, soil in non-size-segregated ambient samples collected on quartz filters can be accounted for by oxidatively removing the carbon and remeasuring the sample. The mineral-based absorbing components from soil will remain on the filter.

A comparison of the integrating plate method and the pyrolysis-oxidation technique at 650°C for Denver low-volume samples collected simultaneously on Nuclepore and Micro-Quartz filters has been reported (24) and is shown in Fig. 3. While there is a significant amount of scatter, there is a correlation coefficient of 0.93, indicating that these methods are comparable despite the problems discussed above.

Method Comparison — The aforementioned work has shown several promising methods which minimize the carbonization of organics during analysis. In order to determine if there are any systematic differences between these methods, they were used to determine elemental carbon on over twenty samples. Denver high-volume filters were selected because they provided sufficient sample size for multiple analysis and because the atmospheric conditions during sample collection had been well characterized (20). The methods included pyrolysis-oxidation at 650°C

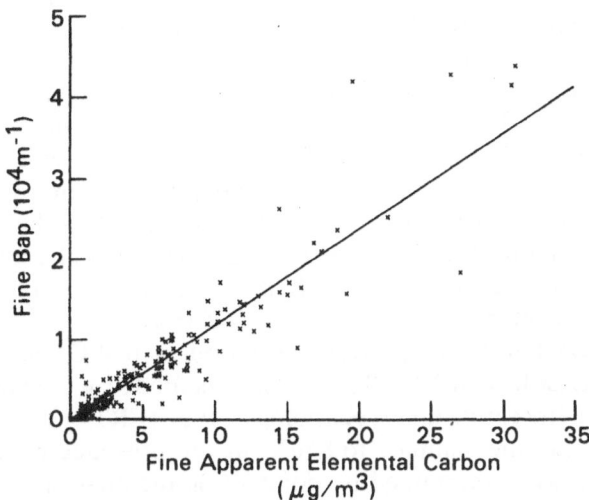

Fig. 3. Comparison between apparent elemental carbon and absorption determined by the integrating plate method.

(apparent elemental carbon); oxidation at 350°C followed by total carbon analysis; oxidation at 350°C followed by pyrolysis-oxidation at 650°C; extraction with benzene-ethanol followed by total carbon analysis; extraction with benzene-ethanol followed by pyrolysis-oxidation at 650°C; 2-hour nitric acid digestion on a steam bath followed by total carbon analysis; and 2-hour nitric acid digestion on a steam bath followed by pyrolysis-oxidation at 650°C. Results of the latter six analyses were compared to the apparent elemental carbon as determined by pyrolysis-oxidation at 650°C. The linear least squares fit and the corresponding correlation coefficients for these comparisons are given in Table 2. Actual plots are given in Appendix B.

TABLE 2

Comparison Between Apparent Elemental Carbon, C_{ae}, and Elemental Carbon Determined by Other Techniques

Analysis Method	Regression Line[a]	Correlation Coefficient
350°C oxidation followed by total carbon analysis	$y = 0.81 C_{ae} + 5.6$	0.95
350°C oxidation followed by 650°C pyrolysis-oxidation	$y = 0.71 C_{ae} - 1.14$	0.99
Extraction followed by total carbon analysis	$y = 0.82 C_{ae} + 7.7$	0.95
Extraction followed by 650°C pyrolysis-oxidation	$y = 0.71 C_{ae} - 2.4$	0.96
Nitric acid digestion followed by total carbon analysis	$y = 0.81 C_{ae} - 0.71$	0.95
Nitric acid digestion followed by 650°C pyrolysis-oxidation	$y = 0.57 C_{ae} - 3.0$	0.95

[a] *y is the value of elemental carbon determined by indicated method.*

References pp. 100-101.

It is apparent from Table 2 that all these methods are highly correlated but give systematically different results. These differences are consistent with the previous discussions which showed oxidation at 350°C, extraction, and nitric acid digestion all remove some of the organic carbon which carbonizes during the pyrolysis step. Thus, analysis after these procedures gives elemental carbon values 20-30% lower for the apparent elemental carbon. This change cannot be used to correct other samples since the amount of carbonization is not constant and will depend on the nature of the organic carbon material in the sample. Of all the methods, nitric acid digestion, followed by 650°C pyrolysis-oxidation, gives the lowest elemental carbon numbers. However, it is not clear that this number is correct since there may be some attack of the elemental carbon by the nitric acid.

A comparison between pyrolysis-oxidation at 650°C and a thermal analysis method using a Dohrman DC-50 analyzer was made by exchanging samples with Environmental Research and Technology, Inc. (ERT). ERT analyzed for total carbon by heating the samples to 850°C in the presence of manganese dioxide. Organic carbon was determined by conducting the analysis at 550°C. Regression analysis showed that the ERT results for total carbon on 151 samples were 8% lower than our results. Results were identical for organic carbon but 20% lower for elemental carbon. The reason for the discrepancy has not been resolved.

DISCUSSION

Our purpose in this work is to find a method which can distinguish between organic carbon and elemental carbon. However, we must remember that the composition of the carbonaceous material emitted from various sources is different. As organic compounds are heated, they lose oxygen and hydrogen and polymerize to form large planar molecular networks. Depending on the industrial process or combustion source involved, the nature and extent of the polymerization differs so that the material emitted can range from coke-like material with a molecular weight greater than 3000, which has large regions of oriented planar molecules, to graphite, which has a highly ordered structure. The material from various sources may also contain heavy molecular weight organic compounds. Thus, it is not entirely clear how the dividing line between organic and elemental carbon should be defined. We conclude that our current state of knowledge only allows us to make operational definitions based on the method used to analyze the material. As we have shown in this study, care must be exercised not to alter the material during analysis in order to minimize errors in the determination. For example, carbonization can cause large errors in the determination of elemental carbon in some samples which are pyrolyzed at 650°C. The amount of error will depend on the type of organic material present and the percent of elemental carbon in the samples. Errors will be small in automotive diesel particulate samples where carbonization of the organics results in small errors since elemental carbon frequently comprises most of the sample. We have previously shown (10) that there are no errors with many types of organic compounds or organic mixtures such as grease, motor oil, and tire tread rubber. However, the error can be large in samples susceptible to carbonization, such as wood-burning particulates and ambient particulate. Obviously, even small amounts of carbonization are intolerable in samples which contain very little elemental carbon. Thus, great care must be taken when applying thermal methods of analysis.

With unknown samples or samples where carbonization may be appreciable, oxidation of the organics at temperatures of 350 to 400°C should be incorporated into the analysis. This can readily be done with a small increase in analysis time by using the automated carbon analyzer previously described (10). One cut from the filter is analyzed for total carbon, a process which takes less than 5 minutes per sample. A second cut is then heated in a furnace for 10 minutes at 350°C and this filter is analyzed using pyrolysis-oxidation at 650°C. Elemental carbon corresponds to the amount of material removed by oxidation at 650°C. Organic carbon then is the difference between total and elemental carbon.

Determining elemental carbon from the difference between temperature programmed heating in helium and in air is not recommended since programming increases errors due to carbonization. This suggests that significant problems may occur in the thermogravimetric analysis of samples run in an inert atmosphere.

While the thermal methods provide rapid, automated analysis, extraction has the advantage of preserving the organics for further analysis. With the Denver particulate, we found that extraction was equivalent to oxidation at 350°C. Thus, thermal analysis of an extracted and unextracted filter determines the extraction efficiency and minimizes the error in the elemental carbon. However, there is still some error since even the most efficient solvents leave some unextracted organics on ambient particulate samples.

Nitric acid digestion followed by total carbon analysis gave results intermediate to the other methods. While this treatment does remove more organics than a 350°C oxidation or extraction, it also appears to convert some elemental carbon into pyrolyzable carbon. Thus, it is not clear that this method has any advantage over oxidative removal in a furnace which is a more rapid technique.

The results from the integrating plate method correlated well with the other methods, and it thus appears to be a viable technique when only elemental carbon determinations are needed. The use of the method as a quantitative tool will require a better knowledge of the absorptvity of elemental carbon. Significant interference by other absorbing compounds may occur when only small amounts of carbon are present in the sample.

The high correlation (Table 2) between the various methods means that any of the methods can be used to obtain a relative ranking of samples in terms of elemental carbon content. However, correction factors become necessary when carbon determinations by different methods are compared.

All of the methods discussed have limitations and must be applied carefully. Until better methods are developed, we recommend the use of a step-wise thermal analysis technique incorporating an oxidation procedure to minimize carbonization. This procedure is rapid, can be automated, and is not excessively expensive. However, samples with little elemental carbon compared to organic carbon will be subject to large errors from carbonization. Carbonization can be monitored with the integrating plate technique.

References pp. 100-101.

ACKNOWLEDGEMENTS

We are grateful for the efforts of Carolina Ang, Pat Mulawa, and David Stroup whose help made this paper possible.

REFERENCES

1. D. Grosjean, Anal. Chem., Vol. 47 (1975), p. 797.
2. B. R. Appel, P. Colodny, and J. J. Wesolowski, Environ. Sci. Technol., Vol. 10 (1976), p. 359.
3. R. McCarthy and C. E. Moore, Anal. Chem., Vol. 24 (1952), p. 411.
4. V. P. Kukreja and J. L. Bove, Environ. Sci. Technol., Vol. 10 (1976), p. 187.
5. J. A. Pimenta and G. R. Wood, Environ. Sci. Technol., Vol. 14 (1980), p. 556
6. H. Puxbaum, Fresenius Z. Anal. Chem., Vol. 298 (1979), p. 250.
7. H. Puxbaum, Fresenius Z. Anal. Chem., Vol. 299 (1979), p. 33.
8. R. L. Johnson and J. J. Huntzicker, In: Proceedings of Carbonaceous Particles in the Atmosphere Conference, T. Novakov, Ed., Lawrence Berkley Laboratory, Report 9037, (1979), pp. 10-13.
9. R. L. Johnson, J. J. Shah, and J. J. Huntzicker, Presented at Conference on Sampling and Analysis of Toxic Organics in the Atmosphere, Boulder, CO, August (1979); ASTM (In Press).
10. S. H. Cadle, P. J. Groblicki, and D. P. Stroup, Anal. Chem., Vol. 52 (1980), p. 2201.
11. S. L. Heister, R. C. Henry, J. G. Watson, and G. M. Hidy, Environmental Research and Technology, Inc., Document P-5417-1, (1980), pp. C7-C8.
12. D. M. Smith, J. J. Griffin, and E. D. Goldberg, Anal. Chem., Vol. 47, (1975), p. 233.
13. H. Rosen, A. D. A. Hansen, L. Gundel, and T. Novakov, App. Optics, Vol. 17 (1978), p. 3859.
14. C. I. Lin, M. Baker, and R. J. Charlson, Appl. Optics, Vol. 12 (1973) p. 1356.
15. R. E. Weiss, A. P. Waggoner, R. J. Charlson, D. L. Thorsell, J. S. Hall, and L. A. Riley, In: Proceedings of Carbonaceous Particles in the Atmosphere Conference, T. Novakov, Ed., Lawrence Berkeley Laboratory, Report 9037, (1979), pp. 257-262.
16. R. G. Delumyea, L.-C. Chu, and E. S. Macias, Atmos. Environ., Vol. 14 (1980), p. 647.
17. T. J. Truex and J. E. Anderson, Atmos. Environ., Vol. 13 (1979), p. 507.
18. F. R. Faxvog and D. M. Roessler, J. Appl. Phys., Vol. 50 (1979), p. 7880.
19. R. L. Dod, H. Rosen, and T. Novakov, Atmospheric Aerosol Research Annual Report for 1977-1978, Lawrence Berkeley Laboratories, Report 8696, pp. 2-10.
20. R. J. Countess, G. T. Wolff, and S. H. Cadle, J. Air Pollut. Control Assoc., Vol. 30 (1980), p. 1194.
21. N. A Kelly, G. T. Wolff, and M. A. Ferman, Atmos Environ. (In Press).
22. G. T. Wolff, N. A. Kelly, and M. A. Ferman, Science, Vol. 211 (1981), p. 703.
23. S. H. Cadle, G. J. Nebel, and R. L. Williams, SAE Transactions, Vol. 87 (1980), pp. 2381-2401.
24. P. J. Groblicki, G. T. Wolff, and R. J. Countess, Atmos. Environ., Vol. 15(1981), p. 2473.
25. D. M. Roessler and F. R. Faxvog, J. Opt. Soc. Am., Vol. 70 (1980), p. 230.
26. S. Ergun, J. T. McCartney, and R. E. Walline, Fuel, Vol. 40, (1961), p. 109.

27. *M. S. Sadler, M. S. Dissertation, University of Washington, Seattle, Washington, (1979).*

28. *W. R. Pierson and P. A. Russell, Atmos. Environ., Vol. 13 (1979), p. 1623.*

29. *R. L. Williams and D. P. Chock, presented at International Symposium on Health Effects of Diesel Engine Emissions, Cincinnati, Ohio, December 1979; also available as General Motors Research Laboratories, Publication GMR-3177.*

30. *G. T. Wolff, R. J. Countess, P. J. Groblicki, M. A. Ferman, S. H. Cadle, and J. L. Muhlbaier, Atmos. Environ., Vol. 15 (1981), p. 2485.*

31. *G. T. Wolff, P. J. Groblicki, S. H. Cadle, and R. J. Countess, These proceedings.*

32. *J. L. Muhlbaier and R. L. Williams, These proceedings, p. 279.*

APPENDIX A — CARBONIZATION

Solvent Extraction — The amount of carbonization occurring during pyrolysis of Denver high-volume filters at 650°C was estimated by analyzing portions of six filters for apparent elemental carbon before and after solvent extraction. Any decrease in apparent elemental carbon caused by solvent extraction was assumed to be due to removal of organic compounds which carbonized during pyrolysis. The solvents used are listed in Table 1 of the text. Extraction by all the solvents except Fluorinert FC-78 reduced the apparent elemental carbon. Since the 4:1 benzene-ethanol removed the most total carbon and caused the largest decrease in apparent elemental carbon, it was used to extract 32 additional Denver high-volume samples. These extracted filters averaged 30% less apparent elemental carbon and 70% less apparent organic carbon than the nonextracted filters. To cross check the data, portions of the extracts were filtered and then dried on pieces of filter paper and analyzed for apparent organic and elemental carbon. An average of 27% of the benzene-ethanol extracts appeared to be elemental carbon. This clearly shows that the extractable material does carbonize. In addition, the amount of apparent elemental carbon produced during analysis of the extracts agreed well with the amount of apparent elemental carbon lost from the filter on extraction. To prove that the results are due to carbonization rather than elemental carbon particles being washed from the filter during extraction, diesel extracts were examined by optical and electron microscopy. No carbon particles were detected; therefore, it was concluded that the decrease in apparent elemental carbon after extraction was a measure of carbonization. We also determined that 50% of the organics removed by water extraction were carbonizable. This information, combined with the fact that the nonpolar Fluorinert FC-78 solvent removed no carbonizable organics, indicates that the polar compounds tend to carbonize.

Integrating Plate Method — The integrating plate method was used to study carbonization of ambient, wood fire, gasoline car, and diesel car samples after heat treatment. Table A-1 lists the different samples studied, the average apparent organic and apparent elemental carbon, as well as the average elemental carbon determined by the integrating plate method. Also given is the average relative standard deviation of the absorbance determined by the integrating plate method from six replicate cuts of each original filter. Some of the error is due to uneven filter loading rather than random variation in the measurement process. Repeat measurements on an individual filter were generally within two percent.

In all but the diesel samples, the elemental carbon determined by the integrating plate method is lower than the apparent elemental carbon which was determined by pyrolysis-oxidation at 650°C. This difference is primarily due to carbonization during pyrolysis. The larger value obtained with diesel samples may be due to an error in the absorptivity or in the total carbon measurements, which were near the lower limit of detection for some samples. The percent of the organic carbon which must have carbonized during pyrolysis to give the difference in the elemental carbon numbers was calculated and is given in column 8 of Table A-1. These numbers should be considered as estimates of the carbonization because large errors in the absolute value of the elemental carbon determined by the integrating plate method may be caused by an uncertain absorption coefficient and an uncertain correction factor for particle concentration in fiber filters.

The integrating plate method was also used to determine the amount of carbonization caused by other thermal treatments, including temperature programming at 25°C/min in helium and air. In these experiments, elemental carbon was determined with the integrating plate method before and after thermal treatment. This permitted the determination of the increase in filter blackness (i.e., elemental carbon) and avoids the uncertainties present in the absolute measurements discussed above. Results are presented in Table A-2. Increases in elemental carbon are due to carbonization. Decreases may be due to the removal of either elemental carbon or other light-absorbing material. While this table does show the amount of error involved in these specific samples, it must be kept in mind that the percent increase depends on the percent elemental carbon initially present. This dependency can be avoided by calculating the amount of the organic carbon that carbonized to produce the observed increase in elemental carbon. Again, this calculation provides estimates rather than absolute values because of uncertainties in the integrating plate method. The results of these calculations are given in Table A-3. It is apparent from data for the ambient samples, which indicate over 100% carbonization of the organics, that the absolute estimates are too high. However, the data is useful in a relative sense to compare different thermal methods. Thus, we see that temperature programming the ambient samples in helium produces a great deal more carbonization than does temperature programming in air. The data on carbonization during

TABLE A-1
Optical Analysis of Carbonization

Sample	Number of Samples	Total Carbon μg/cm²	Average % Apparent Organic Carbon	Average % Apparent Elemental Carbon	Average % Elemental Carbon (Integrating Plate)	Average Relative Standard Deviation (Integrating Plate)	Average Carbonization of Organics
Ambient (Denver)	15	5.7	58	42	32	ND	15
Ambient (SD)	7	7.7	72	28	14	9.9	17
Ambient (LA)	8	8.0	80	20	6.9	10	14
Wood Fire	4	15.9	56	44	39	3.5	8
Gasoline Car	4	12.8	60	40	8.7	14	34
Diesel Car	10	5.8	26	74	106	3.9	—

ND=No Data

TABLE A-2
Average Percent Increase in Elemental Carbon due to Thermal Treatments.

Sample	Program to 400°C in He	Hold at 350°C in air	Program to 350°C in air	Program to 400°C in air	Program to 350°C air 650°C He	Program to 400°C air 650°C He
Ambient (Denver)	300	12	ND	8	ND	0.5
Ambient (SD)	777	192	109	-3	392	a
Ambient (LA)	1343	122	50	38	ND	53
Wood Fire	87	-37	-28	a	-32	-44
Gasoline Car	220	21	25	a	-24	a
Diesel Car	20	-4	-4	-2	-10	-9

ND=No Data

a Most black carbon removed.

TABLE A-3
Average Percent Carbonization of Organics

Sample	Program to 400°C in He	Hold at 350°C in air	Program to 350°C in air	Program to 400°C in air	Program to 350°C air 650°C He	Program to 400°C air 650°C He
Ambient (Denver)	110	12	ND	3	ND	-1
Ambient (SD)	104	-20	10	-3	49	a
Ambient (LA)	99	9	2	2	ND	2
Wood Fire	48	-53	-27	a	-33	-27
Gasoline Car	50	3	2	a	-4	a
Diesel Car	67	-16	-19	-10	-28	-34

ND=No Data

a Most black carbon removed.

temperature programming in helium averaged 104% for the ambient samples. This can be compared to the 15% estimated carbonization which occurs when the samples are quickly heated by dropping them into the hot furnace of the carbon analyzer (Table A-1). Apparently, temperature programming produces much more carbonization than flash heating. The data in Table A-3 also show that the Denver and Louisiana samples can be heated to 400°C without loss of elemental carbon.

Wood samples showed only 8% carbonization during the 650° pyrolysis (Table A-1). This is considerably lower than the 44% obtained from the oxidative removal of organics from three different wood fire samples discussed in the text. Large variations between samples using different woods were observed (see Figure 1 and 2 of text) and may account for some of this difference. More likely, some organics in these samples were highly absorbing and thus were seen as elemental carbon by the integrating plate method. Evidence for this is seen in the large decrease in elemental carbon when the samples were oxidized at 350°C. Indeed, some wood fire samples we have tested are dark brown rather than black, are up to 85% extractable in benzene-ethanol, and can be completely removed from the filter by oxidation at 350°C (32).

Results for gasoline and diesel car particulate are similar to the other samples. Some carbonization occurs during pyrolysis at 650°C. This increases when the samples are programmed in helium but decreases if the samples are oxidized at 350°C. Apparently, oxidation at 400°C can remove the black carbon from gasoline car samples but not from diesel car samples. The decrease in elemental carbon when the diesel car samples were oxidized at 350°C and 400°C suggests the removal of absorbing organic compounds.

One additional set of experiments was conducted to determine the amount of carbonization on Denver samples pyrolyzed at 650°C. Individual samples were pyrolyzed by dropping them into the furnace of the carbon analyzer under a helium atmosphere. The furnace was cooled and the sample recovered for analysis by the integrating plate method. Data from seven samples showed that there was an average of 39% more black carbon after pyrolysis. Assuming black carbon, as determined by the integrating plate method, is identical to oxidizable carbon, this data shows that 22% of the organics carbonized. This is in good agreement with the other measurements of carbonization. Similar experiments were performed by placing samples in a cold, helium-filled tube and inserting the tube into a tube furnace. This procedure, which is equivalent to a rapid temperature program, gave carbonization results intermediate to those of the rapid pyrolysis in the carbon analyzer and the 25°C temperature program. It is apparent from this that temperature programming in helium should be avoided.

The aforementioned experiments have shown that carbonization can be a large error in the thermal analysis of organic and elemental carbon. Oxidation at 350°C minimizes but does not necessarily eliminate this problem. Oxidation at higher temperatures, at least 400°C, can be used with some samples. It appears that the best thermal method is oxidation at 350-400°C, followed by pyrolysis at 650°C and then oxidation at 650°C. Programming the temperature between 350 and 650°C is not recommended since it can increase carbonization errors.

APPENDIX B — COMPARISON OF ELEMENTAL CARBON DETERMINED BY VARIOUS METHODS

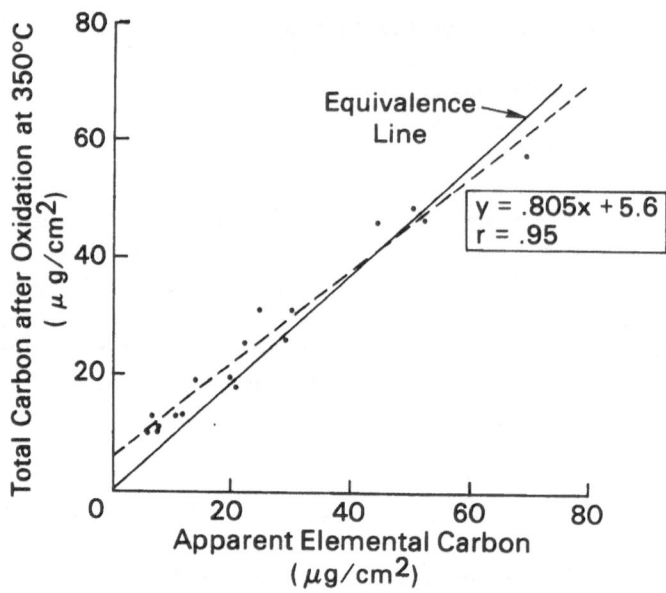

Fig. B-1. Comparison between apparent elemental carbon and total carbon remaining after oxidation at 350°C.

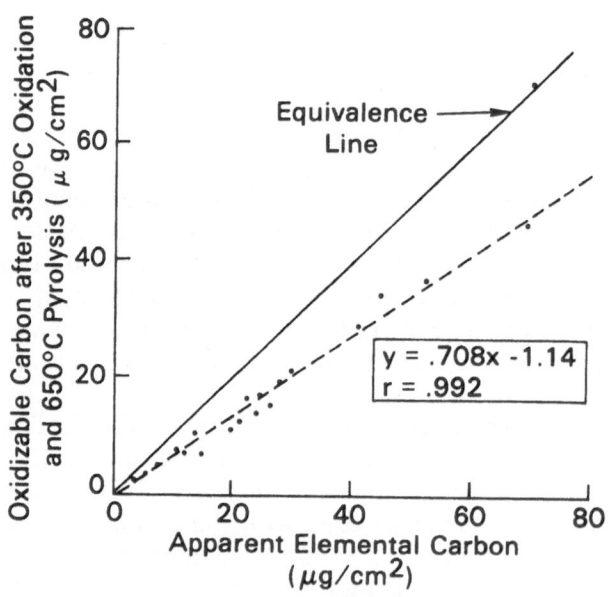

Fig. B-2. Comparison between apparent elemental carbon and elemental carbon remaining after 350°C oxidation.

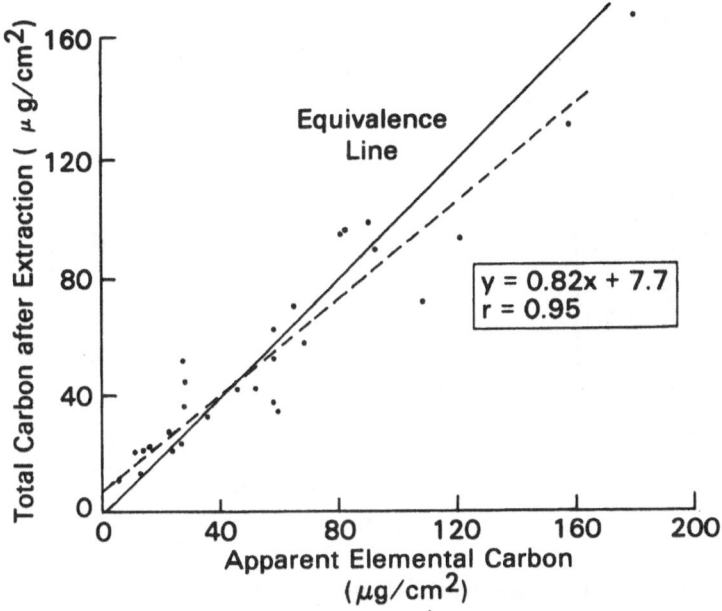

Fig. B-3. Comparison between apparent elemental carbon and total carbon remaining after extraction.

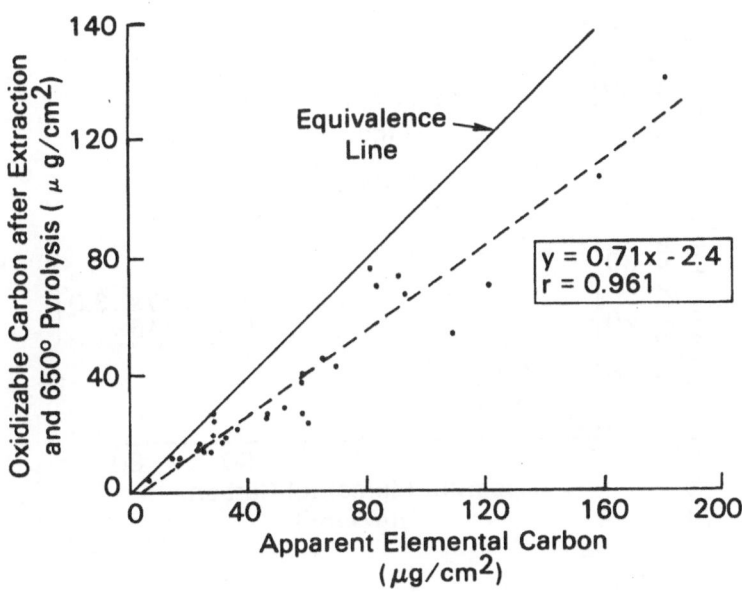

Fig. B-4. Comparison between apparent elemental carbon and elemental carbon remaining after extraction.

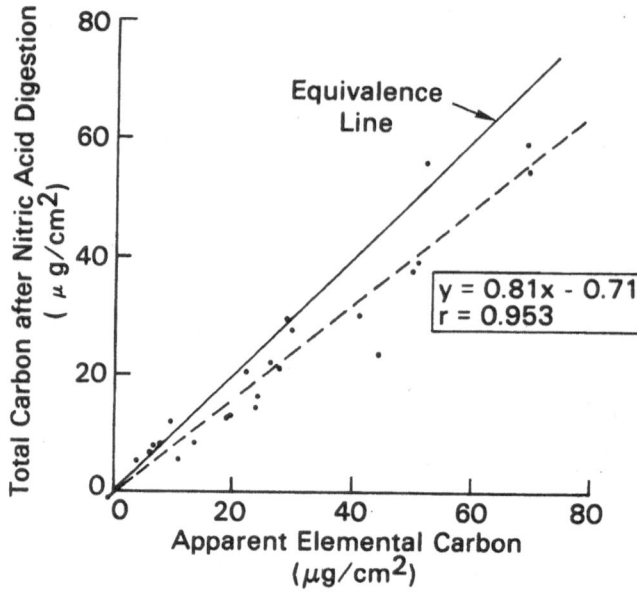

Fig. B-5. Comparison between apparent elemental carbon and total carbon remaining after nitric acid digestion.

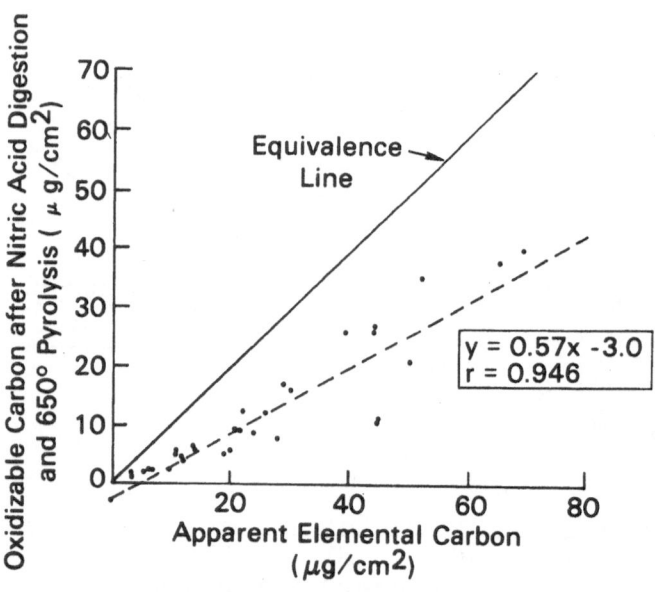

Fig. B-6. Comparison between apparent elemental carbon and elemental carbon remaining after nitric acid digestion.

DISCUSSION

M. Shelef, *(Ford Motor Company)*

Being unfamiliar with this whole question of separating organic carbon from elemental carbon, I will allow myself to make a suggestion. Why not separate the volatile carbon under vacuum, at much lower temperature, thereby avoiding the carbonization of organics?

S. Cadle

The samples that I have tried under vacuum have shown that it is equivalent or less efficient than heating at 350°C in an oven. So, unless you want to collect and preserve the organics, I do not see any advantage. It is easier to put it in an oven.

W. Richards, *(Meteorology Research, Inc.)*

What specific absorption coefficient would bring your integrating plate method in line with the 350°C pretreatment data?

C. Cadle

Around 14 m^2/g.

ANALYTICAL METHODS TO MEASURE THE CARBONACEOUS CONTENT OF AEROSOLS

R. K. STEVENS, W. A. McCLENNY and T. G. DZUBAY

U.S. Environmental Protection Agency
Research Triangle Park, North Carolina

M. A. MASON and W. J. COURTNEY

Northrup Services, Inc.
Research Triangle Park, North Carolina

ABSTRACT

Increased use of diesel fuel and coal over the next decade could lead to increased amounts of carbonaceous material in fine particle aerosols. For this reason, EPA has for the past few years been developing and evaluating a number of analytical methods to characterize ambient concentrations of carbonaceous aerosols. These include methods based either on combustion of carbonaceous material to CO_2 (or CH_4) with subsequent detection, or on nondestructive optical measurements. These methods are designed so that the carbon content of aerosols can be measured in samples collected by dichotomous samplers using a variety of filter media. The combustion method uses a Dohrmann Model DC-50 carbon analyzer with a detection limit of 0.5μg. Optical measurements are made with prototype light transmission and photoacoustic instruments. Comparisons of combustion and optical methods are presented for carbon in aerosols collected in Houston, TX and the Shenandoah Valley of VA.

INTRODUCTION

Carbonaceous material generally is the second largest component in submicron aerosol particles. Fine particle elemental carbon can be a significant factor in visibility degradation since carbonaceous particles both scatter and strongly absorb light. Increased use of wood and fossil fuels over the next decade could lead to increased amounts of carbonaceous material in ambient particles. The potential increase in the emissions of volatile and non-volatile carbon suggests that a need exists for developing methods for comprehensive chemical analyses to determine the types of organics present in ambient aerosols. For these reasons a number of laboratories have recently been developing and evaluating analytical methods to measure carbon. These methods are mainly based on thermal combustion, optical transmittance, photoacoustics, nuclear excitation and wet chemical oxidation principles. Unfortunately, the carbon values obtained are often strongly dependent on the method.

Thermal combustion methods [1-4] are the most widely used and provide measures of the volatile carbon and total carbon or the volatile and graphitic carbon. In the former, the arithmetic difference between volatile carbon and total carbon has been assumed to be the graphitic carbon component. This is clearly an operational definition. One problem with these methods is that the various laboratories using their own versions of thermal combustion analyzers to measure total aerosol carbon and volatile carbon often obtain different values for identical samples. According to Will [5] two independent laboratories analyzed 28 Gelman AE filters that contained fine particle aerosols collected in a forested area in Arkansas between August 17 and September 13, 1979. Table 1 summarizes the results of the study.

TABLE 1
Comparison of Aerosol Carbon Measurements

	Volatile Carbon, $\mu g/m^3$		Elemental Carbon, $\mu g/m^3$	
	Lab I	Lab II	Lab I	Lab II
Average (N = 28)	1.89	3.52	0.85	0.37

Laboratory I used a two step process to measure the volatile and elemental carbon content of the aerosols. The procedure involves heating a portion of the filter to 650°C in helium, oxidizing the evolved organics to CO_2, and detecting the CO_2 with a nondispersive infrared analyzer. Next O_2 is added to the helium stream and the sample is heated at 650°C to convert the remaining carbon to CO_2. It is assumed that the remaining carbon is elemental carbon.

Laboratory II measured carbon in three steps. In step one the sample is oxidized at 340°C in a O_2/He atmosphere, and the CO_2 produced is converted to methane and measured with a flame ionization detector. In step 2 the balance of the volatile organic carbon is volatilized at 580°C in helium. In step three the residual elemental carbon is oxidized at 580°C in O_2/He. Although the two methods appear to be similar, the results of the analyses are somewhat different.

An interesting feature of this intercomparison is the fact that both laboratories found very little elemental carbon. However, optical absorption measurements on filters collected simultaneously with glass fiber filters indicated that the elemental content was substantially greater than the values obtained by combustion methods. This discrepancy may be due to the relatively high temperatures both laboratories used to measure the volatile carbon. Some of the light-absorbing or soot carbon could have been lost during the analysis. Another reason may involve the uncertainties of estimating carbon by optical methods as well as the possible presence of other absorbing aerosol species.

Recently Moyers [6] at the University of Arizona has investigated a wet chemical method to measure the volatile or oxidizable carbon content of fine particles. The method involves digesting a 37 mm dia. Teflon filter in a sealed tube containing H_3PO_4, $K_2S_2O_8$ and water at 100°C in a pressure cooker for 20 minutes. The sealed tube is inserted into an Oceanography International CO_2 nondispersive infrared analyzer; the tube is broken, and CO_2 is release into the analyzer, Results of Moyer's work to date indicate that the carbon oxidized by this technique approaches the amount of volatile carbon measured by a combustion method.

Extraction methods have also been used [7, 8] to differentiate between the elemental and organic fraction of ambient aerosols. These methods require significant quantities of sample and are time consuming and expensive. For this reason, only a limited amount of data has been reported on results from this type of analysis.

Several laboratories [9, 10, 11] measure the optical absorption properties of particles collected on filters. While sampling, it is generally observed that filters become darker as the particle loading increases. The aerosol light absorption coefficient b_{ab} can be estimated from the amount of light transmitted through a filter before and after sampling with the Beer-Lambert relationship:

$$b_{ab} = \frac{1}{L} \ln (I_0/I) \qquad (1)$$

where L = length of the air column sampled and is also the ratio of the volume of air sampled to the deposit area on the filter.

Lin et al. [9] and Weiss et al. [10] have developed what is known as the integrating plate method (IPM) for measuring absorbing particles and have demonstrated that it can be used to measure b_{ab} for aerosol particles collected on Nuclepore filters. Recently, Rosen et al. [11] developed a laser transmission method to measure absorption for particles collected on Millipore filters. In a collaborative study between Lawrence Berkeley Laboratory (LBL) and the University of Washington, Sadler et al. [12] compared b_{ab} results for aerosols collected simultaneously on Millipore and Nuclepore filters. Although the results were highly correlated, the LBL laser transmission method on Millipore filters gave results that were larger by a factor of 2.5 than results by the University of Washington using their integrating plate method on Nuclepore filters.

An empirical relationship obtained for urban areas by Rosen et al. [13] indicated that the ratio of the light absorption coefficient to total carbon concentration is equal to 3.5 m^2/g. On the other hand in a study of both urban areas and pristine areas, Sadler et al. [12] compared total carbon measured by combustion with absorption measured by the integrating plate method on Nuclepore filters and found a non-linear relationship. The ratio of the absorption coefficient to carbon concentration varied from 0.33 to 3 m^2/g.

Recently Yasa et al. [14] used a photoacoustic detector as part of a program to infer that the major absorbing species in urban particles is graphitic carbon. McClenny and Shaw [15] performed a feasibility study at about the same time demonstrating that the photoacoustic technique can detect carbon from a propane flame after collection on a Teflon filter. Based on the feasibility study, EPA established a research program to further develop this technique. As part of this program, Bennett *et al.* [16] of North Carolina State University designed a system to simultaneously measure light transmission (IPM) and photoacoustic response from carbon on Teflon filters. This project has also provided known loadings of carbon on Teflon filters These known loadings have been used in relating optical response to carbon loading in this paper. Howell and Palmer [17] are using a photoacoustic spectrometer to investigate the spectral nature of ambient carbon collected on Teflon filters.

For the past few years our laboratory has been developing improved methods to sample and characterize ambient aerosols. Most of our work has centered on sulfate, nitrate, ammonia, hydrogen, elemental composition and mass of fine and coarse particles collected in dichotomous samplers. In most of our field studies we have used Teflon as a filter medium. Teflon filters are ideal for elemental determinations by x-ray fluorescence (XRF) and for anion-cation determinations, but not for measuring the carbon content of aerosols by combustion procedures. Organic solvent extraction to determine the extractable organics collected on Teflon filters with the dichotomous sampler has not been successful due to the small amount of material collected (generally, less than 800 μg aerosol per filter) and the high blanks resulting from extraction and evaporation. For this reason we have used duplicate dichotomous samplers in most of our field studies to characterize aerosols. One sampler collects on Teflon filters, the other on quartz filters.

During the past two years the following field studies have been conducted to collect aerosols for chemical characterization:

1. Great Smoky Mountains — 6 day study (September 1978)
2. Shenandoah — 21 days (July-August 1980)
3. Houston, Texas — (June-July 1980)

In these studies we measure volatile carbon and total carbon on quartz filters by a combustion method. We compare the difference between total carbon and volatile carbon with the amount of soot which we estimate by using photoacoustic and light transmission methods. The light transmission measurement devices to be discussed are similar to the integrating plate method of Lin *et al.* [9]. Both the integrating plate methods and a new method based on photoacoustics are calibrated in terms of soot mass concentrations by using laboratory generated soot standards. Here we do not report optical data in terms of light absorption coefficients because of a serious dependence on the type of filter media used to collect the particles [12]. Also we have found that light absorption by aerosols collected on Teflon filters depends significantly on whether the aerosol deposit is oriented toward or away from the light source.

EXPERIMENTAL

Sampling — To enable a variety of chemical components to be measured, aerosol samples were collected in at least three separate dichotomous samplers. In the case of the Shenandoah study, four automated dichotomous samplers and one manual dichotomous sampler were used in order to have duplicate samples for certain specialized analyses. The filters used in the dichotomous samplers were 1 μm pore size Teflon filters (Ghia Corporation, Pleasanton, California) and type 2500 QAST quartz filters (Pallflex Products Corporation, Putnam, Connecticut). For the studies in Houston, Teflon, quartz and 0.4 μm pore size Nuclepore filters were used.

To measure daytime and nighttime conditions, filters were changed at about 0700 and 1900 EST. The dichotomous samplers collected particles in the 0-2.5 μm (fine) and 2.5-15 μm (coarse) aerodynamic diameter size ranges. Filters used in acidity measurements were removed daily and stored in an ammonia free atmosphere to prevent neutralization. Table 2 is a summary of the typical sampling conditions used in the three field studies.

TABLE 2
Summary of Aerosol Sampling Conditions[a]

Sampler	Filter Medium	Deposit Diameter	Flow Rate ℓ/min.	Duration Hrs.	Analysis
A	Teflon	30.5	16.7[b]	12	Mass, elemental, b_{ab}, optoacoustic
B	Teflon	29.5	16.7[b]	12[c]	H^+, SO_4^{-2}, NO_3^-, NH_4^+
C	Nuclepore	30.5	16.7	12[d]	b_{ab}
D	Quartz	29.5	16.7	12	Carbon (combustion)

[a] Smoky Mountains sampling site was located near Elkmont, TN at an elevation of 647 m. Shenandoah sampling site was located 12 miles North of Luray, VA. Sampling in Houston was performed on the campus of the University of Houston.
[b] Flow rates were 50 ℓ/min for the Smoky Mountains study.
[c] Sampling time in Smoky Mountains was 24 hours.
[d] Used in Houston study only.

Analysis — *Mass, Elemental, Anion and Cation Measurements* – Mass, sulfate, nitrate, H^+, NH_4^+ and elemental composition for aerosols collected on Teflon filters were measured according to procedures described by Stevens et al. [18]. Mass was determined with a β-gauge; sulfate and nitrates by ion chromatography; H^+ by Grantitration, NH_4^+ by colorimetry and elemental composition by energy dispersive x-ray fluorescence.

Carbon Measurements by Combustion – A Dohrman DC-50 Envirotech Organic Analyzer (Envirotech Corp, Santa Clara, Calif. 95050) is used to measure total and volatile organic carbon collected on Pallflex tissue quartz filters in dichotomous samplers. The DC-50 contains a single quartz tube in which sample pyrolysis, oxidation to CO_2 and reduction to methane take place. Oxygen is provided on demand by MnO_2 in the platinum sample boat. Reduction to methane takes place downstream in a bed of nickel catalyst supplied with hydrogen and maintained at 350°C. The temperature of the combustion area is variable to 1000°C. Carbon is measured by a flame ionization detector (FID) which responds linearly to methane

References pp. 125-126.

throughout the range used for analysis. The instrument is equipped with an integrator which integrates the FID response over the five minute combustion period.

To measure total combustible carbon the combustion zone is adjusted to 850°C. The instrument is calibrated with potassium hydrogen phthalate (KHP). Filter sections of 0.665 cm diameter are placed in the sample boat and injected into the combustion zone. Blanks and duplicates are analyzed to ensure quality control.

Volatile carbon is determined in the same manner, but the temperature of the pyrolysis zone of the DC-50 is adjusted to 450°C. For calibration, an aqueous solution of KHP is injected into the sample boat. Analysis of representative organic compounds at varying temperatures demonstrates that 450°C provides at least 94% recovery for a wide range of organic compounds (See Table 3).

TABLE 3

Recovery Efficiencies (%) for Representative Organic Compounds

| | Temperature (°C) | | | | |
Substance	300	350	400	450	600
Adipic Acid	17	64	103	101	—
Myristic Acid	23	50	90	99	—
Fluoranthene	32	83	100	112	—
Chrysene	10	52	71	94	—
KHP	—	≤50	86	101	100
Acetanilide	58	82	100	100	101
N-Octacosane	81	84	84	94	—

Blank Study – Considerable differences have been noted between the carbon content of blank filters before and after they have been exposed to field conditions. Table 4 illustrates the results obtained using Pallflex tissue quartz filters (lot 4094M). Based upon the results of the filters placed in the sampler, the blank concentration for elemental carbon was calculated to be 1.05 $\mu g/cm^2$ with an estimated standard deviation of 0.46 $\mu g/cm^2$. (Total carbon and volatile carbon con-

TABLE 4

Effect of Filter Handling on Carbon Blanks

| Handling | No. Filters | Total Carbon ($\mu g/cm^2$) | | Volatile Carbon ($\mu g/cm^2$) | |
		Mean	s.d.	Mean	s.d.
As Received	30	1.40	0.18	NA	NA
Field I*	24	2.67	0.48	NA	NA
Field II**	6	5.01	0.46	3.96	0.28

Remained in plastic filter holders in trays.
**Placed in sampler, but not sampled.*

centrations are not independent, therefore the variance of the elemental carbon equals the sum of the variances of the total and volatile carbon less the covariance of total and volatile carbon.) These data suggest that refinements in sample handling and storage are needed.

Duplicate analyses were run on 29 filters for total carbon and 14 filters for volatile carbon. The estimated standard deviation for individual measurements was calculated from these data resulting in an estimated standard deviation of 0.27 μg/cm² total carbon and 0.28 μg/cm² volatile carbon.

Combining the estimation of standard deviation for measurements of samples and blanks, the resulting estimates of the standard deviation for determination of net carbon collected are 0.53 μg/cm² for total carbon, 0.40 μg/cm² for volatile carbon, and 0.61 μg/cm² for elemental carbon. Related to atmospheric carbon this represents a standard deviation of 0.33, 0.24, and 0.37 μg/m³ respectively for total, volatile and elemental carbon samples collected for 12 hours on 6.6 cm² filters at 16.7 ℓ/min.

EPA Light Transmission Method – The light absorption coefficient for fine particle aerosol deposits on Teflon filters was measured using the device shown in Fig. 1. The device is based upon a design described by Lin *et al.* [9]. The light source consists of a light emitting diode having emission at 520 nm wavelength. The Teflon filters were supported by 5 by 5 cm square plastic frames, and their aerosol deposit side faced the light source. An opal glass diffuser plate placed between the light source and the photomultiplier makes this device similar to that used in the integrating plate method [9].

Pacific Photometric
PMT Housing Mod 62, S/N 573

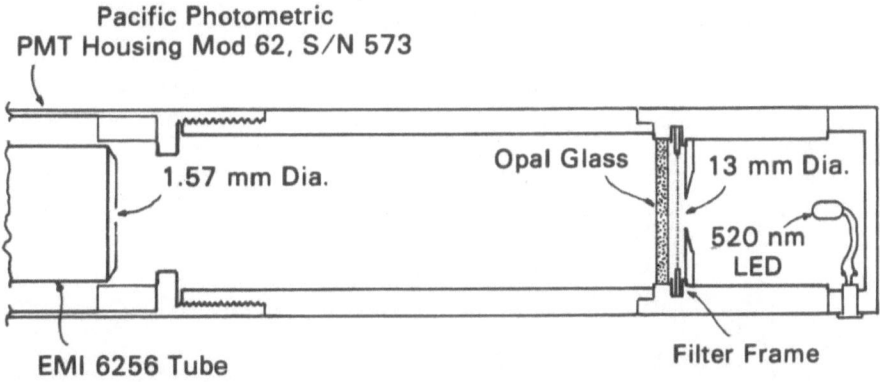

Fig. 1. EPA version of integrating plate method for measuring aerosol absorption.

In making the measurements, a quality assurance procedure was followed which included setting the instrument reponse to a standard value before measuring each filter and setting the instrument zero before measuring each group of 36 filters. Values of I_0 (See Equation 1) were measured in our laboratory in North Carolina before the filters were sent to the field. Later the filters were returned to the laboratory and the quantity I was measured. The data were analyzed using Equation 1. However, the results are expressed in terms of soot concentration using a procedure described later in this paper.

References pp. 125-126.

NCSU Optical System Description – The NCSU system as described by Bennett *et al.* [16] measures aerosol absorption both by a photoacoustic method and by a light transmission method. Fig. 2 is a schematic of the system showing the major components. The NCSU light transmission method uses an opal glass diffuser plate and is similar to the integrating plate method as described by Lin *et al.* [9]. The passband of light, which is effective in causing a photoacoustic signal, is approximately 500-1100 nm as defined by the tungsten iodine light source and the water filter; the passband is 500-600 nm for the light transmission measurements, being limited at the longer wavelengths by the photocathode sensitivity characteristic of the RCA 1P39 photodiode. Measurements were made by placing the sample deposit facing the light source. The light transmission measurements were made in the same way as described above for the EPA system. Photoacoustic measurements required only a measurement subsequent to sample loading since response from a blank Teflon filter is negligible.

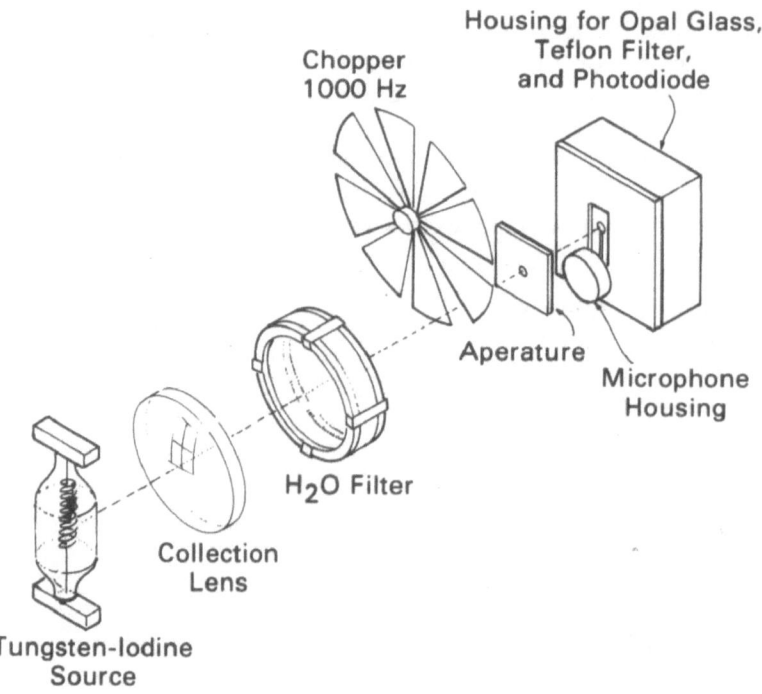

Fig. 2. NCSU system for simultaneous photoacoustic and differential transmission measurements of aerosol absorption.

Calibration of Optical Systems – Calibration standards and data prepared by Bennett *et al.* [16] were used for estimating the elemental carbon content of field loaded filters. The standards were generated by depositing soot from an oxygen-rich propane flame onto Teflon filters held in a manual dichotomous sampler. The mass of soot on the filters was determined gravimetrically to within $\pm$ 25μg per filter. Calibration data obtained by light transmission measurements were plotted as the linear relationship $\ln(I_0/I)$ versus soot loadings in μg per filter. The same calibration curve was used for both the NCSU and the EPA light transmission measurements. Calibration data for the photoacoustic measurements were plotted as photoacoustic signal versus soot loadings per filter.

The carbon loadings used in calibration did not extend below 9.3 $\mu g/cm^2$ (73 $\mu g/$ filter) because of the uncertainty in the gravimetric determinations for the soot standards. The elemental carbon content of the aerosols collected in Houston and in the Shenandoah study were mostly less than 73 μg/filter. For this reason, the extrapolation of the calibration curves towards the origin was crucial to carbon determination.

For light transmission the calibration curve was made to pass through the origin and the point corresponding to the average values of response and of loading. A linear fit to the photoacoustic data was derived by fitting a line from the origin to a point corresponding to the average response and loading of the 6 soot standards of lowest concentration.

Such a calibration of response as a function of carbon loading neglects the presence of ammonium sulfate and other co-collected compounds which are known to alter the response of optical transmission measurements due to their contribution to reflectivity of the sample. [16, 19]. The investigation of these interference effects is currently proceeding at NCSU for both the photoacoustic and differential attenuation measurements of soot on Teflon filters.

DISCUSSION

Background — To compare the various procedures for measuring optical absorption and carbon content of fine particles, it is first necessary to obtain information on the total aerosol composition. Table 5 contains a list of the average concentrations of the components of the fine particle fraction derived from measurements made in the Smoky Mountains, Shenandoah and Houston. Figs. 3 and 4 are time series plots showing how the major constituents in fine particles vary in the Smoky Mountains and Shenandoah, respectively. Fig. 5 is a time series plot of fine particle carbon concentration measured by combustion of samples collected in Houston and Shenandoah.

From these data sets we find that in the Smoky Mountains and Shenandoah, the carbon content is a small part of the total fine particle mass. As will be discussed later, optical absorption measurements of aerosols may be affected by the ratio of absorbing materials to non-absorbing species. Fine particles collected in the Smoky Mountains and Shenandoah were almost always acidic.

Intercomparison of Techniques to Estimate Carbon — The results are shown in Figs. 6 to 8. Fig. 6 and 7 are intercomparisons of optical measurements using a common set of Teflon field loaded filters from the summer 1980 Shenandoah study.

TABLE 5

Average Mass and Elemental Concentrations for Samples Collected in the Great Smoky Mountains, Shenandoah Valley and Houston, TX

	Smoky Mtns. 1978[a] ng/cm³		Shenandoah Valley-1980[b] ng/cm³		Houston TX 1980 ng/cm³	
	F	C	F	C	F	C
Mass	24000	5600	26300	7400	15200	29400
C (Organic)	2200	1200	900	ND*	1200	ND
C (Elemental)	1100	100	1500	ND	1400	ND
Al	20	195	31	315	272	1314
Si	38	580	113	822	660	3483
S	3744	204	4458	255	1086	353
Cl	<10	7	9	161	162	2457
K	40	108	60	129	127	271
Ca	16	322	35	313	186	2106
Ti	<2	10	—	13	18	80
V	—	<1	—	<2	—	—
Fe	28	118	54	161	298	777
Ni	1	1	1	1	1	3
Cu	3	<5	5	6	14	6
Zn	9	<4	11	5	37	34
As	2.2	<1	<1	<1	2	1
Se	1.4	0.2	<1	<1	<1	1
Br	18	5	8	3	26	15
Pb	97	14	51	9	140	44

Averages based on samples collected for the following periods: (a) September 1978; (b) July-August 1980; and, (c) June-July 1980.

*ND = not determined.

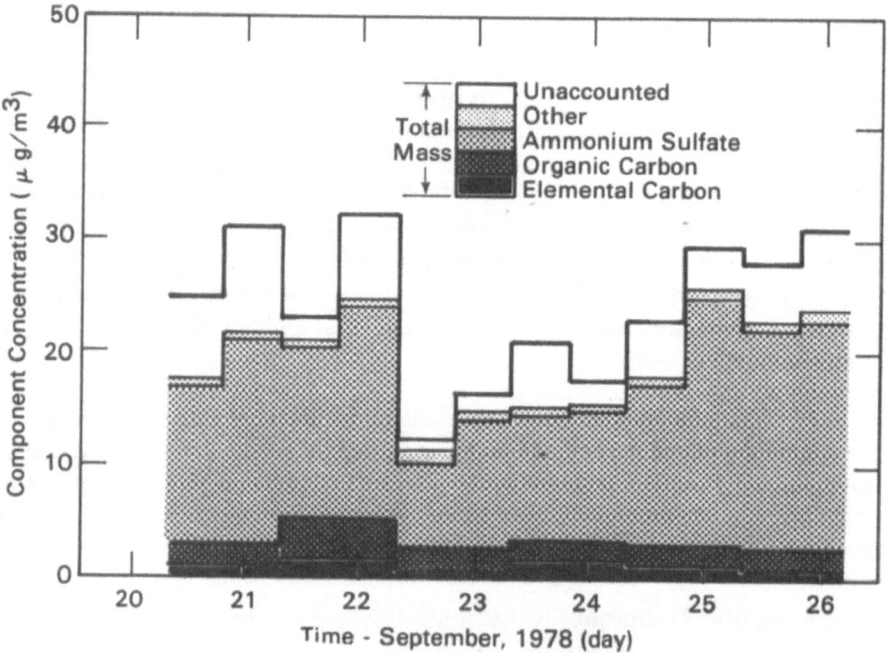

Fig. 3. Time series plot of fine aerosol components in Smoky Mountain study.

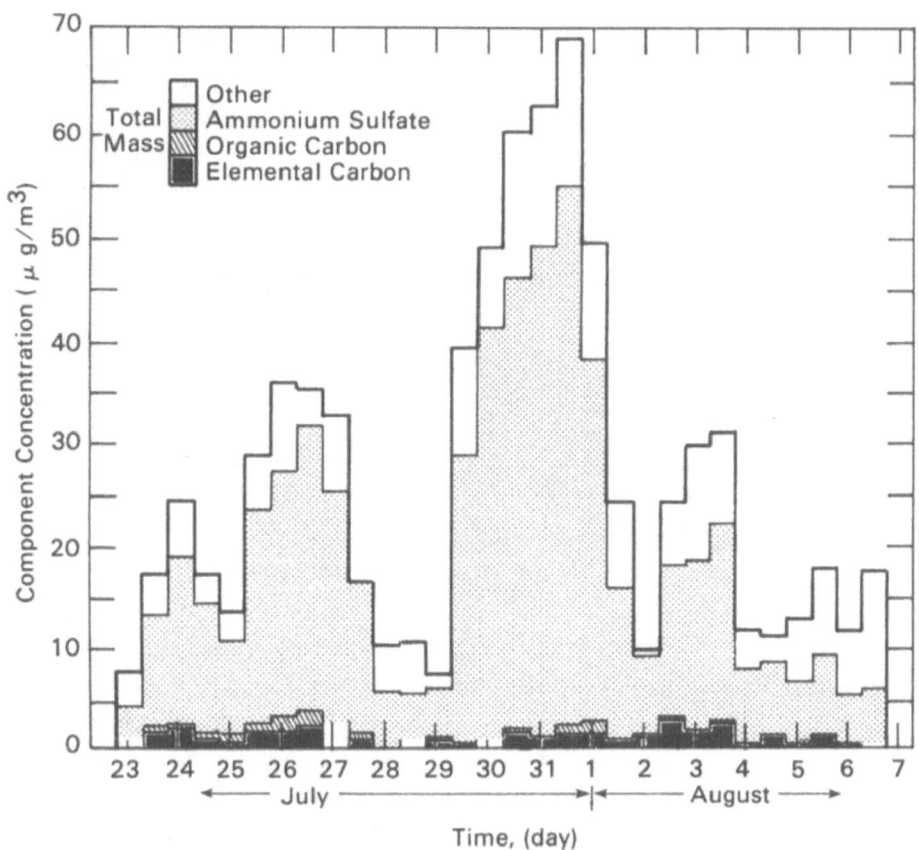

Fig. 4. Time series plot of fine aerosol components in Shenandoah Study. Carbon data are missing for 12-hour periods on the following five days: July 12, 27, 29, 30 and August 6.

Fig. 8 is a comparison between two sets of filters, one set Teflon and the other quartz, sampled simultaneously at a common location in Houston during the period June and July 1980. All samples were of 12 hours duration.

Fig. 6 compares light transmission measurements by the NCSU system (TM #1) and the EPA system (TM #2) in terms of estimated soot concentration. The set of individual comparisons is characterized by a standard linear regression analysis. The two light transmission measurement systems are reading essentially identical values for carbon.

Fig. 7 compares the NCSU photoacoustic and IPM measurements, indicating consistently lower photoacoustic readings and somewhat more scatter than Fig. 6. The lower photoacoustic readings could be due to calibration errors, reflectivity, or multiple scattering by co-collected compounds [16, 19], or to the fact that different passbands of radiation were used in the two techniques. The matter is currently being investigated.

References pp. 125-126.

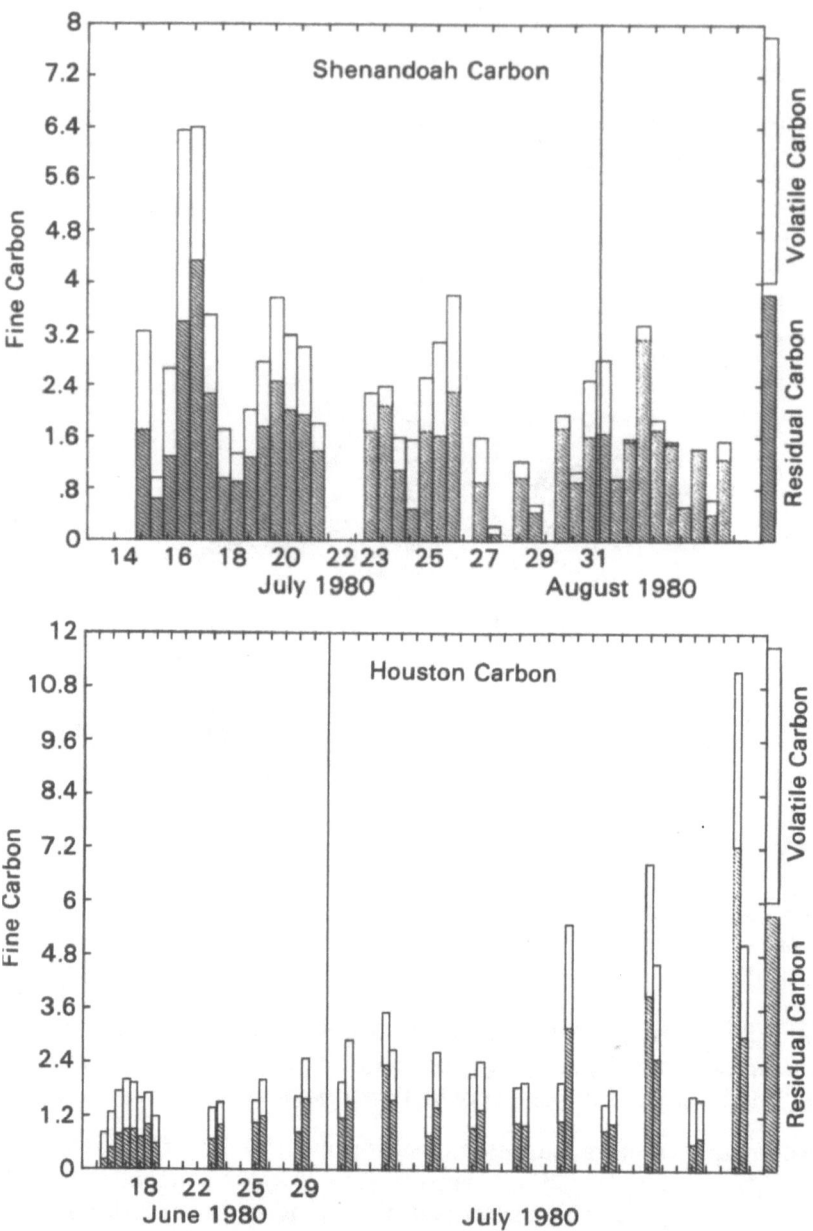

Fig. 5. Time series plot of volatile and residual carbon in Shenandoah and Houston. Carbon measurements in $\mu g/m^3$ were determined by the thermal combustion method.

Fig. 8 compares photoacoustic measurements with combustion measurements for samples collected during the Houston study. The data scatter reflects the combined uncertainty of the two techniques for low carbon values and the fact that two different samplers were used. Similar comparisons between optical and combustion techniques on the Shenandoah filters, for which the range of carbon loadings is

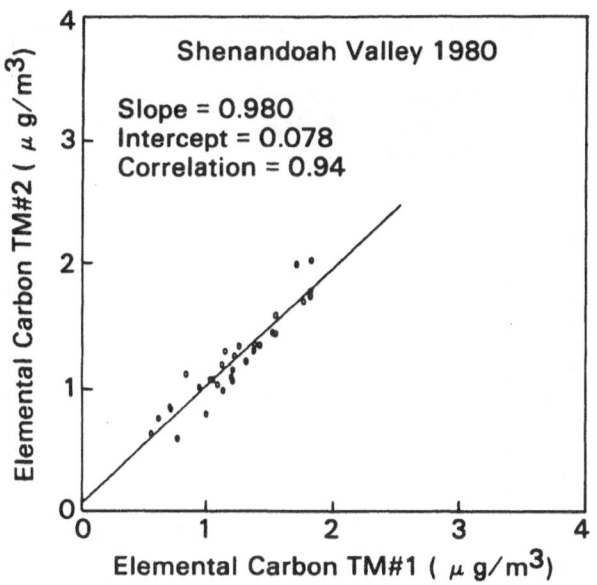

Fig. 6. Comparison of elemental carbon concentrations in micrograms per cubic meter as determined by two different light transmission methods for fine particles (particle dia. <2.5 μm) collected in the Shenandoah Valley during the summer 1980.

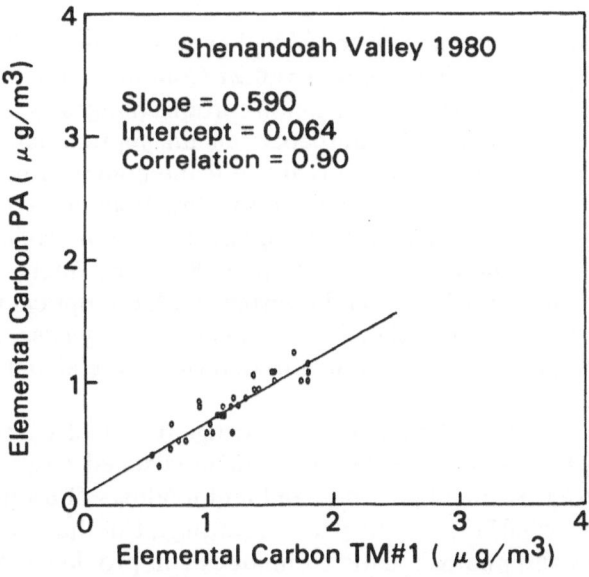

Fig. 7. Comparison of elemental carbon concentrations in micrograms per cubic meter as determined by a photoacoustic method (PA) and an integrating plate method (TM #1) for fine particles (particle dia. <2.5 μm) collected in the Shenandoah Valley during the summer 1980.

References pp. 125-126.

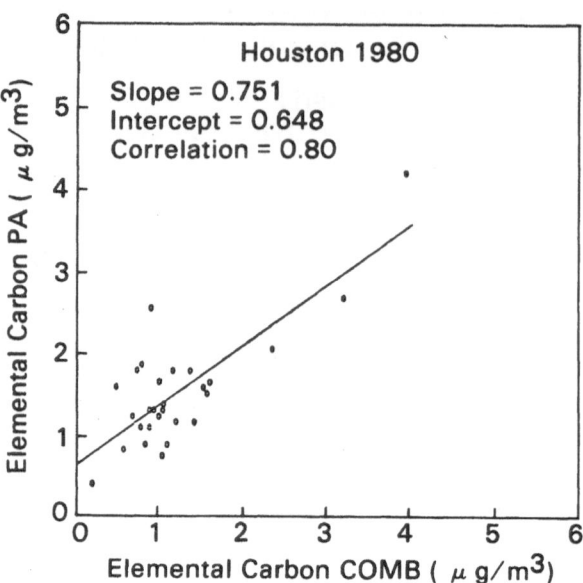

Fig. 8. Comparison of elemental carbon concentrations in micrograms per cubic meter as determined by a Photoacoustic method (PA) and a differential combustion method (COMB) for fine particles (particle dia. ≤ 2.5 μm) collected in Houston, TX during the summer 1980.

even less, show similar characteristics of higher combustion values and of optical technique intercepts, e.g., (TM-NCSU) = 0.21 Combustion + 0.86. The relationship between optical carbon and combustion carbon changes by a factor of 4 for the two sampling locations and the intercepts remain approximately the same. An improvement in agreement would be realized if the combustion estimates of elemental carbon were too high, possibly due to some volatile carbon being read as elemental. Also, some of the samples had significant amounts of iron present in the fine particles. After combustion, even up to 1000°C, the filters were still slightly discolored. The iron, therefore, could interfere with the optical measurements at least for a few of the samples collected in Houston. In any case, more data over a wider dynamic range must be examined to determine the actual comparability of these techniques.

The Shenandoah filter set represents a stringent test of carbon measurement capability for any technique since the values did not exceed 2 μg/m³ (3.6 μg/cm²) for the optical analyses. Elemental carbon ambient loadings of up to 40 μg/cm² have been routinely recorded [13] in typical urban settings. For these urban locations the light transmission and photoacoustic techniques can provide even better nondestructive quantitative analysis of the aerosol absorption. Additional research should eventually clarify the effects of co-collected compounds on estimates of elemental carbon.

SUMMARY AND CONCLUSIONS

- The two light transmission methods are highly correlated for aerosol carbon standards, ($r^2 = 0.99$).
- Photoacoustic measurements of elemental carbon in aerosols collected in the Shenandoah study were about 41% lower than values estimated from light transmission measurements. Explanations for this include: (1) the effects of scattering by co-collected compounds, (2) response differences due to the use of different passbands for the two techniques, (3) uncertainty in calibration of the photoacoustic technique for low elemental carbon loadings.
- Carbon standards deposited on a variety of filter media are needed to calibrate optical and combustion methods used by the various laboratories involved in aerosol characterization.
- Optical methods need to be tested to determine the effects of non-absorbing species such as $(NH_4)_2SO_4$ when measuring the elemental carbon content of ambient aerosols collected on different filter media.
- The thermal combustion measurements were affected by variability in the blanks, variations in the amount of organic vapors adsorbed on the quartz filters during handling and storage as well as the charring of the organics during the combustion process. All of these factors were important since the amount of carbon present in the atmosphere during these studies was relatively low.
- The average elemental carbon content, at least for these studies, was less than 6% of the fine particle mass for samples collected in the Great Smoky Mountains and Shenandoah Valley, and typically less than 5% for samples collected in Houston, Texas.

ACKNOWLEDGEMENTS

The authors gratefully acknowledge the assistance of the following people: C. A. Bennett, Jr. and R. R. Patty of the NCSU Physics Dept. for optical analysis of filters; F. T. Varcoe, J. H. Chance, D. Hern and R. Barbour for collecting samples in the field studies; and G. Gallant and R. Barbour for typing the manuscript.

REFERENCES

1. R. Patterson, Anal. Chem., Vol. 45 (1973), p. 605.
2. S. Gal, F. Paulik, E. Pell and H. Puxbaum, Anal. Chem., Vol. 282 (1976), p. 291.
3. B. R. Appel, P. Colodny and J. C. Wesolowski, "Analysis of Carbonaceous Materials, in Southern California Atmospheric Aerosols." Environ. Sci. Technol., Vol. 10, (1976), pp. 359-363.
4. S. H. Cadle, P. J. Groblicki and D. P. Stroup, Anal. Chem., Vol. 52 (1980), p. 2201.
5. S. Will, "Air Quality In The Rural Ozark Region; Atmospheric Chemical Constituents That Cause Visual Impairment and Their Sources" M.S. Thesis University of Washington, Seattle, WA. 1980.
6. J. Moyers, Private Communication.
7. D. Grosjean, Anal. Chem., Vol. 47 (1975), p. 757.

8. B. R. Appel, E. M. Hoffer, E. L. Kothny, S. M. Wall, M. Haik and R. L. Knights, "Diurnal Variations of Organic Aerosol Constituents in the Los Angeles Basin" Proceedings Carbonaceous Particles In The Atmosphere, LBL Report 9037, Conf. 7803101, UC-11 June 1979.

9. C. I. Lin, M. B. Baker, and R. J. Charlson, Appl. Optics, Vol. 12 (1973), p. 1356.

10. R. E. Weiss, A. P. Waggoner, R. J. Charlson, D. L. Thorsell, J. S. Hall, and L. A. Riley, "Studies of the Optical, Physical and Chemical Properties of Light Absorbing Aerosols" in Carbonaceous Particles in the Atmosphere, Lawrence Berkeley Laboratory, March 20-22, 1978.

11. H. Rosen, A. D. A. Hansen, L. Gundel and T. Novakov, Appl. Optics, Vol. 17, (1978), p. 3859.

12. M. Sadler, R. J. Charlson, H. Rosen and T. Novakov, "An Intercomparison of the Integrating Plate and the Laser Transmission Methods for Determination of Aerosol Absorption Coefficients" LBL Report 11176, July 1980.

13. H. Rosen, A. D. A. Hansen, R. L. Dod, and T. Novakov, Science, Vol. 208 (1980), p. 741.

14. Z. Yasa, N. M. Amer, H. Rosen, A. D. A. Hansen and T. Novakov, Appl. Optics, Vol. 18 (1979), p. 2528.

15. W. A. McClenny and R. W. Shaw, Environmental Sciences Research Laboratory, EPA, Research Triangle Park, N. C., Private Communication.

16. C. A. Bennett, Jr., R. R. Patty, and W. A. McClenny, "Photoacoustic Detection of Carbonaceous Particulates", submitted to Appl. Optics.

17. J. L. Howell and R. A. Palmer, "Photoacoustic Analysis of Atmospheric Aerosol Filters" Paper No. 675, Pittsburgh Conference, Atlantic City, 13 March 1980.

18. R. K. Stevens, T. G. Dzubay, G. Russwurm, D. Rickel, Atmos. Environ., Vol. 12, (1978), p. 55.

19. H. Rosen and T. Novakov, "Optical Attenuation: A Measurement of the Absorbing Properties of Aerosol Particles" in the 1977-1978 Annual Report of the Energy and Environmental Division, Lawrence Berkeley Laboratory, University of California. LBL-8696, UC-11.

DISCUSSION

D. Stedman, *(University of Michigan)*

It is worth pointing out that photoacoustic techniques are by no means all roses. In order to get chopped light converted into sound you have to include a thermal diffusion process. Even if you are working on pure graphite, which, of course, you never are, you are dealing with a crystal with anisotropic thermal diffusion . . . so it is not trivial to get a one-to-one relationship.

R. Stevens

As long as you can empirically calibrate the device against known standards and you understand the optical properties of your photoacoustic cell, many of these problems can be handled by calibration.

V. Mohnen, *(State University of New York)*

Is your figure of 6% elemental carbonaceous material in rural areas consistent with your earlier statement that the visibility reduction that you showed in the first slide is mainly man-made sulfate?

R. Stevens

Yes. In the eastern portion of the United States during the summer, air masses tend to stagnate between the Smoky Mountains and the east coast. This air mass contains aged aerosols in a rather uniform particle size distribution. By the time the air mass moves into the rural areas, sulfate is the dominant species in the fine particle faction. Aerosol elemental carbon generally represents a small fraction of the fine particles. In the 31 day study we did in the Caucusus Mountains of the USSR, sulfate was the dominant species. Looking at 72- and 96-hour trajectories, it appeared that the air mass moved from Eastern Europe to the Caucasus site. Even in these remote areas of the world sulfate still appears to be the major contributor to visibility degradation. This is not to say that we should ignore the contribution of carbon to visibility impairment, but we must keep things in perspective. And, we must develop better procedures to estimate the carbon content of ambient particulate matter.

S. Japar, *(Ford Motor Company)*

I would like to point out that, for something more than a year, photoacoustic techniques have been demonstrated to be a viable method to measure carbon emissions from various combustion processes. We have done it at Ford and Dave Roessler has done it at GM for acetylene flames. In fact, we have demonstrated that for diesel exhaust over a very large concentration range in a dilution tube, the photoacoustic method correlates very well with chemical methods.

R. Stevens

In-situ methods are very valuable tools especially in correlation studies with the collection methods.

J. Ogren, *(University of Washington)*

How did you calculate the optical properties of your laboratory-generated soots for the calibration? And, secondly, I would like to address your observations on the effects of the nonabsorbing aerosols on the integrating plate method. We have been trying to measure absorption in the stratosphere and what appears to be the predominant aerosol up there is sulfur. Impacting the stratospheric aerosol directly on opal glass and in laboratory tests with pure sulfur aerosol, the maximum transmission drop we have observed is on the order of half a precent.

R. Stevens

At the present time, most of our observations with regard to interference of co-collected materials with carbon measurements indicate with our optical transmission system, ammonium sulfate deposited on the teflon filter does produce a significant positive interference in carbon measurements. The interference due to ammonium sulfate appears to be substantially less with the photoacoustic technique compared to optical transmission method.

R. Bradow, *(U.S. Environmental Protection Agency)*

Many of us who have done measurements of organic matter in source measurements have had very little difficulty in separating organic substances thermally from the pyrolizable carbon. Yet in the ambient air, it appears that carbonization of organics occur. Now, we have heard from Dr. Huntzicker that there is a strong possibility that materials of biological origin, simple things like cellulose and the protein material found in plants, are interfering in this method. Might there be three kinds of carbon in which we are interested? In those source samples, which are derived from petroleum combustion, we find two very different kinds of carbon: elemental and the associated organics. In the ambient air there is still a third component, the biological material, which appears to be the principle source of the difficulties in combustion methods.

R. Stevens

Well, I think that is where a combination of techniques plays a very big role. These include: optical microscopy, scanning electron microscopy, X-ray diffraction, and gas chromatography. I think all those will play a big role in aerosol characterization in the next few years.

P. Mueller, *(Electric Power Research Institute)*

If you take the sample through a cyclone or virtual impactor and you end up with two and one half micron particles on your filters, does the biological material interfere?

J. Huntzicker, *(Oregon Graduate Center)*

No, but pyrolysis still occurs on fine particle samples. Charring still happens. Bob, do you know when you remove all the carbon?

R. Stevens

Most of the problems due to biological material seem not to be associated with the fine fraction. Most of the biological material is found in the coarse fraction. For example, we found wings and legs of insects in the coarse fraction when we sampled aerosols at the General Motors site in the Shenandoah Valley.

P. Mueller

We have observed that we do not combust all of the coarse material on filters. It takes more time. So, if you take more time, you might be able to get it. I am not quite sure where the problem really lies.

S. Dattner, *(Texas Air Control Board)*

We have routinely analyzed several summer filters by microscopy and we estimate that there is 20-30 micrograms per cubic meter of pollen in the coarse fraction.

S. Japar

We have recently correlated the in-situ photoacoustic measurements with the integrating plate method using teflon backed filters and found that the correlations are excellent. We found that the integrating plate reading is high by factor of 1-1/2 to 2-1/2 depending on how you orient it in the light beam.

R. Stevens

In the original work of Lin, he oriented the particles in the system away from the light and later reversed this order. I think, however, when the concentrations of carbon on the filter exceed 30 or 40 micrograms per square centimeter, all three methods appear to provide equivalent results. It is at lower concentrations that the problems begin. Carbon measurements in Houston indicate carbon was not the major species in the fine particles. This is typical of air pollution in eastern urban cities. I think California is a special place where the carbon is often the dominant species in the fine particle fraction. In areas where carbon represents a small fraction of the fine particles, optical methods need to be improved in order to account for interference in the carbon measurements by other species co-deposited with carbon.

CARBON ANALYSIS OF ATMOSPHERIC AEROSOLS
USING GRALE AND REFLECTANCE ANALYSIS

E. S. MACIAS and L.-C. CHU

Washington University
St. Louis, Missouri

ABSTRACT

Techniques for total and elemental carbon analysis of atmospheric aerosols are described. Total carbon is determined using the gamma ray analysis of light elements (GRALE) technique which employs the in-beam measurement of γ rays emitted during the inelastic scattering of protons. Elemental carbon is determined by light reflectance. Extensive calibrations of both methods have been carried out. The use of the methods with teflon as well as glass and quartz filters is described.

INTRODUCTION

The importance of atmospheric particulate carbon in visibility reduction, health effects, and other adverse effects of air pollution is being actively discussed in this symposium. Several of the other speakers have described a number of techniques for carbon analysis based on combustion and optical attenuation [1-3]. In this paper we describe two different techniques for total and elemental carbon analysis which we have developed over the past few years. Both methods are totally instrumental, automated, and non-destructive. Total carbon is determined using the gamma ray analysis of light elements (GRALE) technique [4-5]. This method involves the in-beam measurement of γ rays emitted during the inelastic scattering of protons accelerated in a cyclotron. Elemental carbon is determined by a light reflectance method [6]. Much of the analysis has been of aerosols deposited on low carbon glass and quartz filters. The extension of the method to teflon filters is also described [7]. These methods have been used extensively for studies in Charleston, WV [8], St. Louis, MO [5, 9], the southwestern US [10], Los Angeles, CA [11, 12] and China Lake, CA [13].

References p. 143

TOTAL CARBON ANALYSIS BY THE GRALE TECHNIQUE

The total carbon content of the aerosol samples has been determined using the gamma ray analysis of light elements (GRALE) technique [4]. This non-destructive technique is based on the measurement of gamma-ray emissions induced by proton bombardment of aerosol samples. The gamma-ray energy is, in general, unique to a particular nuclide and thus can be used as a signature for the element. Elemental concentrations are obtained in units of $\mu g\,cm^{-2}$ and are not affected by the chemical form of the elements. The GRALE technique has been used for total carbon analysis but the method can also be used for determining the concentration of other elements such as sulfur, nitrogen, and oxygen.

Following collection, the filter samples are mounted in 5 x 5 cm slide mounts after removal of the cellulose backing. The samples are then irradiated with a collimated beam of 7 MeV protons in the external beam facility of the Washington University 135 cm sector focused cyclotron with an automated sample changing chamber maintained in 1 atm of helium. The experimental setup is shown schematically in Fig. 1.

Gamma rays produced in the proton bombardment are detected with a 60 cm^3 lithium drifted germanium Ge(Li) detector (11.7% efficient relative to a 7.5 x 7.5 cm NaI(Tl) detector of 1332 keV gamma rays), with energy resolution of 2.5 keV full width at half maximum for 1332 keV gamma rays. The gamma ray data are stored and processed in a PDP-11 mini-computer with a 28000 word memory (Digital Equipment). The spectra are analyzed immediately after each irradiation with the on-line computer. The intensity of each peak is determined from the integrated peak area after subtraction of background. The peak intensity is corrected for system dead time losses (typically 20%) determined from the area of a pulser peak in the spectrum produced from a 60 Hz tail-pulse generator. These data are normalized to the proton beam intensity and intensities of the filter blanks are subtracted from the aerosol results.

TOTAL CARBON CALIBRATION

Carbon peak intensity is converted into carbon mass by calibrating with standard methionine aerosol ($C_5H_{11}O_2SN$), deposited on the same type of filter. The mass of methionine deposited on the filter is determined using a beta attenuation mass monitor. Filter blanks, atmospheric, laboratory and methionine standard aerosol samples are run under nearly identical conditions which yield nearly equal detector count rates. All samples are analyzed in the same way.

METHOD INTERCOMPARISON

The GRALE technique has been extensively compared to other techniques [4]. In that work the total carbon analysis of the same or duplicate filters was carried out with the GRALE method, the Dohrmann DC-50 carbon analyzer (Envirotech) and the Perkin-Elmer Model 240 Elemental Analyzer (Perkin-Elmer) as shown in Fig. 2.

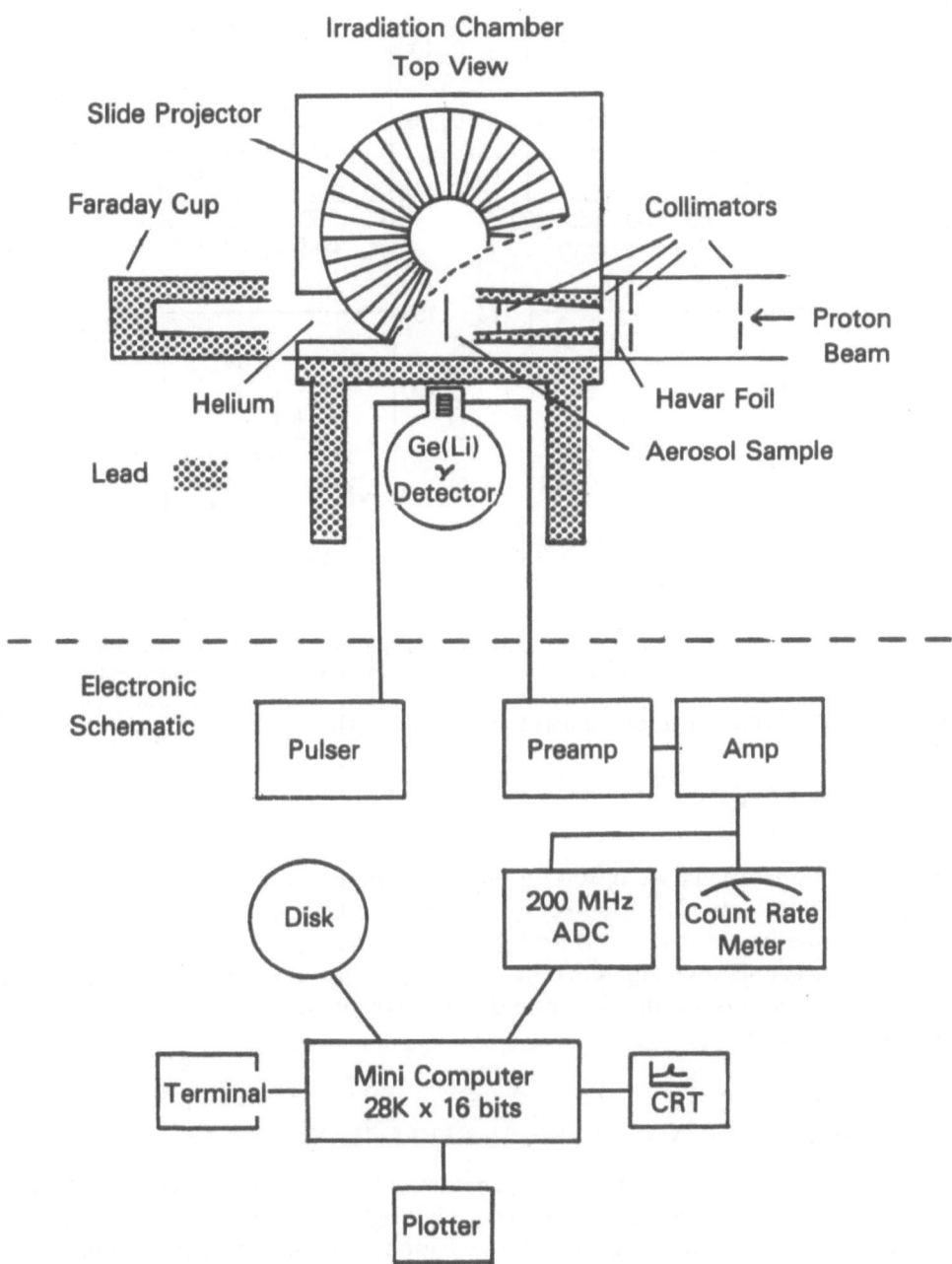

Fig. 1. Schematic diagram of the sample irradiation chamber and electronics used for carbon analysis by the GRALE technique. (From ref. [4]).

References p. 143

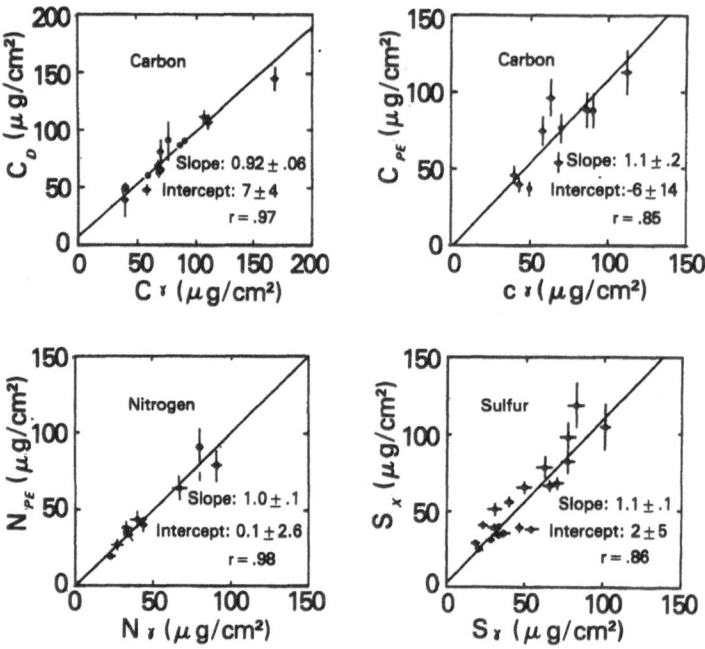

Fig. 2. Method intercomparison of atmospheric fine particle samples. In each plot the data for the γ-ray method is the abscissa. The ordinate is the data for the Dohrmann device in frame a, the Perkin-Elmer device in frames b and c, and x-ray fluorescence in frame d. The line in each pilot is the linear least squares fit to the data [4].

The latter two methods are based on the high temperature combustion of carbon compounds with subsequent detection of CH_4 and CO_2, respectively. This intercomparison indicated that to within 5% there are not systematic discrepancies in any of the methods investigated. The advantages of the GRALE technique over combustion methods are that it is non-destructive and unaffected by the chemical form of the aerosol.

ELEMENTAL CARBON ANALYSIS BY REFLECTANCE

A laboratory reflectometer, shown schematically in Fig. 3 has been constructed for use with fine particle filter samples. This light-tight system consists of two blackened aluminum tubes (5.5 cm dia.) each mounted at 45° to an aluminum sample chamber which holds standard 5 x 5 cm slide mounts. A 12-V tungsten filament lamp powered by a stabilized 12-V supply is used as a light source. Plano-convex lenses (2.3 cm dia., 4.5 cm focal length) are used to focus the incident light onto the sample (0.3 cm²) and reflected light onto a solid state photodetector (N/P silicon solar cell, Centerlab Semiconductor, Inc.). The intensity of the solar cell output is monitored with a digital volt-meter. An unexposed filter, taken from the same filter roll as the samples, is used to determine the initial, or blank, intensity (I_0). Blank values are obtained after the intensity of each sample is determined in order to minimize the effect of any drifts in the electronics.

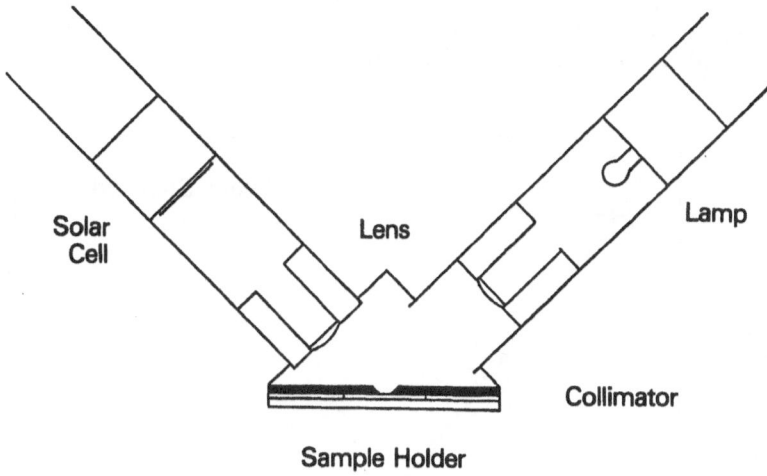

Fig. 3. Reflectance photometer [6].

PREPARATION OF ELEMENTAL CARBON STANDARDS

Aerosol particles of elemental carbon (as soot) were prepared by the combustion of polystyrene, paraffin and butane. The aerosol generated was aspirated into a 5 L container, mixed with filtered air and drawn through a sampler equipped with the on-line reflectometer. The mass of the deposit was determined by beta attenuation [14]. The reflectance of the spot before and after loading was also measured. Since the aerosol could contain appreciable non-elemental carbon (organic) compounds, a method for their removal was developed. A set of butane-generated samples was mounted over a 1.3 cm dia. hole in 5 x 5 cm aluminum plates and total carbon was determined via the GRALE technique. The samples were then placed in the oven of a conventional gas chromatograph equipped with a programmable temperature ramp. Samples were heated at 20°C min^{-1} to a fixed temperature and left at that temperature for 20 min. The reflectance of the sample was measured after cooling to room temperature, then the procedure was repeated at a higher temperature. The reflectance of the sample was nearly constant up to 350°C. Above 350°C the reflectance relative to 25°C increased, indicating loss of elemental carbon. Therefore, a temperature of 300°C was used to remove organic compounds from the soot samples with minimal loss of elemental carbon.

Analysis of butane-generated samples for total carbon content by GRALE after heating to 300°C shows $\sim 25\%$ carbon loss upon heating without appreciable change in reflectance. The after-heating GRALE carbon mass was taken to be the

References p. 143

actual elemental carbon content of the samples which were then considered elemental carbon standards.

A calibration curve, shown in Fig. 4, was prepared by plotting the log of the ratio of the final to initial reflectance of the elemental carbon standards versus the carbon mass determined by GRALE analysis after heating to 300°C. Alternately the total mass of the soot standard after heating to 300°C, determined by beta attenuation, can be used in place of the GRALE Carbon mass.

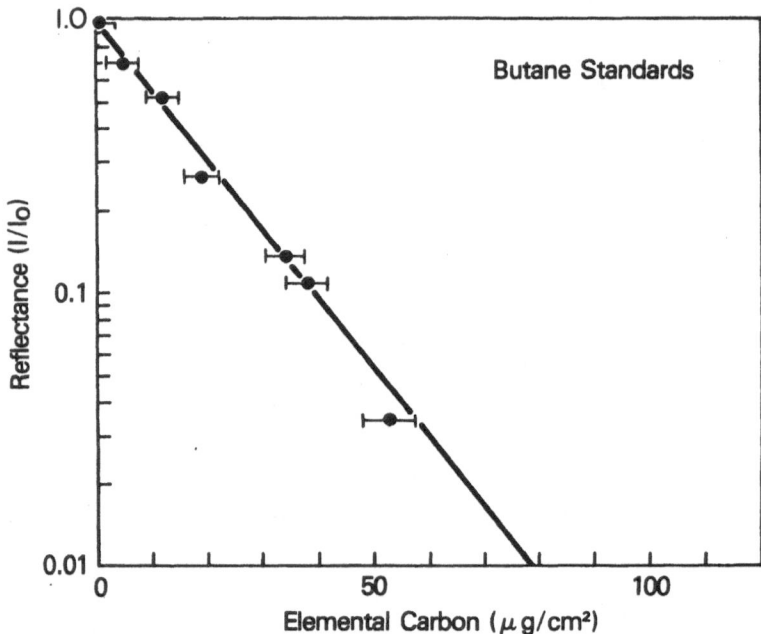

Fig. 4. Calibration curve determined from elemental carbon standards. The log ratio of reflectance is plotted on the ordinate. The elemental carbon mass determined by GRALE analysis after heating to 300°C is plotted on the abscissa [6].

MEASUREMENTS WITH ATMOSPHERIC SAMPLES

Atmospheric fine particle aerosol samples were collected on the roof of the six story Chemistry building of Washington University, located 6 miles west of downtown St. Louis. The immediate area is primarily residential with a major expressway one mile away which is often upwind from the sampling site. Samples were collected on an automated TWOMASS sampler [15] equipped with the on-line reflectometer and β-attenuation system for soot carbon and mass analysis. The samples were analyzed via the GRALE technique for total carbon content. Samples were then heated and the carbon content and reflectance remeasured after heating. Fig. 5 shows the average decrease in total and elemental carbon with temperature for two samples collected on pre-ashed Pallflex Tissue Quartz (2500 QAO) filters. It

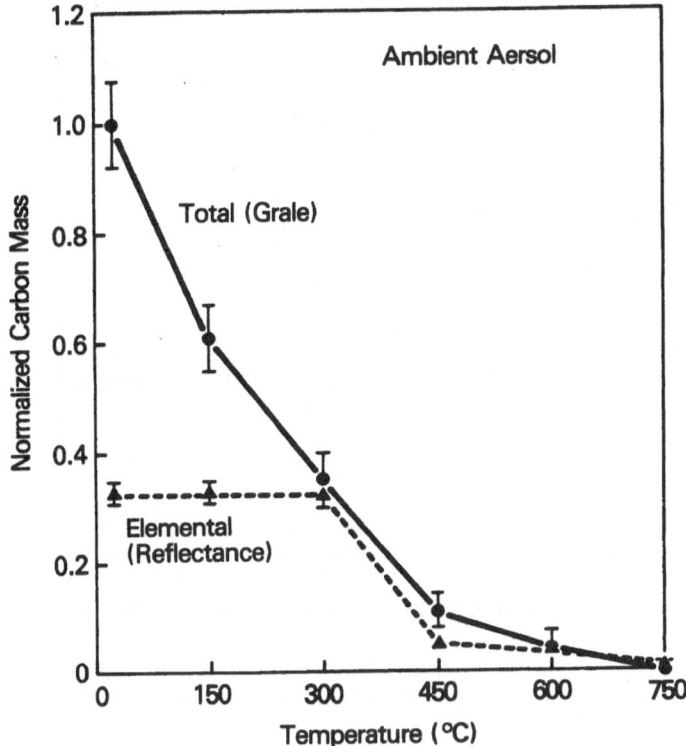

Fig. 5. Average decrease in total and elemental carbon in atmospheric aerosol samples with increasing temperature [6].

can be seen that a significant amount of carbon is lost at temperatures up to 300°C without an appreciable loss of elemental carbon. We assume that this carbon loss at or below 300°C is due to volatile organic compounds. However, we cannot be certain that all organic compounds have been removed at this temperature. The amount of organic carbon loss below 300°C is dependent on the composition of the aerosol. Between 300 and 450°C, the carbon content continues to decrease with a corresponding decrease in reflectance, due primarily to the loss of elemental carbon.

It is instructive to compare the amount of total and elemental carbon determined after heating atmospheric aerosol samples to 300°C in order to estimate the amount of non-elemental carbon remaining at that temperature. This comparison is shown in Fig. 6 for six ambient samples. The data show a high correlation (r = 0.96) indicating a linear relationship between elemental carbon and total mass. The least squares fit to the data has a slope near unity (slope = 0.9 ± 0.2) indicating good agreement between total and elemental carbon. The y intercept = 3 ± 4 μg cm^{-2}, which indicates there may be a small amount of non-elemental carbon remaining at 300°C in these ambient samples.

References p. 143

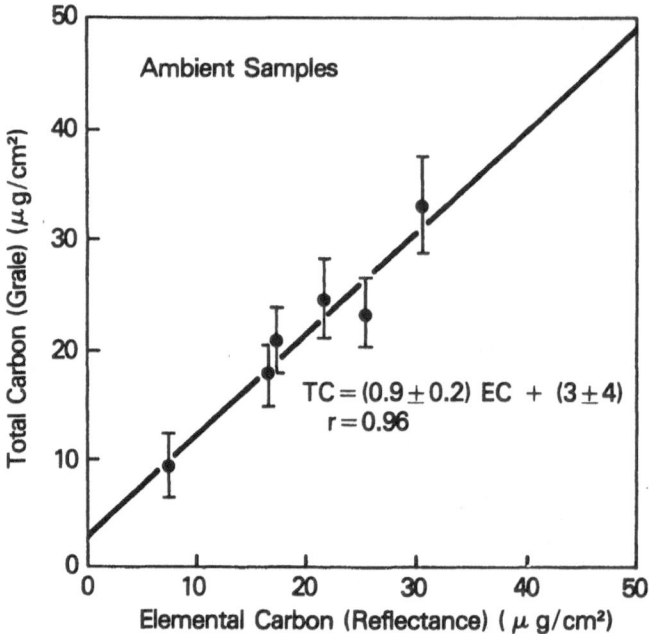

Fig. 6. Comparison of total and elemental carbon determined after heating atmospheric aerosol samples to 300°C. The solid line is the least squares fit to the data [6].

TOTAL AND ELEMENTAL CARBON ANALYSIS ON TEFLON FILTERS

The work described above employed glass or quartz filters. Teflon filters present a different problem because this material is 24% carbon [$(C_2F_4)n$]. Previously no carbon analysis method has been shown to be applicable with this high carbon filter. However, if the carbon to fluorine mass ratio is constant in teflon filters, the fluorine γ-ray can be used as an internal standard in GRALE analysis to indicate the thickness of the filter blanks and to determine the carbon filter blank value. In this section we describe the successful extension of the GRALE and reflectance analysis techniques for total and elemental carbon for use with teflon filters.

Teflon filters with 1 μm pore size mounted on a 37 mm polyolefin ring (Ghia Corp. Pleasanton, CA) and Pallflex E-70 glass fiber filter tape with a detachable cellulose backing (Pallflex Inc. Putnam, CT) were used as filter media.

Ambient aerosols were collected simultaneously with a manual dichotomous virtual impactor sampler [16], and a TWOMASS automated sequential tape sampler from the same aerosol manifold. Both samplers fractionated the aerosol into two size classes; the fine fraction consisting of particles with aerodynamic diameters less than 3μm were examined in this study.

The dichotomous sampler employed teflon filters; the TWOMASS sampler used glass fiber as a filter medium. Six, 12 and 24 hr samples were collected using the dichotomous sampler operated at a flow rate of 14 L/min. Six hr samples were collected with the TWOMASS sampler at a flow rate of 12 L/min. Both samplers were equipped with automatic flow controllers; the flow was also monitored manually.

Standards for total and elemental carbon analyses on teflon filters were prepared with methionine and butane soot aerosols as described previously. GRALE analysis of blank teflon filters yielded a linear relationship between fluorine counts and carbon counts (r = 0.99), as shown in Fig. 7. This indicates that fluorine can be used

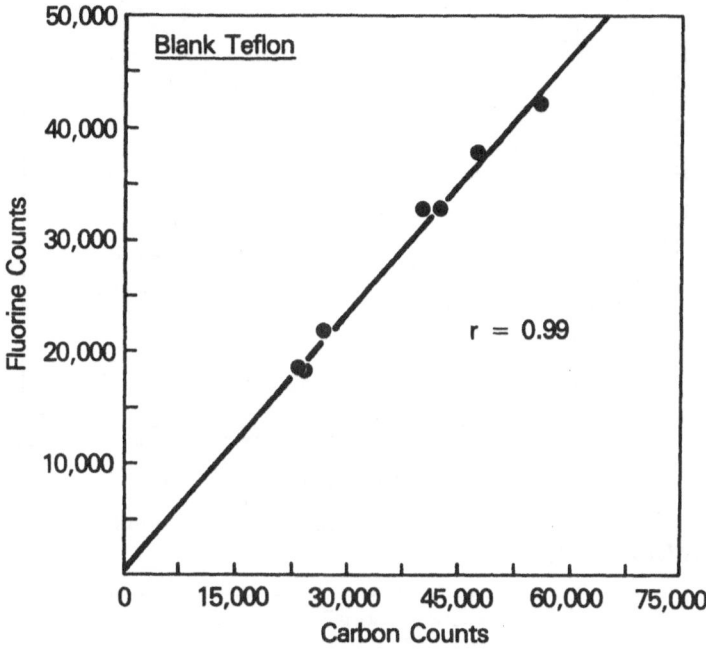

Fig. 7. Correlation between fluorine counts and carbon counts from GRALE analysis for blank teflon filters [7].

for subtraction of carbon background from teflon filters. Teflon blanks gave a C/F count ratio of 0.128 ± 0.004. The large variation in carbon counts on blank filters shows that a reliable calibration can be achieved only when a correction for variations in carbon of the teflon filters is made by using the fluorine peak intensity as an indicator.

The average intensity of the 1.35 MeV fluorine peak from the blank filters was chosen as the standard reference peak for fluorine. The 4.43 MeV carbon peak intensity was obtained using the following equation:

$$C_D = \left[(C/F)_S - \overline{(C/F)}_B \right] \cdot \overline{F_B}$$

where C_D is the total carbon intesity from the deposit, $\overline{F_B}$ is the average fluorine intensity from blank filters, $(C/F)_S$ is the C/F ratio for the sample, and $\overline{(C/F)}_B$ is the average ratio for blank filters.

References p. 143

Using this method, a calibration curve for carbon obtained from methionine standards was constructed as shown in Fig. 8a. As seen in this figure, the propor-

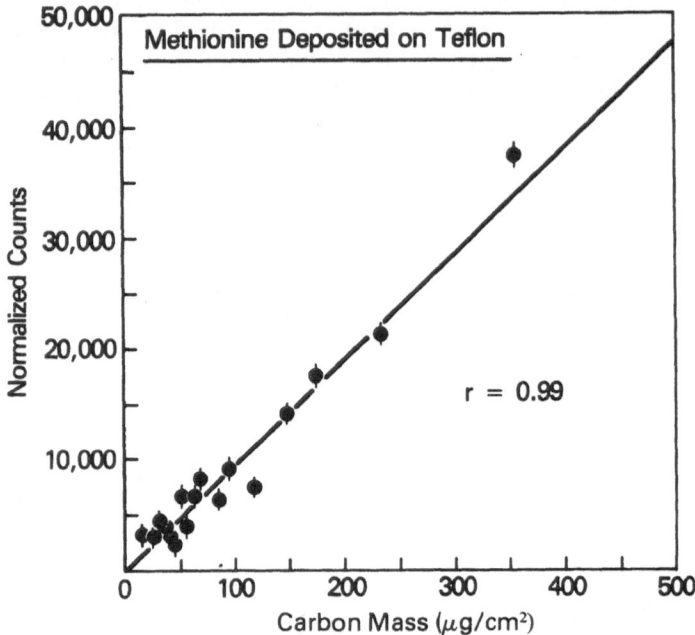

Fig. 8a. Calibration curves for total carbon analysis of methionine aerosols collected on teflon filters by GRALE [7].

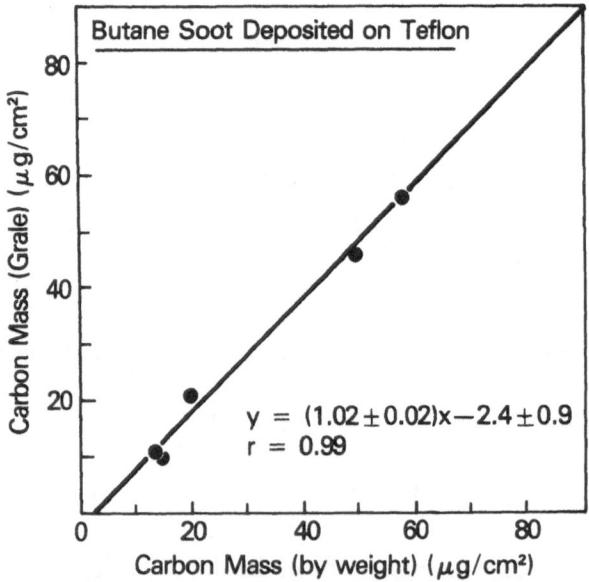

Fig. 8b. Calibration of the laboratory reflectance photometer for carbon generated from a butane flame collected on teflon filters [7].

tionality between the measured γ-ray counts and deposited carbon is excellent (r = 0.99).

A typical calibration curve for elemental carbon, using butane soot deposited on teflon filters, is shown in Fig. 8b. This curve was prepared by plotting the log of the ratio of reflectance of the blank filter (I_O) to the reflectance after addition of the soot (I) versus carbon mass determined by the GRALE technique. The deposited soot mass on these samples was also determined gravimetrically. The comparison of these two methods of calibration is shown in Fig. 9. The linear correlation between the results of two methods is excellent. The slope is 1.02 ± 0.02 and the correlation coefficient is 0.99. The negative intercept indicates that there is a very small amount of impurity in carbon soots which contribute to the weight determined gravimetrically but is not measured as elemental carbon by GRALE.

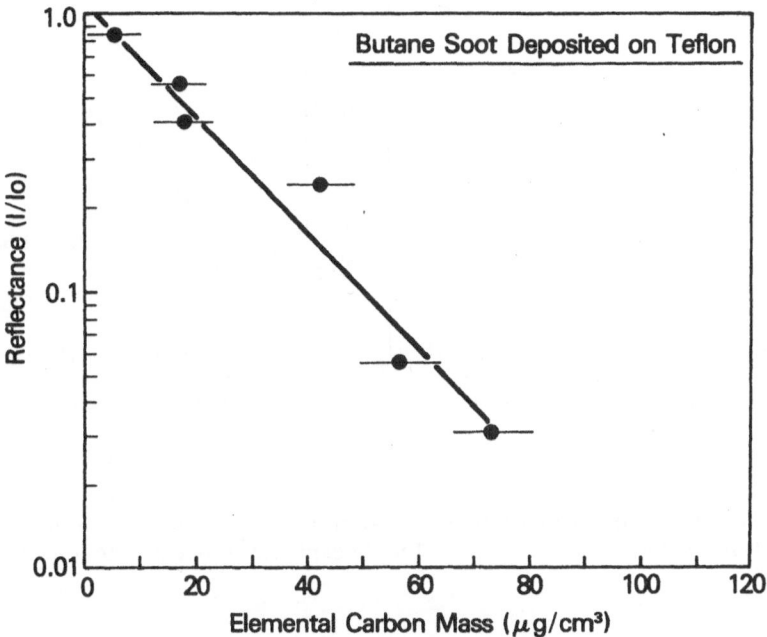

Fig. 9. Comparison of elemental carbon analysis by reflectance using soot standard masses determined gravimetrically and by GRALE [7].

The total carbon detection limit is estimated from the irradiation conditions and counting statistics. The minimum sensitivity (1σ) with teflon filters is 10 $\mu g/cm^2$ which corresponds to 1.5 $\mu g/m^3$ for 6 hr sampling at 14 L/min with a 10 mm diameter sample deposit. This compares to 4.5 $\mu g/cm^2$ on glass fiber filters or 0.7 $\mu g/m^3$ for 6 hr sampling under the same conditions as above.

The reflectance analysis of elemental carbon on teflon filters has a minimum detectable limit of 1.8 $\mu g/cm^2$ or 0.3 $\mu g/m^3$ for six hour sampling as given above. The upper detection limit of the reflectance method due to saturation of the surface is ~70 $\mu g/cm^2$.

References p. 143

COMPARISON OF CARBON ANALYSIS ON TEFLON
AND GLASS FIBER FILTERS

Fine ambient aerosols were collected on both teflon and glass fiber filters during the period of February 16-March 8, 1980, in St. Louis. The comparison of the results on teflon and glass fiber filbers for total carbon analysis by the GRALE technique and for elemental carbon by reflectance analysis during this period is shown in Fig. 10. The results from both filters agree quite well (total carbon slope = 0.99 ± 0.9; r = 0.95; elemental carbon slope = 1.18 ± 0.26; r = 0.93). These data are summarized in Table 1.

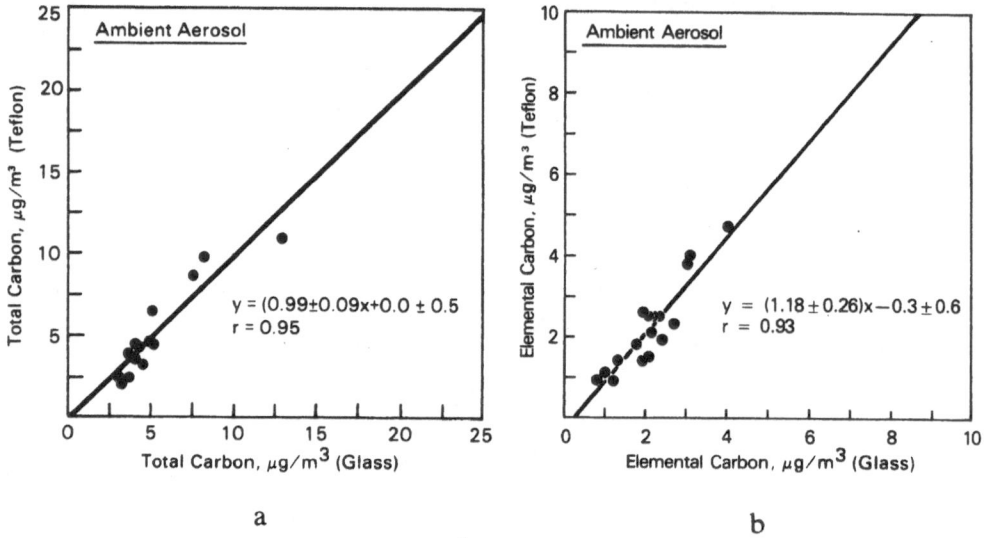

a   b

Fig. 10. Comparisons of (a) total carbon analysis by the GRALE technique and (b) elemental carbon analysis by the reflectance method for atmospheric aerosols on teflon and glass fiber filters [7].

TABLE 1

Average Ambient Aerosol Carbon Concentration Measured on Teflon and Glass Filters. Samples were collected in St. Louis, MO during February and March, 1980.

	All Values Are $\mu g/m^3$	
	Teflon Filters	Glass Fiber Filters
Total Carbon	5.5 ± 2.2	5.5 ± 2.7
Elemental Carbon	1.8 ± 1.1	2.0 ± 0.8
Mass (beta guage)	——	22.1 ± 12.9

ACKNOWLEDGEMENTS

The authors are indebted to the Washington University cyclotron staff for their assistance in sample irradiation. We would also like to thank Charles Lewis, Carole Sawicki, and Robert Stevens of the USEPA for help and encouragement in various stages of this work. This work has been supported in part by the U.S. Environmental Protection Agency, Inorganic Pollutant Analysis Branch under grant R806005.

REFERENCES

1. *J. J. Huntzicker, R. L. Johnson, J. J. Shah and R. A. Cary, These Proceedings.*
2. *R. K. Stevens, T. G. Dzubay, W. A. McClenny and M. A. Mason, These Proceedings.*
3. *H. E. Gerber, These Proceedings.*
4. *E. S. Macias, C. D. Radcliffe, C. W. Lewis and C. R. Sawicki, Analy. Chem., Vol. 50 (1978), p. 1120.*
5. *E. S. Macias, R. Delumyea, L.-C. Chu, H. Appleman, C. D. Radcliffe and L. Staley, "The Determination, Speciation and Behavior of Particulate Carbon," Proceedings of the Conference on Carbonaceous Particles in the Atmosphere, CONF-7803101, Berkeley, California, (1979), pp. 70-78.*
6. *R. G. Delumyea L.-C. Chu and E. S. Macias, Atmos. Envir., Vol. 14 (1980), p. 647.*
7. *L.-C. Chu, R. G. Delumyea and E. S. Macias, Anal. Chem. (Submitted for publication).*
8. *C. W. Lewis and E. S. Macias, Atmos. Envir., Vol. 14 (1980), p. 185.*
9. *L.-C. Chu and E. S. Macias, "Carbonaceous Urban Aerosol-Primary or Secondary." Chemical Composition of Atmospheric Aerosol: Source/Air Quality Relationships, (Edited by E. S. Macias and P. K. Hopke) ACS Books, Washington, DC (in press).*
10. *E. S. Macias, J. O. Zwicker, J. R. Ouimette, S. V. Hering, S. K. Friedlander, T. A. Cahill, G. A. Kuhlmey and L. W. Richards, Atmos. Envir. (Submitted for publication).*
11. *G. R. Cass, P. M. Boone and E. S. Macias, These Proceedings.*
12. *M. H. Conklin, G. R. Cass, L.-C. Chu, and E. S. Macias, "Wintertime carbonaceous aerosols in Los Angeles: an exploration of the role of soot," Chemical Composition of Atmospheric Aerosol: Source/Air Quality Relationships, (Edited by E. S. Macias and P. K. Hopke) ACS Books, Washington, DC (in press).*
13. *J. R. Ouimette, "Aerosol Chemical Species Contributions to the Extinction Coefficient," California Institute of Technology. Ph.D Thesis, 1980.*
14. *E. S. Macias and R. B. Husar, "A review of atmospheric particulate mass measurements via the beta attenuation technique." In Fine Particles: Aerosol Generation, Measurement, Sampling and Analysis, (Edited by B.Y.H. Liu), (1976), pp. 535-564.*
15. *E. S. Macias and R. B Husar, Envir. Sci. Technol., Vol. 10 (1976), p. 904.*
16. *T. G. Dzubay, R. K. Stevens and C. M. Peterson, X Ray Fluorescence Analysis of Environmental Samples, (Edited by T. G. Dzubay) Ann Arbor Science, Ann Arbor, Michigan, 1977.*

DISCUSSION

L. Currie *(National Bureau of Standards)*

What is the typical magnitude of carbon uncertainty in micrograms per square centimeter caused by the nuclear measurement of carbon and fluorine on Teflon filters?

E. Macias

The measurement of total carbon on Teflon filters using the carbon to fluorine ratio has an uncertainty of about 1 to 2 $\mu g/cm^2$ for a 30 mm deposit. This is about a factor of 2 higher than a total carbon measurement on glass filters by GRALE.

L. Currie

I just wanted to note that our laboratory has used the photonuclear activation method to measure carbon eleven.

R. Stevens, *(U.S. Environmental Protection Agency)*

I noticed that the lowest concentration on your graph was 5 $\mu g/m^3$, but the highest concentration we saw in the Smoky Mts. and the Shenandoah Valley was only 2 or 3 $\mu g/m^3$. Would your sensitivity allow you to get down to these levels?

E. Macias

Yes. For sampling times of 6 hours at 18L/ min with a 10 mm diameter sample, the minimum detection limit is about 1 $\mu g/m^3$. Better sensitivity can be obtained by increased the sampling time or irradiation time.

OPTICAL TECHNIQUES FOR THE MEASUREMENT
OF LIGHT ABSORPTION BY PARTICULATES

H. E. GERBER

Naval Research Laboratory
Washington, D. C.

ABSTRACT

A review is presented of the available techniques of measuring particulate light-absorption properties, the imaginary part of the particulates' refractive index and the aerosols' absorption coefficient. Included is a discussion of the applicability, accuracy, and cost of the techniques; and data collected to date with the techniques is summarized. Evidence is presented to answer the question: Does a quantitative relationship exist between the atmospheric light-absorption properties of particulates and elemental carbon? Finally, research areas are identified in which instrumentation development and field work are required to improve the understanding of the role absorbing particulate species play in atmospheric optics.

INTRODUCTION

Light-absorption properties of aerosols and aerosol particles are described in the literature in various ways, but with similar labels, which creates a potential for confusion. Hence, it is worthwhile to begin with a brief summary of the popular descriptions. The absorption properties most often seen are n_2, n_2/ρ_p (m^{3}g^{-1}), b_a (m^{-1}), b_a/M_v (m^{2}g^{-1}), and ω_0, where the first two are properties of the aerosol particles, while the others are properties of the aerosols (aerosol particles and the atmosphere). The imaginary index n_2 of the complex refractive index $n = n_1 - in_2$ is related to the absorption coefficient D (m^{-1}) of bulk material which has the same chemical composition as that of the particle $n_2 = D\lambda/4\pi$, where D is given by the exponential light-attenuation law for the bulk, λ is the wavelength of light, and n_1 is the real part of the refractive index n of the material. Reference to the mass absorption coefficient n_2/ρ_p is often found in the German literature, where ρ_p is the density of the particles. The absorption coefficient, b_a, is the absorption per unit length of the aerosol and b_a/M_v is the specific absorption of the aerosol, where the division by M_v normalizes the absorption to the mass of the particles per unit volume of the atmosphere. The albedo of single scattering ω_0 equals $(b_e - b_a)/b_e$, where b_e (m^{-1}) = $b_a + b_s$ is the atmospheric extinction coefficient, and b_s (m^{-1}) is the scattering coefficient. The specific extinction (or attenuation) b_e/M_v (m^{2}g^{-1}) and specific scattering b_s/M_v (m^{2}g^{-1}) are sometimes also used. The indices and the coefficients are related by Mie theory which describes the interaction between electromagnetic energy and the aerosol particles.

The major groups with interest in aerosol and aerosol-particle light absorption are climatologists, the military, the solar energy community, environmental scientists, and the automobile industry. Generally, each group considers a different set of the absorption properties given above as the most important. For example, the military used n2 with Mie codes, climatologists in addition desire ω_0 for calculating potential climatic change, and environmental scientists use b_a to estimate the deterioration of the optical quality of the atmosphere. Since different absorption properties are often desired and since experimental techniques usually only give one or another of the properties, it becomes desirable to know how the various techniques compare and how the absorption properties are related to each other.

This paper briefly summarizes the existing optical techniques to measure light-absorption properties in the visible and infrared parts of the spectrum. Values of n_2 collected with the techniques in the atmosphere are given for the visible part of the spectrum, and the discussion is also limited to this region. The infrared should not be ignored, however, since certain salt and soil particles absorb significantly in the infrared windows; the infrared data is much more scarcer than data in the visible part of the spectrum, and more difficult to interpret. The relationship between the absorption properties n_2, b_a, ω_0 is explored, and the role of elemental carbon is described. In addition, new developments in light-absorption measurement are listed, and recommendations for future research are given.

EXPERIMENTAL TECHNIQUES

Reported techniques of measuring light-absorption properties of aerosols and aerosol particles can be divided into four categories as shown diagrammatically in Fig. 1. Along with each sketch is a partial list of investigators who have either contributed to the development of a technique or have made use of it.

Bulk Sample — In the bulk-sample category, ambient particles are captured on a filter or on a slide by impaction. This deposit is then subjected to a variety of means usually to extract b_a; although, in some cases [1-3] n_2/ρ_p, and in others [11-12] n_2 are obtained. An approximation formula from Mie theory is used to estimate n_2 from the measured values of b_a (see Role of Elemental Carbon); and if b_s is measured at the same time with a nephelometer (e.g., see Charlson et al.[4]) then ω_0 is obtained.

The means used to extract b_a or n_2/ρ_p depend on the investigator. Fischer [1] measured scattering and extinction through a thick particle deposit located in an integrating sphere. His procedure included measuring ρ_p which was needed to

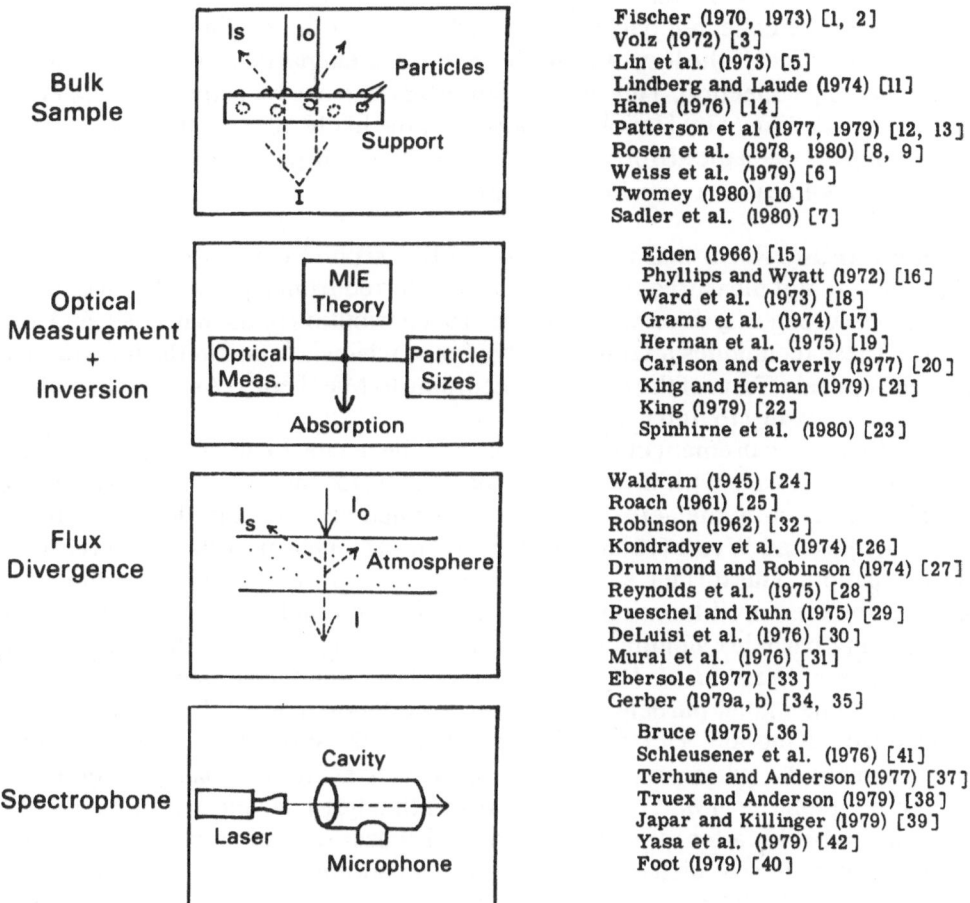

Bulk Sample

Optical Measurement + Inversion

Flux Divergence

Spectrophone

Fischer (1970, 1973) [1, 2]
Volz (1972) [3]
Lin et al. (1973) [5]
Lindberg and Laude (1974) [11]
Hänel (1976) [14]
Patterson et al (1977, 1979) [12, 13]
Rosen et al. (1978, 1980) [8, 9]
Weiss et al. (1979) [6]
Twomey (1980) [10]
Sadler et al. (1980) [7]

Eiden (1966) [15]
Phyllips and Wyatt (1972) [16]
Ward et al. (1973) [18]
Grams et al. (1974) [17]
Herman et al. (1975) [19]
Carlson and Caverly (1977) [20]
King and Herman (1979) [21]
King (1979) [22]
Spinhirne et al. (1980) [23]

Waldram (1945) [24]
Roach (1961) [25]
Robinson (1962) [32]
Kondradyev et al. (1974) [26]
Drummond and Robinson (1974) [27]
Reynolds et al. (1975) [28]
Pueschel and Kuhn (1975) [29]
DeLuisi et al. (1976) [30]
Murai et al. (1976) [31]
Ebersole (1977) [33]
Gerber (1979a, b) [34, 35]

Bruce (1975) [36]
Schleusener et al. (1976) [41]
Terhune and Anderson (1977) [37]
Truex and Anderson (1979) [38]
Japar and Killinger (1979) [39]
Yasa et al. (1979) [42]
Foot (1979) [40]

Fig. 1. Experimental techniques of measuring light absorption by aerosols and aerosol particles. The numbers in brackets refer to the reference list at the end of the paper.

calculate n_2/ρ_p. Volz [3] mixed the particles and other bulk samples with KBr, which upon pressing formed a glassy pellet which was analyzed with infrared spectroscopy. Lin *et al.* [5], Weiss *et al.* [6], and Sadler *et al.* [7] used an "integrating plate method" which was similar to Fischer's [1] technique; however, the later work [5-7] used only a tenuous particle deposit to find b_a. Rosen [8, 9] varied the technique by measuring the transmittance of a laser beam through the filter; and Twomey [10] devised a method which is not influenced by light scattered by the particles. A different approach was used by Lindberg and Laude [11] and Patterson [12, 13] who used measurements of the diffuse light reflected off the particle layer and the Kubelka-Munk theory to infer n_2.

The bulk-sample techniques are attractive due to the generally inexpensive and simple instrumentation needed for the measurements. Thus, these techniques are good for large sampling networks. On the negative side, some uncertainty exists in these measurements, in that it is difficult to establish how well the absorption properties of the collected particles resemble those of the original airborne particles. A major effect is the loss of water from the ambient particles which are usually dried following collection. This error is important and can be estimated when n_2 is desired, as was shown by Hänel [14]; however, the likely alteration of the physical relationship between the absorbing and the nonabsorbing parts of the particles upon deposition could cause errors in both b_a and n_2.

Optical Measurement and Inversion — In this second category the absorption properties are not measured directly, but are inferred from a combination of other optical measurements, information on the size distribution of the particles, and Mie theory. These techniques attempt to solve for n_2 which is found in the integral that sums the optical effects of the particles according to Mie theory. An inversion of the integral is necessary to find n_2.

The optical measurement of the aerosols and the means to determine the particle size distribution vary with the investigator. Eiden [15] measured the ellipticity of sunlight polarized by the aerosols, and he assumed various particle size distributions. Phyllips and Wyatt [16] and Grams et al. [17] measured the scattered-light phase function. The former suspended and observed one particle at a time and matched the measured function to a catalog of such functions, and the latter observed many particles simultaneously and used an impactor for the particle size measurements. Ward et al. [18] used bistatic laser scattering, aureole measurements, and one model particle size distribution. Herman et al. [19], Carlson and Caverly [20], King and Herman [21], and King [22] applied the "diffuse-direct radiation method" where the direct and diffusively-scattered solar radiations are measured; Spinhirne et al. [23] used multi-zenith-angle lidar measurements of the extinction to backscatter ratio of the aerosol. The last several investigators [21-23] estimated particle size distributions by using an additional inversion of the solar spectral optical-depth measurements.

Whereas the category-one techniques and two techniques [16, 17] in the second group give point measurements of the absorption properties, the remainder of the methods have the advantage of giving remote measurements which are the effective mean of the absorption properties over a path in the atmosphere. Precise radiometric measurements are necessary for these techniques. The greatest potential for error comes from the necessity to make a priori assumptions about the physical nature of the aerosol particles. The particles are usually assumed to be spherical (even though it is known that real particles which deviate from that ideal shape can mimic spherical particles with unrealistic absorption properties), the relationship of the absorbing to the nonabsorbing part of the particles is specified (e.g., the particles are often assumed to be homogeneous absorbers), and the dependency of absorption properties with particle size is not considered. Errors introduced by these effects are difficult to evaluate, hence they have generally not been included. Measurement of size distribution is also difficult.

Flux Divergence — In the flux-divergence techniques, measurements are made of the net radiative flux (difference between the upward and downward fluxes) at two levels in an aerosol column through which radiation is passing. The difference in the net flux between the levels results in the total absorption in the layer, and by subtracting the calculated absorption of the gases from the total absorption, a direct measure of b_a is obtained. Radiometers mounted on aircraft were used to make the vertical profiles of the net flux in the atmosphere [24-31], except in one case [32] where measurements on the ground were used. Others [33-35] have tried similar measurements in multipass cells filled with ambient aerosols. In these cases the net loss of radiation by a light beam passing through the cells was measured, the scattering coefficient of the aerosol was measured separately and subtracted from the net loss to obtain b_a.

The flux-divergence techniques have the advantage of providing *in situ* measurements; however, they are prone to large errors, since the absorption value usually results from differencing two large and nearly equal quantities. In some cases reasonable accuracies were obtained, e.g., see Murai *et al.* [31] and Gerber [35].

Spectrophone — The techniques in this category rely on the heating of aerosol particles by their absorption of pulsed light energy. The particles rapidly transfer this energy to the surrounding medium in the form of acoustical energy which a microphone measures to give an indication of b_a. The laser has often been used as the light source [36-39]; a broad band source which simulates solar radiation has also been used [40]. While the preceding techniques provide *in situ* measurements, other approaches have used the spectrophone principle for particles dispersed in KBr [41] and for particles collected on filters [42].

Spectrophones have been especially useful for collecting absorption data in the infrared. Generally large light energies are required to obtain measurements when the particles are weakly absorbing. This raises the possibility that for the *in situ* techniques the volatile components lost by the particles during their exposure to the pulsed light will cause errors in the interpretation of the acoustical signal (see Baker [43]).

MEASUREMENTS OF n_2

Measurements given in the literature of the imaginary index n_2 of atmospheric aerosol particles interacting with visible light near $\lambda = 550$ nm are shown in Fig. 2. This wavelength region was chosen, since it permitted the comparison of a large fraction of the reported values of n_2.

The data in Fig. 2 shows a variability in n_2 in excess of two orders of magnitude. Thus it becomes a difficult choice when, for instance, a value of n_2 is desired for modeling with Mie theory. Two factors influence the variability. The errors inherent in each of the categories of instrumentation described in the preceding contribute to the spread of the values, and so does the natural variability in the absorption properties of the ambient particles. An instrumentation intercomparison is an appropriate means to help separate these two causes of variability (see NEW DEVELOPMENTS). An inspection of the notes in Fig. 2 permits some conclusions to be drawn on the variability of n_2 as a function of geographical region, although there

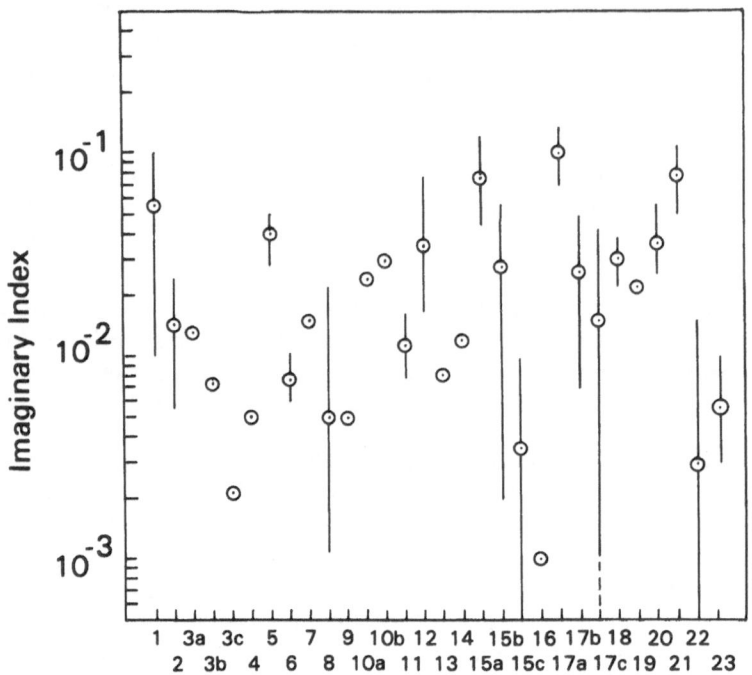

Reference	Index	Notes
1 Eiden (1966) [15]	.055	urban, relative humidity (RH) 70%
2 Volz (1972) [3]	.014	water insoluble rain residue
3a Fischer (1973) [2]	.013	urban, W. Germany, RH=35%
3b Fischer (1973) [2]	.0074	rural, W. Germany, RH=35%
3c Fischer (1973) [2]	.0021	Atlantic Ocean, RH=35%
4 Ward et al. (1973) [18]	.005	urban, Florida
5 Lin et al. (1973) [5]	.04	urban, NYC
6 Lindberg and Laude (1974) [11]	.0077	desert, Southwest US
7 Kondratyev et al. (1974) [26]	.015	desert, Central Asia*
8 Grams et al. (1974) [17]	.005	soil derived aerosol, Southwest US
9 Drummond and Robinson (1974) [27]	.005	measured during BOMEX*
10a Hänel (1976) [14]	.024	Atlantic Ocean, RH=70%
10b Hänel (1976) [14]	.030	urban, W. Germany, RH=70%
11 DeLuisi et al. (1976) [30]	.011	rural, Southwest US
12 Murai et al. (1976) [31]	.035	urban, Tokyo*
13 Patterson et al. (1977) [12]	.008	Saharan aerosol
14 Carlson and Caverly (1977) [20]	.012	Saharan aerosol
15a Weiss et al. (1979) [6]	.076	urban, US
15b Weiss et al. (1979) [6]	.027	rural, US
15c Weiss et al. (1979) [6]	.0035	remote, Southwest US, Mauna Loa*
16 Foot (1979) [40]	.001	rural, England*
17a Gerber (1979) [35]	.10	industrialized harbors, Europe
17b Gerber (1979) [35]	.026	Atlantic Ocean, east of Azores
17c Gerber (1979) [35]	.015	Mediterranean Sea
18 King (1979) [22]	.0306	urban, Southwest US
19 Patterson (1979) [13]	.022	urban, Denver "brown cloud"
20 Rosen et al. (1980) [9] +	.036	urban, Anaheim
21 Sadler et al. (1980) [7] +	.078	urban, U. of Washington
22 Spinhirne et al. (1980) [23]	.003	urban, Southwest US
23 Rahn and Shaw (1980) [44]	.0055	remote, Arctic*

Fig. 2. The imaginary part n_2 of the refractive index of atmospheric aerosol particles for visible light with a wavelength near 550 nm. The points give the mean, and the vertical lines give the range of the measurements; except in one case [23] where they give the measurement uncertaintiy (*n_2 was found by inverting measurements of the single scattering albedo with Fig. 3; †n_2 was estimated with Eq. 1. The ratio of particulate carbon mass to total particulate mass is from Macias *et al.* [47], and the mass measured on the quartz and millipore filters was assumed to be correct.)

is much overlap in the data. The largest values of n_2, up to about 0.1, are found in urban areas. Rural areas generally show smaller values, soil derived particles give a n_2 of about 0.01 and smaller, and the few measurements in remote areas show the smallest values. Those trends are suggested especially by the work of Fischer [2] and Weiss *et al.* [6]. In comparing n_2 values by instrumentation category, the optical-measurement-and-inversion category generally gives the smallest values of n_2 [16-18, 23]. Foot [40] gives an unusually small value of n_2 for a rural area, and Spinhirne *et al.* [23] for an urban area.

ROLE OF ELEMENTAL CARBON

It has been suggested for some time that the absorption of visible light is dominated by the soot-like particles in the combustion residues of various fuels, e.g., see Twitty and Weinman [45] and Fischer [2]. Not until recently, however, have chemical identifications of the dominant absorber been made, and have quantitative relationships between this absorber and some of its optical absorption properties been specified. Rosen *et al.* [8] found that the Raman spectra of radiation backscattered from ambient particles collected on a filter correlated strongly with light absorption by particles for the case where the spectra corresponded to a form of graphitic carbon (elemental carbon). The resistance of the absorber to high temperatures also indicated graphitic carbon [46]. Weiss *et al.* [6] also concluded that a graphitic-type of carbon was the predominant absorber in urban and rural areas. They reached this conclusion by measuring the light absorption of ambient particle deposits washed with strong solvents. The quantitative relationship between the absorber and its absorption properties is found in the good correlation reported between the measured values of b_a and the mass of total particulate carbon [7, 9]. About 15% of the total carbon mass is suggested to consist of elemental carbon [47].

These important findings suggest that the variability of n_2 in Fig. 2 may be largely due to the elemental carbon content of the aerosol particles. This explanation may not always hold for very clean or remote areas as suggested by work of some [6, 7, 26]; although, in certain remote areas such as the Arctic, long-distance advection of pollutants cause absorbing carbonaceous particles to play an important role [48-50].

The role of elemental particulate carbon as the predominant absorber in the visible simplifies matters, in that only one instead of many materials must be considered for that role. However, because elemental carbon is a strong absorber and consists of particles with complex geometry and optical properties [51, 52], and is usually a minor constituent of largely nonabsorbing ambient particles, the complexity of the optical characteristics of atmospheric aerosols is increased so that new uncertainties must be resolved.

RELATIONSHIP OF n_2, b_a, b_s, ω_0

It is well known [53, 54] that Mie theory predicts an approximately linear proportionality between the absorption cross section of the ambient particles and the

References pp. 155-157.

product of n_2 times the volume of the particles. This relationship depends on λ and to some degree on n_1 and the shape of the particles, and it is valid for $\alpha \lesssim 10$ [55] ($\alpha = 2\pi r/\lambda$, r = radius of the particle). By summing the cross sections of all particles in typical size distributions, the relationship becomes

$$b_a \approx \frac{n_2 M_v K_1}{\lambda \rho_p} \quad , \tag{1}$$

where K_1 has been given values of 6.86 ($\lambda = 500$ nm) [5], $12\pi/(n_1^2+2)$ [55], and 4π [56]. Equation (1) shows that with measurements of M_v and ρ_p it is possible to relate b_a and n_2; b_a is predicted by equation (1) with an accuracy of better than 40% ($\lambda = 550$ nm) as compared with the complete Mie calculations [55]. Generally equation (1) is used to estimate b_a. It is important to point out that M_v and ρ_p refer to properties of ambient particles. Hence, a correction must be applied for water lost by hygroscopic particles which are dried and then weighed on filter samples.

The relationship between ω_0 and n_2, shown in Fig. 3, was calculated with Mie theory ($\lambda = 632.8$ nm) for about 90 different ambient particle size distributions measured in the maritime atmosphere near Europe [35]. In nearly all cases the aerosols contained a large continental component. The curves in Fig. 3 demonstrate that $1-\omega_0$ is a strong function of n_2 and relatively independent of n_1 and the particle size distribution, i.e.,

$$1 - \omega_0 \approx f(n_2) \quad , \tag{2}$$

where $f(n_2) = 1.47 (n_2 + 0.001)^{0.152} - 0.627$ $(0.005 < n_2 < 0.2)$ for the solid curve in Fig. 3. Jennings et al. [57] reached similar conclusions concerning ω_0 and n_2 by using model size distributions in their calculations. Fig. 3 provides the opportunity to test values of n_2 determined with equation (1) for specified values of b_a, b_s, M_v, and ρ_p. The data points from Weiss et al. [6] are average values for various urban and rural sampling sites, those of Lin et al. [5] are from measurements in New York City and Hañel [14] gives one point for North-Atlantic maritime aerosols and another for an urban location in Europe. As shown, these data points agree well with the calculations which suggests that equation (1) has been properly applied.

Unfortunately the agreement in Fig. 3 does not mean that we have arrived at correct values of n_2 for aerosol particles. The approximation formulas (1) and (2) hold for Mie theory used with the usual assumption of spherical homogeneous particles. Since aerosol particles are highly inhomogeneous, especially in their absorption properties as was related in the preceding section, n_2 given in Fig. 3 is at best an *effective imaginary index* of the ambient particles. In other words, n_2 in Fig. 3 corresponds to spherical homogeneous particles which behave like the ambient inhomogeneous particles in their measured optical property, b_a. The relationship between the *effective imaginary index* and the real n_2 of ambient particles is not presently known, and it deserves attention.

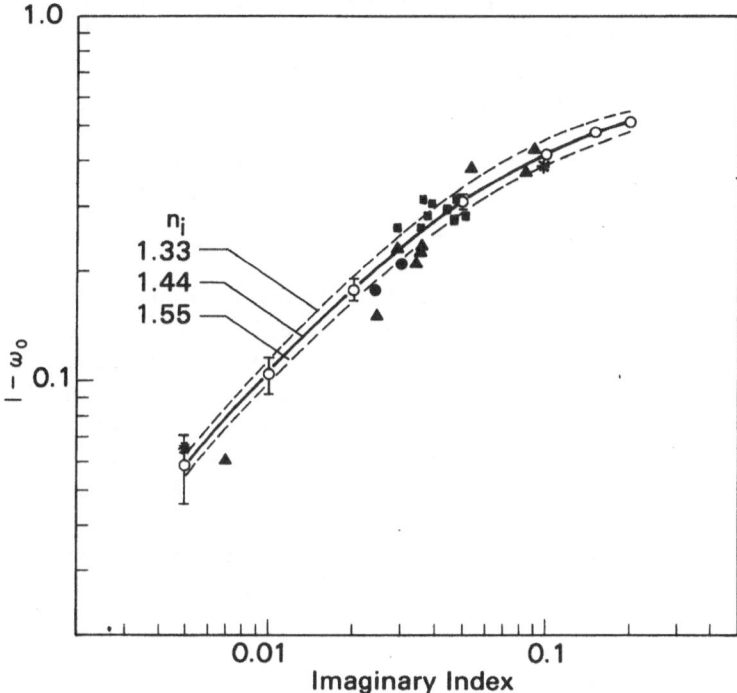

Fig. 3. The single scattering albedo ω_0 of ambient aerosols as a function of their imaginary index n_2, real index n_1, and particle size distribution. The vertical lines/correspond to twice the standard deviation of the variation due to a large sample of ambient size distributions. Other data points are from Lin *et al.* [5] (■), Weiss *et al.* [6] (▲), Hanel [14] (●), and Gillespie *et al.* [58] (*).

The inhomogeneity of the particles also causes large uncertainties in attempts to find b_a in equation (1) from estimates of n_2. To help illustrate this point, Gillespie *et al.* [58] calculated b_a and ω_0 with Mie theory given a hypothetical atmospheric aerosol consisting of a small mass fraction of soot particles ($n_2 = 0.5$) and a large fraction of nonabsorbing salt particles ($n_2 = 0$) with realistic size distributions. In case 1 they calculated ω_0 by summing the optical effects of each type of particle, while in case 2, ω_0 was again found by summing, but this time with a volume-averaged values of $n_2 = 0.005$ used for all the particles. The results of their calculations, given in Fig. 3, show a large difference in the calculated values of ω_0 for the two cases. This result emphasizes the point that when assumptions must be made concerning the physical relationship between the absorbing and nonabsorbing parts of ambient particles in order to arrive at absorption properties of the particles, large uncertainties can arise. This point applies to inversion techniques (see Optical Measurement and Inversion) where it must be shown that the uncertainty in the physical particle models chosen for the calculations does not translate to the results.

References pp. 155-157.

The curves in Fig. 3 also give information on the relationship between b_s and M_v. This relationship is significant, because both quantities are easily measured, and the ratio b_s/M_v for ambient aerosols has been shown to be relatively constant [59-62]. The ratio is found by combining $b_e = b_a + b_s$ with (1) and (2);

$$\frac{b_s}{M_v} \approx \frac{n_2 K_1}{\lambda \rho_p} \left[\frac{1}{f(n_2)} - 1 \right] \tag{3}$$

Evaluating (3) for various values of n_2 shows that b_s/M_v is relatively insensitive to most values of n_2 expected in the atmosphere. However, in areas significantly polluted with elemental carbon, light absorption by those particles should noticeably decrease b_s/M_v. For a reasonably clean atmosphere, equation (3) predicts $b_s/M_v = 1.05\ m^2g^{-1}$ ($n_2 = 0.02$, $n_1 = 1.44$, $\lambda = 550\ nm$, $\rho_p = 1.5\ g\ cm^{-3}$, K_1 [55]). This value is less than the nearly constant value of $b_s/M_v = 3.13\ m^2g^{-1}$ found by Waggoner and Weiss [62] at many different sites, because b_s/M_v in equation (3) corresponds to ambient aerosols, while their measurements were made on the dried small particle fraction. It may be possible to use equation (3) for estimating n_2 from measurements of b_s/M_v from the small particle fraction.

NEW DEVELOPMENTS

On July 28-August 8, 1980 the First International Workshop on Light Absorption by Aerosol Particles was held to compare various techniques of measuring aerosol light absorption. The workshop was organized by H. Gerber of Naval Research Laboratory and E. Hindman of Colorado State University under the auspices of the Radiation Commission of IAMAP, the American Meterorological Society, and the Optical Society of America. It was held at the Aerosol and Cloud Simulation Laboratory of CSU, and the primary sponsor was NSF. Eighteen scientific groups attended the workshop, and a similar number of instruments were compared by exposing them simultaneously to the same well characterized aerosols. At least one technique in each of the categories described in the section on Experimental Techniques was present. The aerosols which were used contained $(NH_4)_2SO_4$, Methylene Blue, and soot particles, as well as various mixtures thereof, ambient particles, and soil particles. The results of the workshop are expected early in 1981.

Some noteworthy developments in instrumentation are underway. HSS Inc. of Bedford, MA, and PMS Inc. of Boulder, CO are developing compact instruments capable of measuring transmittance through ambient aerosols. If successful, these devices used in conjunction with nephelometer measurements will give *in situ* values of b_a. Work is in progress on spectrophone techniques for filter samples (W. McClenny, EPA; Palmer, Duke University; R. Patty and C. Bennet of N. Carolina State University); of particular interest is the spectral evaluation of particle samples at Duke University. G. Hañel of Goethe-Universitat, Frankfurt, W. Germany has developed a differential spectrophone which operates with a simulated solar light source and a superior sensitivity.

RECOMMENDATIONS FOR RESEARCH

The following areas of study are recommended to properly establish the role that absorbing particles play in atmospheric optics:

a. Establish the meaning of the *effective imaginary index* for atmospheric particles. This is a necessary step before confidence can be acquired in the use of Mie theory for modelling absorption. Suggested courses of action are to support ambient measurements of b_a with other optical measurements (e.g., scattering phase function [17], backward-hemisphere scattering ratio [63]) and with particle-size-distribution measurements used with Mie calculations. Also, the physical appearance of the ambient elemental carbon particles and their relationship to the non-absorbing particles in the atmosphere need to be determined.

b. It is recommended that measurement of the absorption properties are at least accompanied by measurements of b_s, M_v, and relative humidity.

c. Study the particle-size dependence of the absorption properties, which appears to be significant [64]. The wavelength dependence of the particles also needs study, especially in the infrared windows.

d. Conduct an expanded and coordinated measurement program to determine the global area over which elemental carbon predominates; identify other absorbing species and make vertical measurements in the atmosphere.

e. Plan for a second instrumentation workshop to stress the comparison of remote-sensing and point-measurement techniques which on the average have produced significantly different adsorption properties. Vertical profiling in the atmosphere with point techniques in the area of remote sensing is necessary.

f. Desirable goals in instrumentation development are to arrive at simple *in situ* and sattelite techniques for measuring aerosol light absorption.

REFERENCES

1. K. Fischer, Beitr. Phys. Atmos., Vol. 43 (1970), p. 244.
2. K. Fischer, Beitr. Phys. Atmos., Vol. 46 (1973), p. 89.
3. F. Volz, J. Geophys. Res., Vol. 77, (1972), p. 1017.
4. R. J. Charlson, H. Horvath, and R. F. Pueschel, Atmos. Environ., Vol. 1 (1967), p. 469.
5. C. I. Lin, M. Baker, and R. J. Charlson, Appl. Opt. Vol. 12, (1973), p. 1356.
6. R. E. Weiss, A. P. Waggoner, R. J. Charlson, D. L. Thorsell, J. S. Hall, and L. A. Riley, "Studies of the optical physical and chemical properties of light absorbing aerosols," Proc. Conf. Carbonaceous Particles in the Atmosphere, Report LDL-9037, Lawrence Berkeley Laboratory, Berkeley, CA (1979).
7. M. Sadler, R. J. Charlson, H. Rosen, and T. Novakov, Atmos. Environ., (In press).
8. H. Rosen, A. D. A. Hansen, L. Gundel, and T. Novakov, Appl. Opt., Vol. 17 (1978), p. 3859.
9. H. Rosen, A. D. A. Hansen, R. L. Dod, and T. Novakov, Science, Vol. 208 (1980), p. 741.
10. S. Twomey, Appl. Opt., Vol. 19 (1980), p. 1740.
11. J. D. Lindberg, and L. S. Laude, Appl. Opt., Vol. 13 (1974), p. 1923.
12. E. M. Patterson, D. A. Gillette, B. H. Stockton, J. Geophys. Res., Vol. 82 (1977), p. 3153.

13. E. M. Patterson, "Optical properties of urban aerosols containing carbonaceous material," Proc. Conf. Carbonaceous Particles in the Atmosphere, Report LDL-9037, Lawrence Berkeley Laboratory, Berkeley, CA (1979).
14. G. Hänel, Adv. Geophys., Vol. 19 (1976), p. 73.
15. R. Eiden, Appl. Opt., Vol. 5 (1966), p. 569.
16. D. T. Phillips, and P. J. Wyatt, Appl. Opt., Vol. 11 (1972), p. 2082.
17. G. W. Grams, I. H. Blifford, D. A. Gillette, and P. R. Russell, J. Atmos. Sci., Vol. 13 (1974), p. 459.
18. G. Ward, K. M. Cushing, R. D. McPeters, and A. E. S. Green, Appl. Opt., Vol. 12 (1973), p. 2585.
19. B. M. Herman, R. S. Browning, and J. J. DeLuisi, J. Atmos. Sci., Vol. 32 (1975), p. 918.
20. T. H. Carlson, and R. S. Caverly, J. Geophys. Res., Vol. 82 (1977), p. 3141.
21. M. D. King, and B. M. Herman, J. Atmos. Sci., Vol. 36 (1979), p. 163.
22. M. D. King, J. Atmos. Sci., Vol. 36 (1979), p. 1072.
23. J. D. Spinhirne, J. A. Reagan, and B. M. Herman, J. Appl. Meteor., Vol. 19 (1980), p. 426.
24. J. M. Waldram, Quart. J. Roy. Met. Soc., Vol. 71 (1945), p. 319.
25. W. T. Roach, Quart. J. Roy. Met. Soc., Vol. 87, p. 346.
26. K. Y. Kondratyev, O. B. Vassilyev, S. V. Greshechkin, and L. S. Ivlev, Appl. Opt., Vol. 13 (1976), p. 478.
27. A. J. Drummond, and G. D. Robinson, Appl. Opt., Vol. 13 (1974), p. 478.
28. D. W. Reynolds, T. H. Vonder Haar, S. T. Cox, J. Appl. Meteor., Vol. 14 (1975), p. 433.
29. R. F. Pueschel, and P. M. Kuhn, J. Geophys. Res., Vol. 80 (1975), p. 2960.
30. J. J. DeLuisi, P. M. Furukawa, D. A. Gillette, B. G. Schuster, R. J. Charlson, W. M. Porch, R. W. Fegley, B. M. Herman, R. A. Rabinoff, J. T. Twitty, and J. A. Weinman, J. Appl. Meteor., Vol. 15 (1976), p. 455.
31. K. Murai, M. Kobayashi, T. Yamauchi, and R. Goto, Papers in Meteor. and Geophys., Vol. 27 (1976), p. 21.
32. G. D. Robinson, Arch. of Meteor. Geophys. and Bioclimat., Vol. 12 (1962), p. 19.
33. J. F. Ebersole, "Experimental Investigation of Battlefield Aerosols Pertinent to High Energy Laser Propagation," Report ARI-RR-115, Aerodyne Research, Inc., Burlington, MA (1977).
34. H. E. Gerber, Appl. Opt., Vol. 18 (1979), p. 110.
35. H. E. Gerber, J. Atmos. Sci., Vol. 36 (1979), p. 2502.
36. C. W. Bruce, J. Opt. Soc. Am., Vol. 66 (1976), p. 1071A.
37. R. W. Terhune, and J. E. Anderson, Optics Letters, Vol. 1 (1977), p. 70.
38. T. J. Truex, and J. E. Anderson, Atmos. Environ., Vol. 13 (1979), p. 507.
39. S. M. Japar, and D. K. Killinger, Chem. Phys. Letters, Vol. 66 (1979), p. 207.
40. J. S. Foot, Quart. J. Roy. Met. Soc., Vol. 105 (1979), p. 275.
41. S. S. Schleusener, J. D. Lindberg, K. O. White, and R. L. Johnson, Appl. Opt., Vol. 15 (1976), p. 2546.
42. Z. Yasa, N. M. Amer, H. Rosen, A. D. A. Hansen, and T. Novakov, Appl. Opt., Vol. 18 (1979), p. 2528.
43. M. B. Baker, Atmos. Environ., Vol. 10 (1976), p. 241.
44. K. A. Rahn, and G. E. Shaw, (1980). Personal communication.
45. J. T. Twitty, and J. A. Weinman, J. Appl. Meteor., Vol. 10 (1971), p. 725.
46. H. Rosen, A. D. A. Hansen, L. Gundel, and T. Novakov, "Identification of the graphitic component of source and ambient particulates by raman spectroscopy and an optical attenuation technique," Proc. Conf. Carbonaceous Particles in the Atmosphere, Report LDL-9037, Lawrence Berkeley Laboratory, Berkeley, CA (1979).
47. E. S. Macias, R. Delumyea, L. Chu, H. R. Appleman, C. D. Radcliffe, L. Staley, "The determination, specification, and behavior of particulate carbon," Proc. Conf. Carbonaceous Particles in the Atmosphere, Report LDL-9037, Lawrence Berkeley Laboratory, Berkeley, CA (1979).
48. G. E. Shaw, These Proceedings, p. 53.

49. *K. A. Rahn, C. Brosset, and B. Ottar, These Proceedings.*

50. *H. Rosen, A. D. A. Hansen, R. L. Dod, L. Gundel and T. Novakov, These Proceedings.*

51. *D. M. Roessler, and F. R. Faxvog, Appl. Opt., Vol. 18 (1979), p. 1399.*

52. *D. M. Roessler, and F. R. Faxvog, J. Opt. Soc. Am., Vol. 70 (1980), p. 230.*

53. *Van de Hulst, Light Scattering by Small Particles, Wiley, New York (1957).*

54. *A. P. Waggoner, M. B. Baker, and R. J. Charlson, Appl. Opt., Vol. 12 (1973), p. 896.*

55. *G. Hänel, and R. Dlugi, Tellus, Vol. 29 (1977), p. 75.*

56. *J. L. Nolan, Measurements of light absorbing aerosols from combustion sources, Proce. Conf. Carbonaceous Particles in the Atmosphere, Report LDL-9037, Lawrence Berkeley Laboratory, Berkeley, CA (1979).*

57. *S. G. Jennings, R. G. Pinnick, and J. J. Auvermann, Appl. Opt., Vol. 17 (1978), p. 3922.*

58. *J. B. Gillespie, S. G. Jennings, and J. D. Lindberg, Appl. Opt., Vol. 17 (1978), p. 989.*

59. *R. J. Charlson, Environ. Sci. and Technol., Vol. 3 (1969), p. 913.*

60. *H. J. Ettinger, G. W. Roger, J. Air Pollut. Control Assoc., Vol. 22 (1972), p. 108.*

61. *J. G. Kretzschwar, Atmos. Environ., Vol. 9 (1975), p. 931.*

62. *A. P. Waggoner, and R. E. Weiss, Atmos. Environ., Vol. 14 (1980), p. 623.*

63. *P. S. Bhardwaja, J. Herbert, and R. J. Charlson, Appl. Opt., Vol. 13 (1974), p. 731.*

64. *J. D. Lindberg, J. B. Gillespie, Appl. Opt., Vol. 74 (1977), p. 2628.*

DISCUSSION

G. Sverdrup, *(Battelle Columbus)*

You showed that there was a great deal of spread in the imaginary refractive indices and yet that was without regard to the specific materials in the particles. Later you indicated that, on the other hand, when specific materials were considered, it became too complex. On the one hand I thought you said it is too complex to consider, but in the real world the results are no good if you take too simplistic a view. So, I wonder what you are really arguing for?

H. Gerber

The large spread in the reported values of the imaginary index for atmospheric particles could possibly have been explained if the concentration of carbonaceous material in the particles had been measured. However, due to the optical complexity of these inhomogeneous particles which contain a small fraction of highly-absorbing elemental carbon, it is not entirely clear yet how the material content of the particles can be related to their imaginary index. Determining the physical relationship between the absorbing and non-absorbing parts of the atmospheric particles and modeling with Mie theory should improve this situation. The imaginary index can be estimated. Measurements are made of the total fine particle mass and the aerosol absorption coefficient, and one of the simple formulas that I mentioned is used. This index is, however, an effective index which relates to the absorption coefficient but which is of uncertain validity for other Mie calculations.

R. Stevens, *(U.S. Environmental Protection Agency)*

What do you mean by small particles and what particle size range are you referring to?

H. Gerber

It is the small particle fraction that you get from the dichotomous sampler. This fraction usually dominates aerosol optical effects in the visible spectrum, and it avoids misleading results by eliminating the few very large particles which may strongly influence the total particle mass.

One additional comment I would like to add here. There is considerable interest in the relationship between the absorption coefficient of aerosols and the visual range. Surprisingly, given a constant number and size distribution of atmospheric aerosol particles, the visual range is not strongly dependent on the absorption coefficient. This comes about, because the total extinction coefficient due to the particles is nearly independent of their absorption properties as was shown by Jennings et al.* Thus the visual range according to the Koschmeider relationship should also be relatively insensitive to absorption since the relationship shows an inverse proportionality to the total extinction coefficient.

D. Roessler, *(General Motors Research Laboratories)*

The Koschmeider relationship as described in Middleton's book** was derived under the assumption that the particles scatter but do not absorb light. We have found that in the case of absorbing aerosol, Koschmeider's relationship must be modified in a way where both the aerosol scattering coefficient and total extinction coefficient are included***.

*Jennings, S.G., Pinnick, R.G., and Auvermann, J. J., Appl. Opt. 17, (1978), p. 3922.
**Middleton, W.E.K., Vision Through the Atmosphere, Univ. of Toronto Press, Toronto, Ontario, 1963.
***Roessler, D.M., and Faxvog, F.R., Atmos. Environ. 15, (1981), p. 151.

CHEMICAL AND CATALYTIC PROPERTIES
OF ELEMENTAL CARBON

S. G. CHANG, R. BRODZINSKY, L. A. GUNDEL
and T. NOVAKOV

Lawrence Berkeley Laboratories
University of California
Berkeley, California

ABSTRACT

Elemental carbon particles contain many defects — dislocations in their graphitic structure — which constitute the active sites. Carbon atoms located at these sites show strong tendencies to react with other molecules because of residual valencies. The interaction of oxygen and water occurs in air, resulting in the incorporation of oxygen and hydrogen with carbon particles. Nearly every type of oxygen-containing functional group known in organic chemistry has been postulated in the carbon particle. The nature of these oxygen functional groups can affect the chemical reactivity of carbon particles. Very little is known about the surface nitrogen functional groups on carbon particles, however. We will discuss the interaction of elemental carbon with NH_3, leading to the formation of nitrogenous functional groups. The possible environmental influence of these nitrogenous functional groups will be presented.

Carbon particles can also play an important role as catalyst for many chemical reactions in the atmosphere. The assessment of the impact of these carbon-catalyzed reactions on air quality is difficult because the kinetics and mechanisms of these reactions have not been well studied. We have recently completed a kinetic study on the catalytic oxidation of SO_2 on carbon particles in an aqueous suspension. A rate law and a mechanism for this reaction will be presented. We will assess the relative importance of this carbon-catalyzed oxidation of SO_2 in the atmosphere with respect to competing reactions in liquid water.

INTRODUCTION

Elemental carbon particles resulting from incomplete combustion of fossil fuel are one of the major constituents of airborne particulate matter [1, 2]. These particles are chemically and catalytically active and can be effective carriers for toxic air

References pp. 178-179.

pollutants through their adsorptive capability. The chemical, adsorptive and catalytic behaviors of carbon particles depend very much on their crystalline structure, surface composition, and electronic properties. This paper discusses these properties and examines their relevance to atmospheric chemistry.

STRUCTURE

The diameter of elemental carbon particles varies from 50Å or even smaller to several thousand Å. The results of X-ray diffraction [3] have shown that each particle is made up of a large number of crystallites 20 to 30 Å in diameter. Each crystallite consists of several carbon layers with a graphitic hexagonal structure, having defects, dislocations, and discontinuities in the layer planes, and thus containing high concentrations of unpaired electrons which constitute the active sites. Carbon atoms located at these sites show strong tendencies to react with other molecules because of residual valencies. During particle formation, interactions of air, water, flue gas, etc., with carbon particles occur, resulting in the incorporation of oxygen, hydrogen, and nitrogen into the structure. Therefore, elemental carbon particles may be regarded as a complex three-dimensional organic polymer with the capability of transferring electrons, rather than merely as an amorphous form of elemental carbon.

SURFACE COMPLEXES

Surface complexes may determine the adsorptive, electrical and catalytic properties of carbon particles [4]. It is therefore important to determine the structure of surface species. Many studies have been carried out in the past on the determination of surface oxygen species [5]. Nearly every type of oxygen-containing functional group known in organic chemistry has been postulated to exist on the carbon surface (Fig. 1). The functional groups most often suggested are carboxyl groups, phenolic hydroxyl groups, and quinone carbonyl groups [6-10]. Less often suggested are ether, peroxide, and ester groups in the forms of normal and fluorescein-like lactones [11], carboxylic acid anhydrides [12], and cyclic peroxides [13]. The relative amounts of these complexes and their structure depends on the thermal history of carbon particles [14-17].

Little is known about the structure of surface nitrogen species, although the capability of fixation of nitrogen [18] in carbon particles and the promoting effect of the catalytic activity of nitrogenous carbon [19] have been observed. We have investigated the structure of surface nitrogen complexes produced as a result of the reaction between carbon particles and NH_3 in both an oxidizing [1] and a reducing atmosphere.

The first set of experiments involves the exposure of combustion-produced soot with NH_3 in air. The nature of surface nitrogen species, thus formed, was studied with the aid of ESCA. Soot particles for these experiments were generated by a premixed propane-oxygen flame. The exposure of soot particles to NH_3 was done

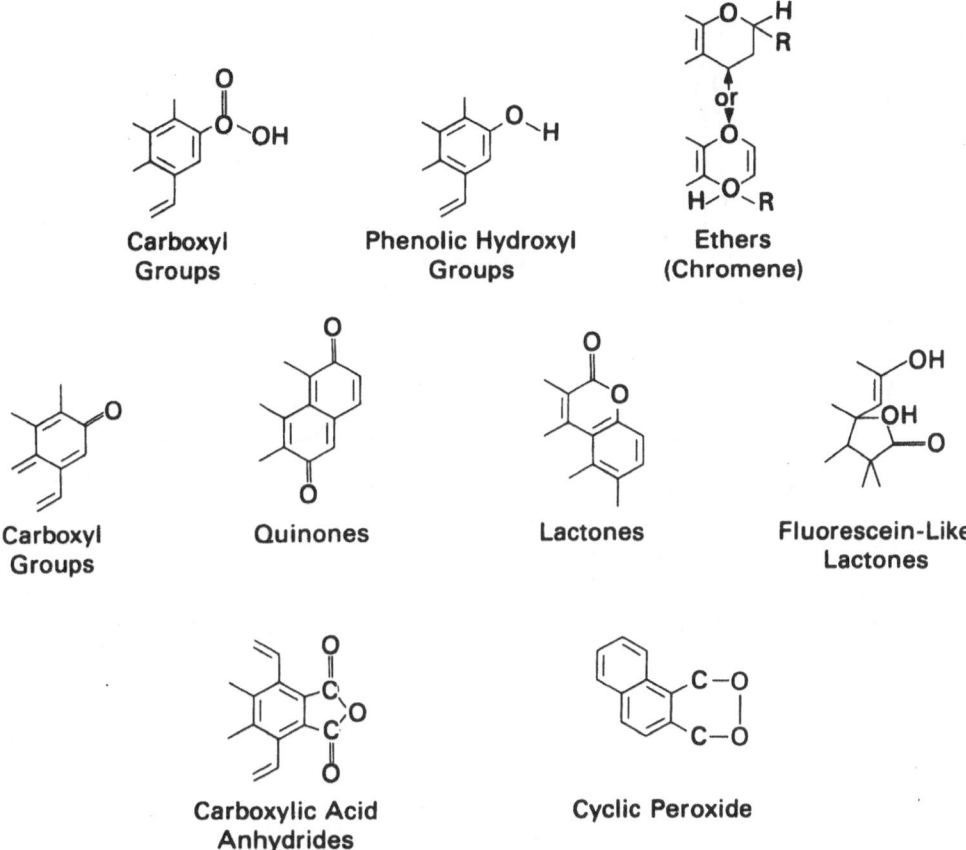

Fig. 1. Oxygen-containing functional groups on elemental carbon particle surfaces.

under two different experimental conditions: in a static regime, with propane soot precollected on a silver membrane filter subsequently exposed to the reactant gas at ambient temperature; and in a flow system, by introducing the reactant gas downstream from the propane-oxygen flame, i.e., while the soot particles are still at high temperature.

ESCA spectra of the nitrogen (ls) region of soot samples prepared in these ways are shown in Figs. 2 and 3. It is evident from Fig. 2 that interaction of NH_3 with "cold" soot particles results in ammonium-like species. However, as seen from Fig. 3, NH_3 interacting with "hot" soot particles produces species with ESCA peaks of binding energies designated as N_x species. Ammonium in these samples is probably produced on soot particles after they have been collected on the filter and cooled down.

Using ESCA, Novakov *et al.* [20] have observed, in addition to commonly occurring nitrate and ammonium, two reduced nitrogen species with N (ls) binding energy corresponding to N_x surface species produced under laboratory conditions. Chemical equivalency of ambient and synthetic N_x species is demonstrated by their

References pp. 178-179.

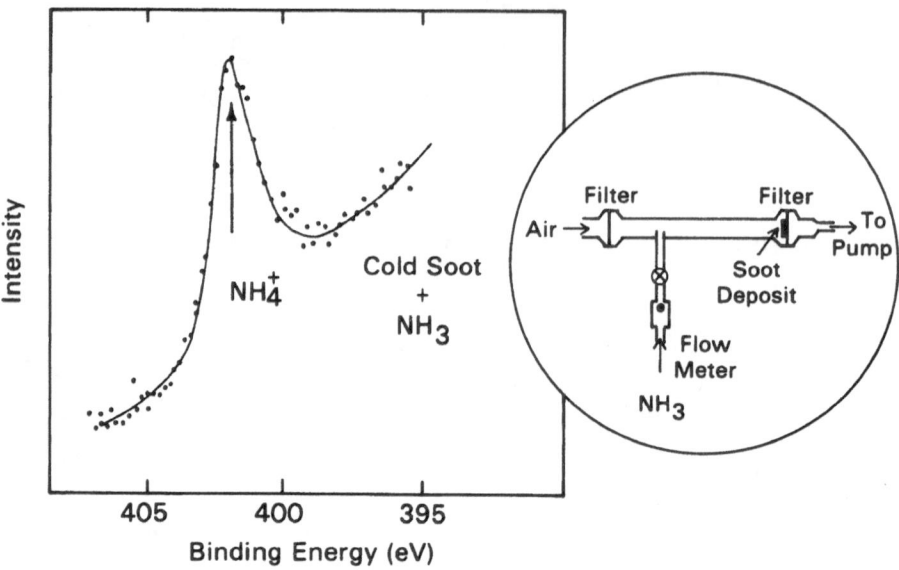

Fig. 2. Nitrogen (ls) ESCA spectrum of cold soot particles exposed to NH_3. The setup used for exposure is also shown.

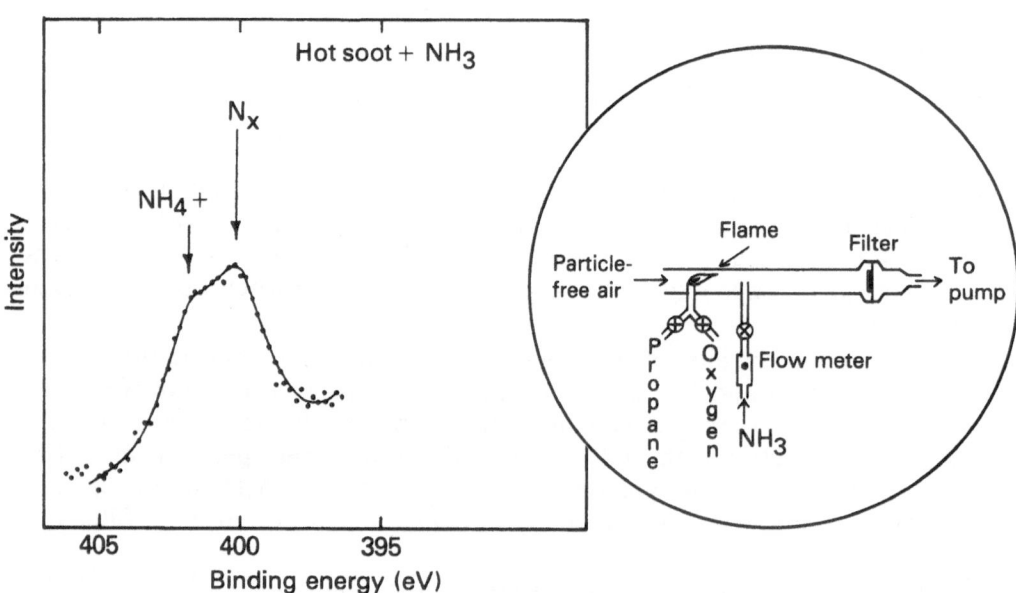

Fig. 3. Nitrogen (ls) ESCA spectrum of hot soot particles exposed to NH_3. The experimental arrangement used for sample preparation is also shown.

thermal behavior. The experimental procedure is to measure ESCA spectra at gradually increasing sample temperatures. The results of such measurements for one ambient particulate sample, collected in Pomona, California, during a moderate smog episode (24 October 1972) and for a sample prepared by NH_3-hot soot interaction are shown in Figs. 4 and 5.

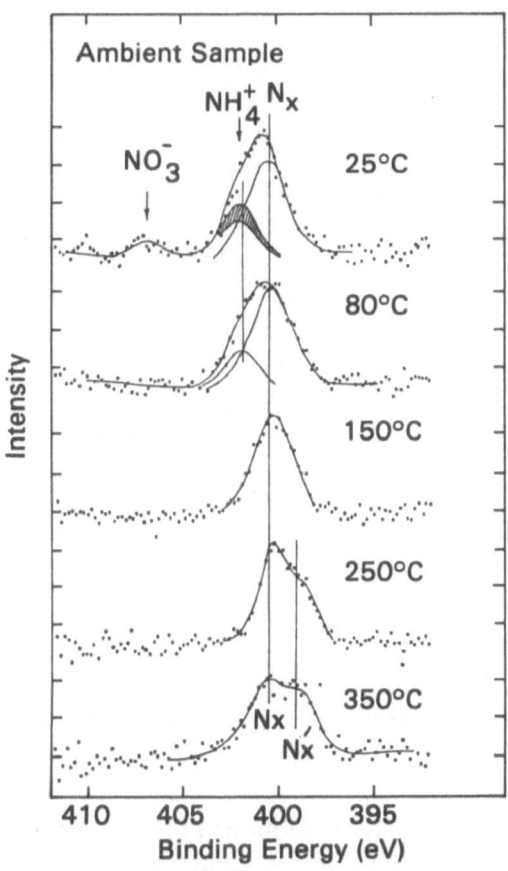

Fig. 4. Nitrogen (ls) ESCA spectrum of an ambient sample as measured at 25, 80, 150, 250, and 350°C.

The spectrum of the ambient sample (Fig. 4) at 25°C shows the presence of NO_3^-, NH_4^+, and N_X. At 80°C the entire nitrate peak is lost, accompanied by a similar loss in ammonium peak intensity. The shaded portion of the ammonium peak in the 25°C spectrum represents the ammonium fraction volatilized between 25 and 80°C. It appears therefore that the nitrate in this sample is mainly in the form of ammonium nitrate. At 150°C the only nitrogen species remaining is N_X. The ammonium fraction still present at 80°C, but absent at 150°C, is associated with some ammonium compound more stable than NH_4NO_3, possibly ammonium sulfate. At 250°C the appearance of another peak, labeled N_X', is seen. This peak continues to increase at 350°C. The total peak areas of spectra recorded at 150, 250, and 350°C remain constant, indicating the N_X component is transformed into N_X' by heating in vacuum.

References pp. 178-179.

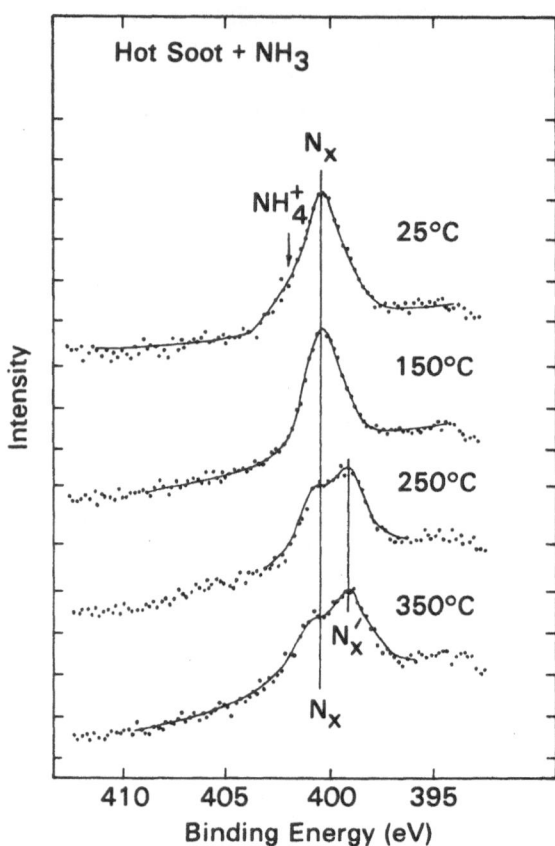

Fig. 5. Nitrogen (ls) spectrum of (hot) soot sample exposed to NH_3 as measured at 25, 150, 250, and 350°C.

N_x species produced by surface reactions of hot soot with NH₃ have the same kind of temperature dependence as the ambient samples. This is illustrated in Fig. 5. The spectrum taken at room temperature shows that most nitrogen species in this sample are of the N_x type. Heating the sample in vacuum to 150°C does not influence the line shape or intensity. At 250°C, however, the formation of N_x' is evident. Further transformation of N_x to N_x' occurred at 350°C.

Both ambient and synthetic N_x' species will remain unaltered even if the temperature is lowered back to room temperature with the sample in vacuum. However, if the sample is taken out of vacuum and exposed to moisture, N_x' will be transformed back to the original N_x compound. It can be concluded that N_x' species are produced by dehydration of N_x.

The outlined results indicate that reduced nitrogen species of the N_x and N_x' type observed in ambient pollution particulates are chemically analogous to the reduced nitrogen species produced by surface reactions of ammonia with finely divided carbon or soot at elevated temperature. The same reactants at room temperature produce surface ammonium compounds.

Based on these experimental results, N_X was assigned to a mixture of amines and amides, and N_X' to nitrile. Since prior to the interaction with NH_X, the soot particle surface was in contact with air and flue gas, it therefore should be covered with surface oxygen complexes. By using the most often-mentioned surface oxygen-carbon functional groups (i.e., carboxyl groups and phenolic hydroxyl groups) and in analogy with organic chemistry, we can describe some possible reactions of NH_3 and soot leading to the formation of amides, amines, nitriles, and ammonium-salt-like compounds associated with soot particle surfaces.

At low temperatures soot particles covered with surface carboxyl or phenolic groups may act as a Bronsted acid when interacting with NH_3. Carboxyl ammonium or phenolic ammonium salts will be formed as the result of proton exchange. Ammonia may also be physically adsorbed by hydrogen bonding to surface OH or COOH groups. At elevated temperatures the carboxyl group carbon is electrophilic and has the tendency to accept an electron pair from the basic species in the process of coordination. The nucleophilic substitution reaction of NH_3 with carboxylic acid yields an amide which may dehydrate and become a nitrile upon further heating. Carboxyl and phenolic hydroxyl ammonium salts may dehydrate at elevated temperature to produce amides and/or nitriles and amines respectively.

The photoelectron spectroscopic results indicate that the amides and amines correspond to the N_X species. These appear as broad peaks indicating the presence of more than one chemical species. Nitriles formed from amides by dehydration on

$$N_X \underset{+H_2O}{\overset{-H_2O}{\rightleftharpoons}} N_X{}'$$

heating correspond to the N_X' species. We have established the reversibility of the process. The carboxyl ammonium and phenolic hydroxyl ammonium salts produced by NH_3 chemisorption correspond to the volatile ambient ammonium species.

We have studied the stability of ambient particulate nitrogen in water by combining ESCA measurements with determination of total nitrogen by proton activation

References pp. 178-179.

[21]. Our results with samples from several locations (Berkeley, Los Angeles, and St. Louis) indicate that 1) a large fraction of N_x (85%) originally present in ambient particulate matter can be removed by water extraction; and 2) more NH_4^+ appears in the extract than was present on the untreated sample and less N_x appears in the extract than was present on the untreated sample. The N_x deficiency in the extract matches the surplus in NH_4^+. The former behavior could be attributed to water-soluble stoichiometric compounds such as amines and surface species such as amides and nitrile which can undergo hydrolysis. The latter could be attributed to the hydrolysis of amide and nitrile groups. This behavior may be responsible for the disagreement of chemical composition observed in black episodes with those predicted from the equilibrium phase diagram as constructed by Brosset [22].

The other set of experiments [23] involves the grinding of a purified grade POCO graphite in NH_3 in the absence of air at room temperature. The concentration of nitrogen with respect to carbon was determined by ESCA. The information on the structure of surface nitrogen species was obtained with the aid of Fourier transform infrared spectroscopy. To help in the assignment of vibrational frequencies, infrared spectra of graphite particles after reaction with deuterated ammonia were also obtained.

The grinding reduces particle sizes and creates fresh surfaces. Surface carbon atoms of graphite particles show a strong chemical reactivity because of unsaturation in valency. Figs. 6a and 6c show infrared spectra of the graphite particles after extensive grinding in an atmosphere of NH_3 and ND_3, with expansions of these spectra in Figs. 6b and 6d. These infrared spectra suggest the occurrence of dissociative chemisorption of NH_3 on the carbon particle surface. Vibrational frequencies associated with the surface groups C-NH , C=N-H, C≡N, and C-H are observed in Figs. 6a and 6b. The isotope shifts shown in Figs. 6c and 6d support these assignments. Surface CNH_2 groups give rise to two bands near 3400 cm^{-1} that

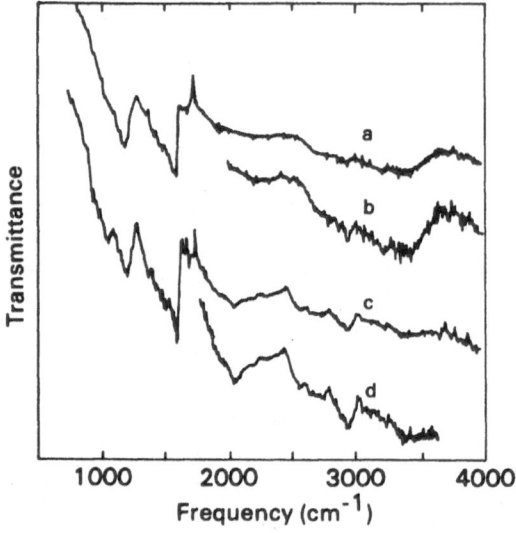

Fig. 6. Infrared spectra of the graphite particles after extensive grinding in an atmosphere of NH_3 (a) and ND_3 (c). (b) and (d) are 2X expansions along the ordinate of (a) and (c) respectively.

are attributed to symmetric and antisymmetric N-H stretching modes. These two bands should shift to 2500 cm^{-1} for CND$_2$. This shift is shown in Figs. 6c and 6d. A NH$_2$ bending mode near 1580 cm^{-1} should shift to about 1200 cm^{-1} for the ND$_2$ groups. However, a strong band due to the k=0,E$_{2g}$ phonon mode of the graphite lattice [24] and/or a vibrational mode of the aromatic structure of graphite [25] also occurs at about 1580 cm^{-1}. Likewise, the C-N stretching mode vibrates at approximately 1200 cm^{-1} and appears in both the C-NH$_2$ and the C-ND$_2$ surface groups.

We have detected surface nitrogen groups indicating the dissociation of more than one bond in a molecule of ammonia. A band between 1600 and 1700 cm^{-1} could be assigned to immines (C=NH and C=N-C), a weak band at 2300 cm^{-1} to nitrile (C≡N), and one at 2180 cm^{-1} to isocyanide (-N$^+$≡C$^-$).

The evidence of the dissociative chemisorption of ammonia on carbon particle surfaces is also supported by the appearance of the C-D stretching band at 2050 cm^{-1}. The assignment of the C-H stretching is ambiguous because the C-H stretching is near 2900 cm^{-1} where a vibrational band appears on both NH$_3$ and ND$_3$ samples. This band could be the overtone and/or combination band resulting from the strong absorption band between 1300 and 1600 cm^{-1}. There is also a band, possibly of the same nature, at 2700 cm^{-1} in both samples.

ACIDITY

Depending on the thermal history or the condition of activation, elemental carbon particles may possess either acidic or basic character. Because of this property, soot may influence the pH of atmospheric water droplets [26]. It has been shown that activation of elemental carbon by exposure to O$_2$ at temperatures between 200 and 400°C produces an acidic type. By contrast, activation of carbon at high temperatures either in pure CO$_2$ or under vacuum, followed by exposure to oxygen at room temperature, results in a basic type.

The acidic character can be explained by the dissociation of acidic surface oxygen functional groups such as carboxyl and hydroxyl in solution. The nature of the basic character has been a topic of considerable discussion and controversy [27-32]. The idea of the presence of basic sites in the form of surface oxides has been proposed by Shilov and Chmutov [27] as well as Garten and Weiss [28], among others, to account for the chemisorption of acids. The latter suggested that the oxides were in the form of a chromene-like structure after neutralization with acid, which would result in the formation of carbonium ion. The presence of carbonium cationic sites on the surface of acid-treated carbon was confirmed by Rivin [29], but it could not be established whether the basic sites are due to the presence of the chromene-like surface oxides or the inherent property of the polynuclear aromatic structures of the carbon particles. Frumkin *et al.* [30] proposed an electrochemical theory in which the adsorption of the acids by carbon is determined by the electrical potential at the carbon solution interface and by the capacity of the double layer. According to Steenberg [31], adsorption of acids involved primary adsorption of protons by physical force and secondary adsorption of anions in the diffuse double layer. On the contrary, Mattson and Mark [32] attributed the neutralization of acids at high acid concentrations to the primary adsorption of the anions and secondary adsorption of the protons.

References pp. 178-179.

CATALYSIS

Elemental carbon particles are effective catalysts [33] for many different types of reactions including oxidation-reduction, halogenation, hydrogenation-dehydrogenation, dehydration, polymerization, and isomerization. Table 1 lists a

TABLE 1
Some Reactions Catalyzed by Carbon.

Reactions	References
1. $SO_2 + 1/2\,O_2 \rightarrow SO_3$	Novakov et al. [34], Chang et al. [25], Brodzinsky et al. [36], Chang et al. [37]
2. $SO_2 + NO_2 \rightarrow SO_3 + NO$	Cofer et al. [38], Britton and Clarke [39]
3. $SO_2 + O_3 \rightarrow SO_3 + O_2$	Cofer et al. [40]
4. $NO + 1/2\,O_2 \rightarrow NO_2$	Rao and Hougen [41]
5. $2H_2O_2 \rightarrow 2H_2O + O_2$	Bente and Walton [42]
6. $CO + Cl_2 \rightarrow COCl_2$	Dulou [43]
7. $SO_2 + Cl_2 \rightarrow SO_2Cl_2$	Dulou [43]
8. HCOOH $\nearrow H_2 + CO$ $\searrow H_2 + CO_2$	Stumpp [44]
9. ⬡—CHO → ⬡—COOH	Gundel [45]
10. Hydroquinone → quinhydrone → quinone	Bente and Walton [42]

few reactions catalyzed by carbon that could have direct bearing on atmospheric chemistry. At this time it is difficult to assess the importance of all these reactions in the atmosphere because useful rate equations have not been determined. We have recently performed a study [35, 36] of the kinetics and mechanism for the catalytic oxidation of SO_2 on carbon in aqueous suspensions and have obtained a rate equation applicable to atmospheric conditions. The reaction was studied by batch (flask) experiments, from which a rate law was derived. This rate law has been confirmed by fog chamber studies [46]. The flask experiments were performed using suspensions of commercially available activated carbons as well as suspensions of combustion-produced soots.

Fig. 7 shows the typical reaction curves for the oxidation of S(IV) in aqueous suspensions of soot particles collected from acetylene and natural gas flames. The reaction occurs in two steps. The initial disapperance of S(IV) is so fast that its rate could not be followed by the analytical techniques used. The second step is characterized by a much slower reduction of S(IV). The results obtained with these combustion-produced soots were reproduced (Fig. 8) with suspensions of similar concentrations of one of the commercially available activated carbons (Nuchar C-190, a trademark of West Virginia Pulp and Paper Co.). Fig. 8 also shows a mass balance between the S(IV) consumed and the sulfate produced. At a constant temperature, the amount of S(IV) oxidized by the rapid first step was found to be proportional to the carbon particle concentration.

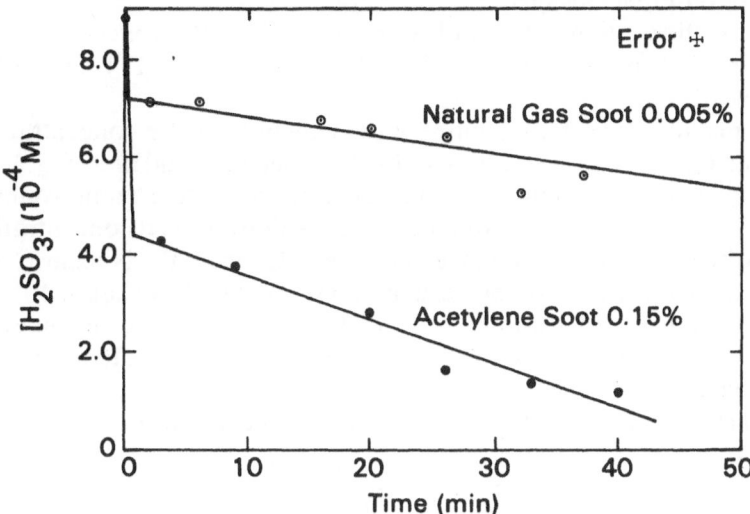

Fig. 7. S(IV) concentration as a function of time for acetylene and natural gas soot suspensions.

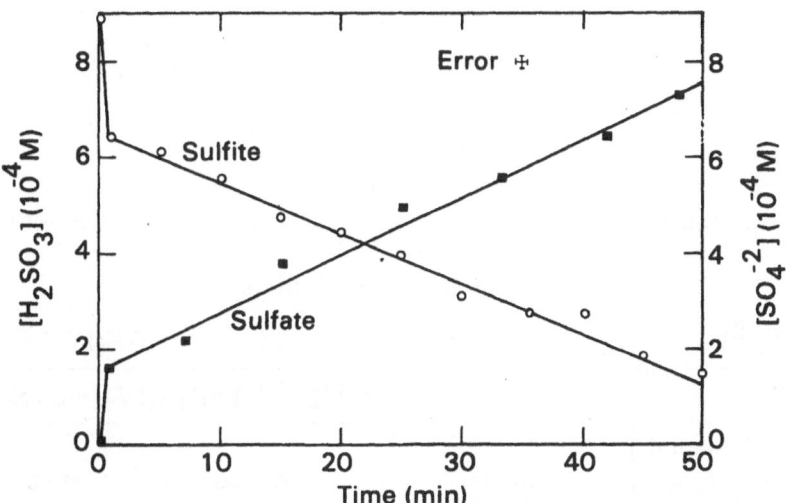

Fig. 8. S(IV) and SO_4^{-2} concentrations as a function of time for a 0.16 weight % activated carbon suspension.

The second step of the reaction has the following characteristics:
1. The reaction rate is independent of pH (pH < 7.6), and therefore $SO_2 \cdot H_2O$, HSO_3^- and SO_3^{-2} are indistinguishable in terms of oxidation on the carbon surfaces.

References pp. 178-179.

2. The reaction is first order with respect to the concentration of carbon particles.
3. The activation energy of the reaction is ~ 8.5 kcal/mole and is slightly different for different carbons.
4. The reaction rate has a complex dependence on the concentration of S(IV), ranging between second and zeroth order as the S(IV) concentration increases.
5. The reaction rate has a complex dependence on the concentration of dissolved O_2, with the order of reaction between zero and first. Fig. 9 shows the effective rate of reaction (normalized carbon concentration, room temperature — 20°C, and air) as a function of the sulfurous acid concentration for the activated carbons studied (see also Table 2). From the Nuchar C-190 curve, the rate of reaction is second order with respect to S(IV) below 10^{-7} M, moves through a first order around 5×10^{-6} M, and becomes independent of S(IV) concentrations above 10^{-4} M. The other curves are seen to be similar in their behavior.

Based on the experimental results, we propose the following reaction mechanism

$$C_X + O_2\,(\ell) \underset{k_{-1}}{\overset{k_1}{\rightleftharpoons}} C_X \cdot O_2\,(\ell) \qquad (1)$$

$$C_X \cdot O_2\,(\ell) + S(IV) \underset{k_{-2}}{\overset{k_2}{\rightleftharpoons}} C_X \cdot O_2\,(\ell) \cdot S(IV) \qquad (2)$$

TABLE 2

Summary of Kinetic Data for the Catalytic Oxidation of SO₂ by
Various Elemental Carbon Particles in Aqueous Suspension.
Reaction rate equation:

$$\frac{d[SO_4^{-2}]}{dt} = Ae^{-Ea/RT}\,[C_X]\left\{\frac{K_1[O_2]}{1+K_1[O_2]}\right\}\left\{\frac{\alpha[S(IV)]^2}{1+\beta[S(IV)]+\alpha[S(IV)]^2}\right\}$$

Kinetic data \ Elemental carbon	Nuchar C-190 (WESVACO)	Nuchar SN (WESVACO)	Aktivkohle (MERCK)
A (moles/g-sec)	0.874	0.158	2.473
E_a (kcal/mole)	8.8	8.1	8.8
K_1 (ℓ/mole)	2.103×10^3	7.427×10^3	4.372×10^3
α (ℓ^2/mole²)	2.404×10^{12}	4.915×10^8	9.519×10^{11}
β (ℓ/mole)	1.219×10^7	2.956×10^5	3.738×10^7

$$C_\chi \cdot O_2 (\ell) \cdot S(IV) + S(IV) \underset{k_{-3}}{\overset{k_3}{\rightleftarrows}} C_\chi \cdot O_2 (\ell) \cdot 2S(IV) \tag{3}$$

$$C_\chi \cdot O_2 (\ell) \cdot 2S(IV) \overset{k_4}{\rightarrow} C_\chi + 2S(VI), \tag{4}$$

where C_χ = soot, $O_2(\ell)$ = dissolved oxygen molecule, $S(IV)$ = sulfite species, and $S(VI)$ = sulfate species.

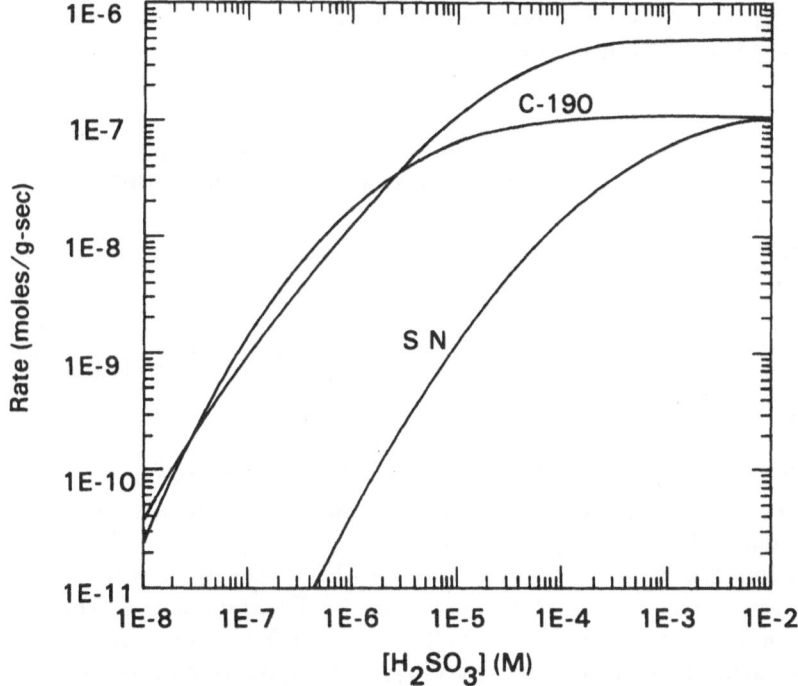

Fig. 9 Effective rate of oxidation of S(IV) catalyzed on various activated carbon particles vs. S(IV) concentration.

Equation (1) indicates that dissolved oxygen is adsorbed on the soot particle surface to form an activated complex. This adsorbed oxygen complex then oxidizes the S(IV) to form sulfate according to Equations (2)-(4). If one assumes that the reaction follows the Langmuir adsorption equilibrium [47], the rate of acid formation is

$$\frac{d[S(VI)]}{dt} = 2k_4 [C_\chi] \left(\frac{K_1[O_2]}{1 + K_1[O_2]} \right) \left(\frac{K_2[S(IV)]}{1 + K_2[S(IV)]} \right) \left(\frac{K_3[S(IV)]}{1 + K_3[S(IV)]} \right) \tag{I}$$

where $K_1 = k_1/k_{-1}$, $K_2 = k_2/k_{-2}$, $K_3 = k_3/k_{-3}$.

References pp. 178-179.

The experimental results yield the following rate law for this reaction:

$$\frac{d[S(VI)]}{dt} = k[C_x] \left(\frac{K[O_2]}{1 + K_1[O_2]} \right) \quad f[S(IV)] \qquad \text{(IIa)}$$

$$\text{where } f[S(IV)] = \left(\frac{\alpha[S(IV)]^2}{1 + \beta[S(IV)] + \alpha[S(IV)]^2} \right),$$

$[C_x]$ = grams of carbon particles per liter,
$[O_2]$ = moles of dissolved oxygen per liter, and
$[S(IV)]$ = total moles of S(IV) per liter.

For Nuchar C-190 the following constants were determined: $k = 0.874\, e^{-4428/T}$ moles/g·sec ($T = °K$); $K_1 = 2.103 \times 10^3$ L/mole, $\alpha = 2.404 \times 10^{12}$ L²/mole², and $\beta = 1.219 \times 10^7$ L/mole.

We have previously reported [36] the experimental rate law for Nuchar C-190 in the form

$$\frac{d[S(VI)]}{dt} = k[C_x] [O_2]^{0.69} f[S(IV)] \, , \qquad \text{(IIb)}$$

where $k = 9.04 \times 10^3\, e^{-5888/T}$ moles $\cdot^{31} \cdot L^{.69}$/g·sec, and the numerical value for α and β were slightly different. The current values for the activation energy, α, and β were calculated with our most recent data.

Equation (IIb) can be changed simply to Equation (IIa). A fractional order adsorption reaction is achieved by substituting the Freundlich isotherm, $\theta^{27} k[X]^n$, for the Langmuir isotherm [47], $\theta = K[X]/(1 + KF24X)$, where θ is the fraction of the surface covered by the adsorbed species X. Multiplication of the S(IV) terms in Eq. (I) yields the experimentally derived expression, where $K_2 K_3 = \alpha$ and $K_2 + K_3 = \beta$.

The dependence of the rate of formation of sulfate on the partial pressure of SO_2 (P_{SO_2}) in the atmosphere can be obtained from Eq. (II). Because the effect of P_{SO_2} on the rate is contained in f[S(IV)], we illustrate the dependence of f[S(IV)] on P_{SO_2} and on the pH of the aqueous droplets in Figs. 10 and 11. The term, f[S(IV)], or the rate of production of sulfate (because the rate is linearly proportional to f[S(IV)], decreases as the pH decreases at a given P_{SO_2}. The magnitude of the change in f[S(IV)] per unit of pH change is much larger at lower values of P_{SO_2}. Also, f[S(IV)] (or the rate) depends only slightly on P_{SO_2} under most atmospheric conditions, i.e. when P_{SO_2} is between 1 and 10 ppb and the pH ranges between 5 and 6. The term, f[S(IV)], increases only 10% and twofold, respectively, at pH of 6 and 5 for a P_{SO_2} increase from 1 to 10 pbb. However, f[S(IV)] does depend strongly on P_{SO_2} when the pH is low.

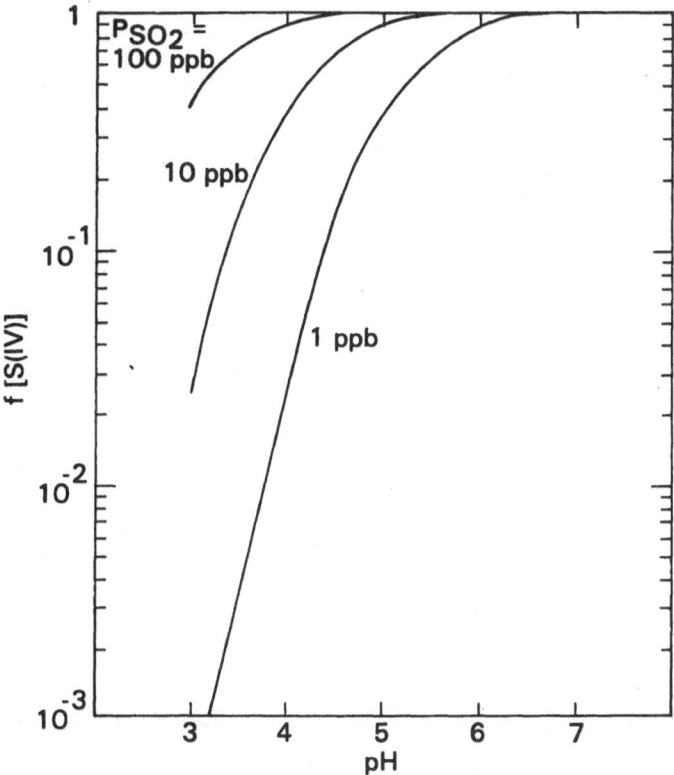

Fig. 10. The effect of pH of aqueous droplets on f [S(IV)] at P_{SO_2} = 100, 10, 1 ppb.

The catalytic oxidation of sulfurous acid on carbon particles of different origins shows the same type of kinetic behavior. However, the rate constants of several different types of carbon particles were studied and found to differ from type to type. In principle, the reaction rate should be proportional to the concentration of active sites on the carbon particles, rather than to the concentration of carbon particles. The number of active sites per unit mass of carbon particles is different from type to type and is not necessarily proportional to the surface area. Siedlewski [48] has shown, by means of the electron paramagnetic resonance method, that free electrons on carbon particles can serve as active centers for the adsorption of oxygen molecules and for the oxidation of SO_2. The concentration of free electrons. is related to the origin and thermal history of the carbon particles.

It is therefore impractical to formulate a generally applicable rate constant for atmospheric soot particles because these particles may arise from the combustion of different types of fossil fuel under different combustion conditions and thus possess a different catalytic activity.

We have carried out a box-type calculation [35, 37] to compare the relative importance of sulfate production by soot particles with other mechanisms involving liquid water. The systems considered in the batch reactor include SO_2-CO_2-$H_2O(\ell)$-air and any of the oxidizing agents such as O_2, O_3, HNO_2, or catalysts such as Fe^{+++}, Mn^{++}, and soot. The role of NH_3 was also investigated in these

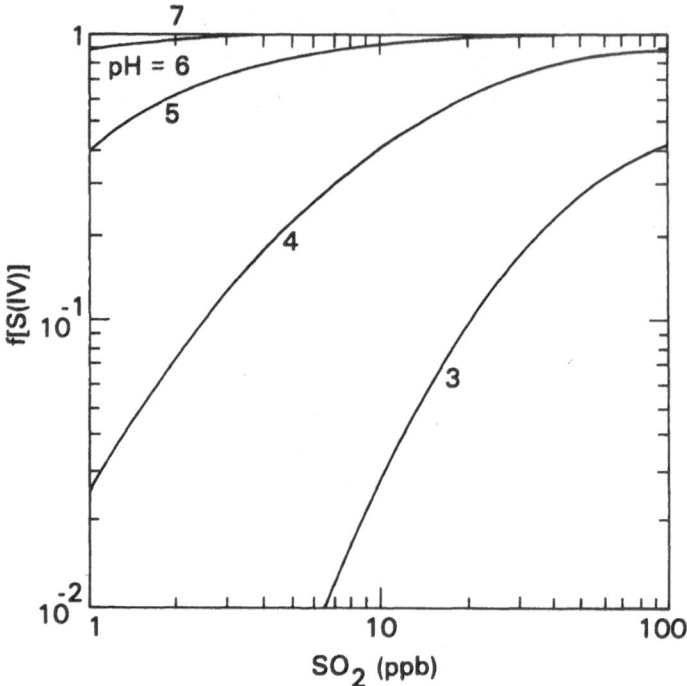

Fig. 11. The effect of partial pressure of SO_2 on f[S(IV)] at pH of 7, 6, 5, 4, and 3.

reactions. The kinetic results of Chang *et al.* [37, 49], Beilke *et al.* [50], Erickson *et al.* [51], Freiberg [52], and Matteson *et al.* [53] for nitrous acid, oxygen, ozone, iron, and manganese systems respectively were used in these calculations.

All the oxidation mechanisms considered except Mn^{++} are pH dependent. Most of the mechanisms have lower oxidation rates at a lower pH, but some are more sensitive to change in pH than others. The HNO_2 mechanism shows a larger oxidation rate when the solution is more acidic, however. The following initial conditions were used in the calculation: liquid water, 0.05 g/m³; SO_2, 0.01 ppm; O_3, 0.05 ppm; and CO_2, 0.000311 atm. Concentrations of particulate Fe and Mn of 250 ng/m³ and 20 ng/m³ respectively were assumed. However, only 0.13% of the total iron and 0.25% of the manganese are water soluble, according to Gordon *et al.* [54]. The concentration of soot and HNO_2 were taken as 10μg/m³ and 8 ppb respectively. The latter corresponds to 25 ppb of NO and 50 ppb of NO_2 at equilibrium conditions. For NH_3 a concentration of 5 ppb was used, which is higher than the highest equilibrium pressure of NH_3 in the United States as calculated by Lau and Charlson [55]. Tables 3 and 4 list the equilibrium equations and oxidation rate equation used for this comparative study.

The following assumptions were made in the calculations:

1. The size of liquid water drops suspended inside the box is so small that the absorption rate of gaseous species (SO_2, NH_3, and HNO_2) is governed by chemical reaction rates.

TABLE 3
Chemical Equilibrium Constants at 25°C.[a]

$$H_2O \overset{K_w}{\rightleftharpoons} H^+ + OH^- \qquad\qquad K_2 = 1.0008 \times 10^{-14}$$

$$CO_2(g) + H_2O(\ell) \overset{H_c}{\rightleftharpoons} CO_2 \cdot H_2O \qquad H_c = 3.4 \times 10^{-2}$$

$$CO_2 \cdot H_2O \overset{K_{1c}}{\rightleftharpoons} HCO_3^- + H^+ \qquad K_{1c} = 4.45 \times 10^{-7}$$

$$HCO_3^- \overset{K_{2c}}{\rightleftharpoons} CO_3^{-2} + H^+ \qquad\qquad K_{2c} = 4.68 \times 10^{-11}$$

$$NH_3(g) + H_2O(\ell) \overset{H_a}{\rightleftharpoons} NH_3 \cdot H_2O \qquad H_a = 57$$

$$NH_3 \cdot H_2O \overset{K_a}{\rightleftharpoons} NH_4^+ + OH^- \qquad K_a = 1.774 \times 10^{-5}$$

$$SO_2(g) + H_2O(\ell) \overset{H_s}{\rightleftharpoons} SO_2 \cdot H_2O \qquad H_s = 1.24$$

$$SO_2 \cdot H_2O \overset{K_{1s}}{\rightleftharpoons} HSO_3^- + H^+ \qquad K_{1s} = 1.7 \times 10^{-2}$$

$$HSO_3^- \overset{K_{2s}}{\rightleftharpoons} SO_3^{-2} + H^+ \qquad\qquad K_{2s} = 6.24 \times 10^{-8}$$

$$HSO_4^- \overset{K_{3s}}{\rightleftharpoons} H^+ + SO_4^{-2} \qquad\qquad K_{3s} = 1.2 \times 10^{-2}$$

$$HNO_2(g) + H_2O \overset{H_N}{\rightleftharpoons} HNO_2 \cdot H_2O \qquad H_N = 5.1 \times 10^{-4}$$

$$HNO_2 \cdot H_2O \overset{K_N}{\rightleftharpoons} H^+ + NO_2^- \qquad K_N = 5.1 \times 10^{-4}$$

$$O_2 + H_2O \overset{H_{O_2}}{\rightleftharpoons} O_2 \cdot H_2O \qquad\qquad H_{O_2} = 1.08 \times 10^{-3}$$

$$O_3 + H_2O \overset{H_{O_3}}{\rightleftharpoons} O_3 \cdot H_2O \qquad\qquad H_{O_3} = 1.23 \times 10^{-2}$$

[a]*Concentrations in moles/ℓ* and gas pressure in atm.

References pp. 178-179.

TABLE 4
Rate of SO_2 Oxidation by Various Mechanisms in Aqueous Droplets.

O_2 $Rate = \dfrac{H_S\{k_2 + k_{10}K_W/[H^+]\}\,K_{2S}k_3}{k_{-2}[H^+]^2 + k_{-10}[H^+] + K_{2S}k_3}\,P_{SO_2}$

O_3 $Rate = \{k_4[HSO_3^-] + k_5[SO_3^{-2}]\}\,[O_3 \cdot H_2O]$

Fe^{+++} $Rate = \dfrac{k_O k_S^2 H_S^2 P_{SO_2}^2\,[Fe^{+++}]}{[H^+]^3}$

Mn^{++} $Rate = 3.67\times10^{-3}[\chi] - 1.17\,\{[HSO_4^-] + [SO_4^{-2}]\}^2\,\{[Mn^{++}] - [\chi]\}\,x[H_2O(\ell)]^{-2}$

where $\chi = \dfrac{k_1 H_S P_{SO2}[Mn^{++}]}{k_1\{H_S P_{SO_2} + [H_2O(\ell)]\,[Mn^{++}]\} + 0.17}$

Soot $Rate = k_6[C_\chi]\,[O_2]^{.69}\,\dfrac{\alpha[S^{+4}]^2}{1 + \beta[S^{+4}] + \alpha[S^{+4}]^2}$

HNO_2 $Rate = k_7[H^+]^2[NO_2^-] + k_8[H^+]\,[NO_2^-]\,[HSO_3^-] + k_9[NO_2^-]\,[HSO_3^-]^2$

$^a[H_2O(\ell)]$ in cc/m³, concentration in mole/ℓ, gas pressure in atm, and time in sec, $k_O = 151.69$ ℓ/mole-sec, $k_S = 1.84\times10^{-2}$ mole/ℓ, $k^1 = 8.12\times10^4$ ℓ/mole-sec, $k_2 = 3.4\times10^6$ sec^{-1}, $k_{-2} = 2\times10^8$ ℓ/mole-sec, $k_{10} = 2.9\times10^5$ ℓ/mole-sec $k_{-10} = 2.3\times10^{-7}$ sec^{-1}, $k_3 = 1.7\times10^-$ sec^{-1}, $k_4 = 1.1\times10^5$ ℓ/mole-sec, $k_5 = 7.4\times10^8$ ℓ/mole-sec, $k_6 = 1.2\times10^{-4}$ mole$^{.3}\cdot\ell^{.7}$/g·sec, $\alpha = 1.5\times10^{12}$ ℓ^2/mole², $\beta = 3.06\times10^6 \times 10^6$/mole, $k_7 = 8\times10^5$ ℓ^2/mole²-sec, $k_8 = 3.8\times10^3$ ℓ^2/mole²-sec, $k_9 = 9\times10^{-4}$ ℓ^2/mole²-sec.

2. There is no mass transfer of any species across the box during the reaction; therefore, the SO_2 (and NH_3 or HNO_2) in each box is depleted with time. The mass balance of the SO_2, CO_2, NH_3, and HNO_2 is always maintained, i.e.,

$$\Delta[SO_2]g = \Delta[SO_2 \cdot H_2O] + \Delta[HSO_3^-] + \Delta[SO_3^{-2}] +$$

$$\Delta[HSO_4^-] + \Delta[SO_4^{-2}]; \ \Delta[CO_2]g = \Delta[CO_2 \cdot H_2O] + \Delta[HCO_3^-] + \Delta[CO_3^{-2}];$$

$$\Delta[NH_3]g = \Delta[NH_3 \cdot H_2O] + \Delta[NH_4^+]; \text{ and } \Delta[HNO_2]g =$$

$$\Delta[HNO_2] + \Delta[NO_2^-] + 2\Delta[N_2O]g; \text{ all units are in moles.}$$

3. The growth of liquid water droplets due to the vapor pressure lowering effect of the sulfuric acid formed in the droplets is neglected.

The rate of sulfate production is determined by a calculation scheme involving a combination of equilibrium and kinetic steps. Equilibrium between SO_2 in the gas phase and sulfur (IV) in the droplet is several orders of magnitude faster than oxidation of sulfur (IV) to sulfate [56]. Similar assumptions were made regarding NH_3 and CO_2 gases. Therefore, initially gases are taken to be in equilibrium with the aerosols. Then the formation of sulfate proceeds by the given time-dependent production rate. The increase in the sulfate level in the small time step Δt causes the reduction in pH of the solution, which in turn disturbs the equilibrium between the aerosol and its surrounding gaseous environment. More gases are dissolved in the aerosol to maintain the equilibrium. At the same time, these gases are depleted in the surrounding atmosphere. After each calculation, the time step is adjusted and the process is repeated until a 24-hour period is completed. The results are shown in Fig. 12.

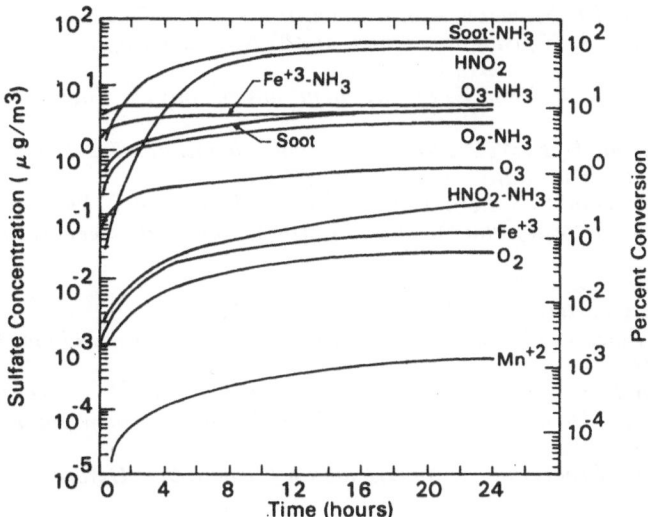

Fig. 12. Comparison of the relative importance of various sulfate production mechanisms involving liquid water based on box-type calculations. The following initial conditions were used in the calculation: $P_{SO_2} = 0.01$ ppm; $P_{CO_2} = 0.000311$ atm; $P_{NH_3} = 5$ ppb; $P_{O_3} = 0.05$ ppm; $P_{HNO_2} =$ ppb; $[Fe^{+3}] = 1.2 \times 10^{-7}$ mole/ℓ; $[Mn^{+2}] = 1.8 \times 10^{-8}$ mole/ℓ; soot $= 10\,\mu g/m^3$; and liquid water $= 0.05$ g/m³.

References pp. 178-179.

Fig. 12 indicates that mechanisms involving O_3, soot, and HNO_2 can be important for sulfate aerosol formation. In general the O_3 mechanism is more important under high pH and/or conditions which lead to photochemical activity and high concentrations of O_3, whereas both the soot and the HNO_2 processes are more important when the lifetime of fog or clouds is long and the pH of the droplets is low. Both the soot and the HNO_2 processes can be dominant close to sources and in heavily polluted urban areas where the concentrations of soot and NO_x are high and the pH of aqueous droplets is low.

The rate constant for atmospheric soot particles depends on the nature and history of particle production as discussed previously. In a fog chamber study, Benner et al. [46] have recently found that the reaction rate of soot particles from a natural gas diffusion flame can be considerably faster than the reaction rate reported here. More studies of rate constants involving soot from different types of fuel is therefore warranted.

ACKNOWLEDGEMENTS

This work was supported by the Biomedical and Environmental Research Division of the U.S. Department of Energy under contract no. W-7405-ENG-48 and by the National Sicence Foundation.

REFERENCES

1. S. G. Chang and T. Novakov, Atmos. Environ., Vol. 9 (1975), p. 495.
2. H. Rosen, A. D. A. Hansen, L. Gundel, and T. Novakov, Appl. Opt., Vol. 17 (1978), p. 3859.
3. U. Hofmann and D. Wilm, Z. Elektrochem. angew. physik. Chem., Vol. 42 (1936), p. 504.
4. V. A. Garten and D. E. Weiss, Rev. Pure Appl. Chem., Vol. 7 (1957), p. 69.
5. H. P. Boehm, Advan. Catalysis, Vol. 16 (1966), p. 179.
6. R. W. Coughlin and F. S. Ezra, Environ. Sci. Technol., Vol. 2 (1968), p. 291.
7. B. R. Puri, in Chemistry and Physics of Carbon, New York, Dekker, Vol. VI (1970), p. 191.
8. B. R. Puri, Carbon, Vol. 4 (1966), p. 391.
9. R. N. Smith, Quarterly Rev., Vol. 13 (1959), p. 287.
10. Y. A. Zary'yanz, V. F. Kiselev, N. N. Lezhnev, and D. V. Nikitina, Carbon, Vol. 5 (1967), p. 127.
11. V. A. Garten, D. E. Weiss, and J. B. Willis, Aust. J. Chemi., Vol. 10 (1957), p. 295.
12. H. P. Boehm, E. Diehl, W. Heck, and R. Sappok, Angew. Chem. Int. Ed., Engl., Vol. 3 (1964), p. 669.
13. B. R. Puri, in, Proceedings, Conference on Carbon, 5th, Oxford, Pergamon, Vol. 1 (1962), p. 165.
14. P. J. Hart, F. J. Vastola, and P. L. Walker, Jr., Carbon, Vol. 5 (1967), p. 363.
15. N. R. Laine, F. J. Vastola, and P. L. Walker, Jr., J. Phys. Chem., Vol. 67 (1963), p. 2030.
16. H. B. Palmer and C. F. Cullis, in Chemistry and Physics of Carbon, New York, Dekker, Vol. 1 (19   ), p. 265.
17. S. W. Weller and T. F. Young, J. Am. Chem. Soc., Vol. 70, (1948), p. 4155.

18. P. H. Emmett, Chem. Rev., Vol. 43 (1948), p. 69.
19. E. C. Larsen and J. H. Walton, J. Phys. Chem., Vol. 44 (1940), p. 70.
20. T. Novakov, P. Mueller, A. E. Alcocer, and J. W. Otvos, J. Colloid Interface Sci. Vol. 39, (1972), p. 225.
21. L. A. Gundel, S. G. Chang, M. S. Clemenson, S. S. Markowitz, and T. Novakov, In: Nitrogenous Air Pollutants, Ann Arbor, Ann Arbor Science, (1979), p. 211.
22. C. Brosset, paper presented at the Chemical Institute of Canada Annual Meeting, Ottawa, Ontario, June 8-11, 1980, sessions on acid precipitation.
23. T. Novakov and S. G. Chang, Lawrence Berkeley Laboratory Report LBL-6323, (1977).
24. F. Tuinstra and J. L. Koenig, J. Chemi. Phys., Vol. 53 (1970), p. 1126.
25. R. A. Friedel and L. J. E. Hofer, J. Phys. Chem., Vol. 74 (1970), p. 2921.
26. S. G. Chang and T. Novakov, Lawrence Berkeley Laboratory Report LBL-4446, (1975).
27. N. Shilov and K. Chmutov, Z. Phys. Chem., (Leipzig), Vol. A149 (1930), p. 211.
28. V. A. Garten and D. E. Weiss, Aust. J. Appl. Sci. Vol. 7, (1956), p. 148.
29. D. Rivin, in Proceedings of the 5th Conference on Carbon, Vol. II New York, Pergamon, (1963), p. 199.
30. A. Frumkin, R. Burstein, and P. Lewin, Z. Phys. Chem., Vol. A157 (1931), p. 442.
31. B. Steenberg, Adsorption and Exchange of Ions on Activiated Charcoal, (Uppsala, Almquist and Wiksells, 1944).
32. J. S. Mattson and H. B. Mark, Jr., Activated Carbon, Ch. 6 New York, Marcel Dekker, (1971).
33. R. W. Coughlin, I&EC Product Res. Develop. Vol. 8 (1969), p. 12.
34. T. Novakov, S. G. Chang, and A. B. Harker, Science, Vol. 186 (1974), p. 259.
35. S. G. Chang, R. Brodzinsky, R. Toossi, S. S. Markowitz, and T. Novakov, in Proceedings, Conference on Carbonaceous Particles in the Atmosphere, Lawrence Berkeley Laboratory Report LBL-9037, (1979), p. 122.
36. R. Brodzinsky, S. G. Chang, S. S. Markowitz, and T. Novakov, J. Phys. Chem. Vol. 84 (1980), p. 3354.
37. S. G. Chang, R. Toossi, and T. Novakov, Lawrence Berkeley Laboratory Report LBL-11380; accepted for publication in Atmos. Environ., (1980).
38. W. R. Cofer III, D. R. Schryer, and R. S. Rogowski, Atmos. Environ., Vol. 14 (1980), p. 571.
39. L. G. Britton and A. G. Clarke, Atmos. Environ. Vol. 14 (1980), p. 829.
40. W. R. Cofer III, D. R. Schryer, and R. S. Rogowski, submitted for publication in Atmos. Environ., (1980).
41. M. N. Rao and O. H. Hougen, Chem. Eng. Progr. Symp. Series, Vol. 48, (1952), p. 110.
42. P. F. Bente and J. H. Walton, J. Phys. Chem. Vol. 47 (1943), p. 133.
43. R. Dulou, Chim. Ind. (Paris), Vol. 54 (1945), p. 396.
44. E. Z. Stumpp, Anorg. allgem. Chem., Vl. 337 (1965), p. 292.
45. L. Gundel, private communication, (1979).
46. W. H. Benner, private communication, (1980).
47. A. Clark, The Theory of Adsorption and Catalysis, New York, Academic, (1970).
48. J. Siedlewski, Int. Chem. Eng., Vol. 5 (1965), p. 297.
49. S. B. Oblath, S. S. Markowitz, T. Novakov, and S. G. Chang, Lawrence Berkeley Laboratory Report LBL-10504; accepted for publication in J. Phys. Chem. (1980).
50. S. Beilke, D. Lamb, and J. Müller, Atmos. Environ., Vol. 9 (1975), p. 1083.
51. R. E. Erickson, L. M. Yates, R. L. Clark, and D. McEwen, Atmos. Environ., Vol. 11 (1977), p. 813.
52. J. Freiberg, Atmos. Environ., Vol. 9 (1975), p. 661.
53. M. J. Matteson, W. Stober, and H. Luther, Ind. & Eng. Chem. Fundam., Vol. 8 (1969), p. 677.
54. G. E. Gordon, D. D. Davis, G. W. Israel, H. E. Landserberg, and T. C. O'Haver, Report NSF/RA/E-73/189 (1975); available from NTIS as PB 262 574).
55. N. C. Lau and R. J. Charlson, Atmos. Environ., Vol. 11 (1977), p. 475.
56. S. Beilke and G. Gravenhorst, Atmos. Environ., Vol. 14 (1978), p. 463.

DISCUSSION

V. Mohnen, *(State University of New York)*

You said that one particular clean graphite, when you ground it up, did show adsorption sites or ammonia. We have done the same experiment using graphite from the General Electric Company. We were told that this graphite was taken out of the oven at about 3000°C and it is as clean as one can get it. When we tried the same experiment, it did not work. This tells me the degree of cleanliness is important and not every graphite will do what you say.

S.-G. Chang

It is very difficult to obtain an infrared spectrum for black carbon species because of their high absorptive nature. We have observed infrared structures of nitrogen complexes in carbon particles by grinding a high purity POCO graphite in an ammonia atmosphere for about a hundred hours. Whether or not an infrared spectrum can be obtained depends on how effectively and how long the graphite has been ground. With the aids of an ESCA, infrared structures can be observed with a FT-infrared spectrophotometer if the nitrogen to carbon ratio after grinding reaches 0.1.

R. Stevens, *(U.S. Environmental Protection Agency)*

In the past three or four years we have been analyzing samples from many urban areas and we have been saying that measuring fine particle sulfur was equivalent to measuring sulfate. In the last two years, we have run across several examples where we measure the total sulfur and we measure the sulfate and they do not agree. This indicates that we may not extract all the sulfur from the aerosol. In every case that this happened so far, there has been a substantial quantity of carbon present. Now, I wonder if your laboratory has made any investigation of the interaction of the soot or graphitic-like material with sulfuric acid creating some kind of stable nonextractable combination of sulfuric acid in the polymer such that the sulfur no longer is soluble even in acid.

S.-G. Chang

The reaction between the aromatic hydrocarbons and sulfuric acid can produce sulfonic acid. Soot particles, which have aromatic structure, might form surface sulfonate groups as a result of reaction with sulfuric acid. Such sulfuric acid groups can undergo hydrolysis at high temperature, however.

R. Stevens

Maybe it is polymeric?

S.-G. Chang

It is possible. We did examine the nature of the interaction between carbon and sulfuric acid and we found that the sulfuric acid is neutralized to form a sulfate. This is based on an infrared spectrum.

C. Brosset, *(Swedish Air & Water Pollution Institute)*

You are quite correct, you can get sulfate buried in the lattice of graphite.

M. Shelef, *(Ford Motor Company)*

In your introduction you have said that the density of electrons is related to the catalytic activity. At the end of your talk, you have shown that you have eight different carbons that have widely different catalytic activity. It is easy to measure the density of unpaired electrons. Have you done this?

S.-G. Chang

That has been done by Siedlewski.* He found that the oxidize SO_2 is proportional to the epr intensity for carbon particles.

J. Siedlewski, Int. Chem. Eng., Vol. 5 (1965), p. 297.

SESSION III
SOURCES OF CARBON

Session Chairman

A. P. WAGGONER

University of Washington
Seattle, Washington

FIREPLACES, FURNACES AND VEHICLES
AS EMISSION SOURCES OF PARTICULATE CARBON

J. L. MUHLBAIER and R. L. WILLIAMS

General Motors Research Laboratories
Warren, Michigan

ABSTRACT

Particulate carbon in the atmosphere results from the incomplete combustion of fossil and contemporary fuels. Fuel-specific emission rates of elemental and organic carbon are presented for residential wood-burning fireplaces and gas furnaces as well as precatalyst, catalyst, and diesel automobiles. The fireplace emissions from several types of wood were collected from residential chimneys using standard EPA source-sampling procedures. Furnace emissions were collected with a laboratory setup and the effects of variations in stoichiometry and cycling rate were evaluated. Automotive emissions were collected from a dilution tunnel using two driving cycles both at sea level and high altitude. Carbon analysis was performed by several methods, including temperature programming, combustion analysis, and solvent extraction. Organic carbon associated with the particulate emissions was also tested for benzo(a)pyrene content. The emission rates and annual fuel consumption data were used to estimate the relative strength of these particulate carbon sources on a national basis.

INTRODUCTION

Particulate carbon is an atmospheric pollutant which can cause visibility reduction [1], promote chemical reactions [2], and possibly disturb the global heat balance [3]. Analytical methods for separating and measuring elemental and organic carbon [4, 5] have only recently become available. As a result, emission sources of carbon have not been adequately characterized. Since all combustion sources which burn carbonaceous fuel emit particulate elemental and organic carbon [6], many different sources need to be evaluated.

References p. 198

In this study, we have examined three types of sources: residential natural gas furnaces, wood-burning fireplaces, and automobiles. The automobile tests included precatalyst and catalyst-equipped gasoline cars as well as diesel cars. Based on their emissions, it is possible to estimate the relative contribution of each source to nationwide ambient carbon levels.

SAMPLE COLLECTION

Fireplace — Sampling was conducted at two residences, both of which had brick fireplaces built on an outside wall of the house. An aluminum chimney extension with an 8 cm port was placed on the chimney and the sample probe was inserted through the port. Sampling was accomplished with source-sampling equipment which complied with EPA Method-5 procedures [7]. A flue sample was drawn through a 1.3 cm diameter nozzle and through a heated probe to a cyclone and to a glass fiber filter held at about 110°C. The cyclone removed particles greater than 10 micrometers in diameter. After the filter, the sample passed through two water-filled impingers, an empty impinger, a dessicant, and then to a dry gas meter.

Due to the very low flow rate in the chimney, the pitot tube readings were too low to obtain an accurate flow measurement. Therefore, the flow rate was determined at frequent intervals during sampling using a thermal anemometer. All samples were taken from the center of the chimney. A study by Snowden, *et al.* [8] has indicated that there is a reasonably flat velocity profile across fireplace chimneys. In addition, traversing the chimney is not appropriate in a situation in which the flow changes considerably during sampling. For this reason and the fact that the 'stack' diameter was small, traverses of the stack were not performed. Due to the lack of continuous monitoring of the flow in the chimney, the sampling flow rate was simply held constant throughout the tests. Since the flow and particulate loading in the chimney decrease with time, the effect of constant-flow sampling is to underestimate the early particulate loadings and overestimate the later loadings.

A single test consisted of burning preweighed split logs which were placed in three lots of approximately 0.5 kg each on hot coals. The test continued until the quantity of coals was judged to equal the original amount. A single test took approximately 45 minutes. Nine tests were made on five varieties of hardwood and six tests were made on two softwoods. In this context, hardwood is used to designate deciduous trees and softwood designates coniferous trees. The moisture content varied from 4 to 16%. In addition, two brands of synthetic 'logs' were tested. One had a sawdust base and the other a peanut by-product base. The base material is held together with wax, and copper is added for color effects.

Particulate was collected on Whatman GF/A glass fiber filters. The front catch (nozzle, probe, and cyclone) was rinsed with deionized water and acetone into a preweighed beaker, taken to dryness, and reweighed. The impinger water and rinsings, which made up the condensable catch, were treated similarly.

Furnace — A Lennox 116 x 10^6 J/h (110,000 BTU) furnace and a KGA 150 L, 42 x 10^6 J/h (40,000 BTU) hot water heater were used to study emissions from gas appliances. The furnace contained four parallel ribbon jets, each with an air shutter for controlling air to the flame. Since a natural gas supply was not available, cylinders of 93 % methane were used. This corresponds to the composition of natural gas in the Detroit area, which is 93 % methane and has a gross heating value of 3.7×10^7 J/m^3. Liquid propane with a heating value of 9.4×10^7 J/m^3 was also used as a fuel. The fuel was passed through a dry gas meter to monitor usage, which was approximately 0.046 m^3/min of methane or 0.019 m^3/min of propane. The hot water heater could only be run on methane and no regulation of air to the flame was possible. The fuel usage for the hot water heater was 0.017 m^3/min methane. The exhaust gases were drawn through a 20 cm diameter flow tube to a floor exhaust. The air flow was controlled by the floor exhaust rate rather than by natural draft. The particle collection devices were mounted at the far downstream end of the flow tube. There were three particle collection devices: a 47 mm filter, a 142 mm filter, and an 8-stage Andersen impactor. Gelman A-E glass fiber filters were used for mass measurement of particles. Each collection device was equipped with a rotameter and a manometer to monitor the air flow and pressure drop across the filter. The temperature at the sampling point was about 90°C. The filters were not heated but quickly reached the temperature of the exhaust gas.

The air velocity through the tube was measured to determine the fraction of exhaust air sampled. The flow changed according to the load on the exhaust system. The air flow was monitored by measuring the CO_2 concentration at the sampling point. From the volume of fuel burned, the expected emission rate of CO_2 was calculated. Carbon dioxide was the only major carbon combustion product with CO present only as a trace constituent. From the CO_2 concentration in the tube, the dilution could be calculated to arrive at a flow rate. The volume flow through the tube varied from 2-6 m^3/min, which agrees with the average flow of 4 m^3/min we found in two residential chimneys. The air velocity through the sampling probes was two to four times the velocity of the exhaust gas. A high sampling rate was considered necessary due to the low emissions. This super-isokinetic condition is not expected to compromise the results since the particles are sub-micron and are expected to follow the air stream.

Vehicles — Exhaust particulate samples were collected by filtration using standard dilution tunnel techniques. For the tests at the Warren, Michigan, laboratory, a 0.3 m diameter tunnel was used with a flow rate of 17 m^3/min. For the high altitude tests at the Denver laboratory, a smaller (0.2 m diameter) tunnel was used with a flow rate of 8.8 m^3/min. In both cases, the entire exhaust was transferred to the tunnel where it was mixed with prefiltered air to cool and dilute the exhaust in a manner similar to dilution in open air.

In both facilities, the test vehicle was driven through a prescribed cycle on a chassis dynamometer. The particulate material was removed from a proportional sidestream and collected on Dexiglas #225 filters. The collected particulate and the transmitted gases were measured to determine the mass emission rates of the exhaust components. Vehicle descriptions and other test details will be given below. The driving conditions used were the Federal Test Procedure (FTP) urban cycle, the Highway Fuel Economy Test (HFET) cycle, and constant-speed cruise.

ANALYTICAL METHODS

Carbon Analysis — *Carbon Fractionation* – Samples were collected for carbon analysis on glass or quartz fiber filters which were preheated overnight at 500°C. To obtain a qualitative idea of the type of carbon present in each sample, the filters were heated slowly in air (25°C/min) and the continuous evolution of CO_2 was monitored. Carbon thermograms from four types of samples are shown in Fig. 1. The four thermograms are all dominated by a single large peak. Most of the carbon is burned off above 500°C in the case of the furnace, softwood burn, and diesel thermograms. This fact, combined with the small extractable fraction of these samples, indicates that the material is elemental carbon. Since the peaks occur at different temperatures, the elemental carbon may be present in different forms. On the other hand, the major peak from the hardwood burn is at a somewhat lower temperature and 75% of the total particulate is extractable. These two results indicate that the carbon from the hardwood burn is primarily organic.

Standard Method – The filters were analyzed for both organic and elemental carbon by the method of Cadle *et al.* [5]. The analysis consisted of dropping a 1 cm² filter section into a 650°C zone in helium to volatilize the organics, which were then catalytically oxidized to CO_2. During the second step of the analysis, air entered the system and the elemental carbon was oxidized to CO_2. The amount of carbon dioxide evolved during the first and second steps is proportional to the organic and elemental carbon concentrations, respectively. A problem inherent in this technique is that the organic carbon may carbonize or char in the first step of analysis and incorrectly be measured as elemental carbon in the second step.

Modification to Reduce Charring – To estimate the amount of charring, the method was modified in the following manner: one sample was run using the original method. A second sample of each filter was heated in a tube furnace to 350°C in air in an attempt to remove organics without charring. After the organics were removed, the sample was then analyzed for elemental carbon by the original method. The difference between the total carbon determined by the first method and the elemental carbon determined by the second method represented the organic carbon. To illustrate the extent of the charring problem for natural products such as wood, the apparent elemental carbon content of the hardwood emissions decreased on the average from 41% to 5% after charring was minimized by the modified procedure. On the other hand, we do not expect a significant amount of charring during the carbon analysis of fossil-fuel emissions [5, 9]. Therefore, the carbon analysis results for furnace and vehicle emission samples were obtained using the standard method, while the modified method was used for the fireplace emission samples.

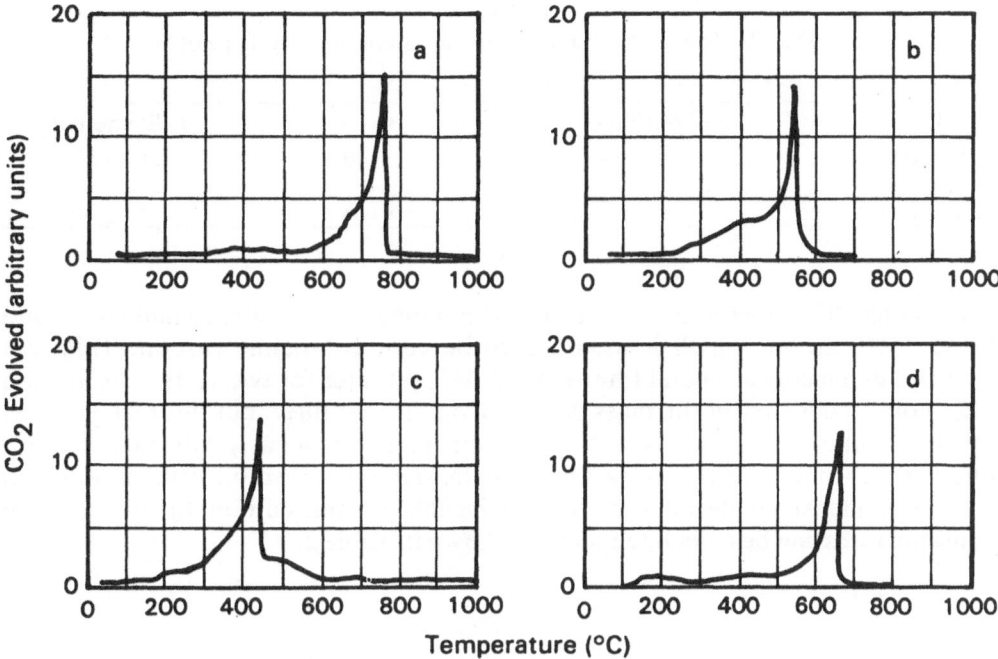

Fig. 1. Thermograms of particulate from the following sources: a) furnace operating rich, b) softwood burned in fireplace, c) hardwood burned in fireplace, and d) diesel automobile.

% **Extractable and Benzo(a)pyrene** — The extractable percentage and benzo(a)pyrene concentrations were determined by the method of Swarin and Williams [10]. A portion of the particle-laden filter was extracted in benzene-ethanol and taken to dryness to determine the extractable fraction. This extract was then dissolved in hexane-methylene chloride, dried, and dissolved in acetonitrile. An aliquot was injected onto a Zorbax-ODS column in a liquid chromatograph and the BaP was detected by fluorescence.

RESULTS

Carbon Emissions from Fireplaces — The results of the carbon analysis from the fireplace filter samples are shown in Table 1. The total carbon content of the filterable emissions was about 55% for the hardwoods and 80% for the softwoods and synthetic woods. The five hardwoods showed surprising similarities in the fraction of organic and elemental carbon despite the fact that total emissions varied from 2-18 g/kg of wood burned. The three synthetic log samples showed wide variations. Emissions from a split sawdust log were 5% organic/71% elemental, but the unsplit sawdust log emissions were 52% organic/14% elemental carbon. The extractable fraction ranged from 40 to 90% from the natural woods.

References p. 198

TABLE 1
Carbon Distribution of Filterable Particulate Emissions by Type of Wood.

Type	No. of Filters Analyzed	% Organic C	% Elemental C
Softwood	4	40 ± 7	36 ±5
Hardwood	6	50 ± 9	5.1 ± 2.5
Synthetic Log	3	26 ± 24	55 ± 36

A sample of the material collected in the impingers was analyzed and was found to be 45% carbon, which is assumed to be entirely organic carbon. This was checked by placing a second filter behind the first filter for two tests. The second filter collected 5-10% of the mass collected on the first filter, but this material was totally organic carbon indicating that no elemental carbon passed through the first filter. If the front catch is assumed to have the same composition as the filter catch and the condensed material is assumed to be 45% organic carbon, the final carbon emission rates can be calculated and are shown in Table 2.

TABLE 2
Carbon Emission Rates by Type of Wood.

Type	Organic C	Elemental C g/kg wood	Remainder
Softwood	2.8 ± 1.5 (45%)	1.3 ± 0.5 (21%)	2.1 ± 0.6 (34%)
Hardwood	4.7 ± 3.4 (47%)	0.39 ± 0.34 (4%)	4.9 ± 4.2 (49%)
Synthetic Log	1.7 ± 1.1 (24%)	3.5 ± 3.3 (49%)	2.0 ± 1.3 (28%)

The softwood and hardwood particulate emissions are both about 45% organic carbon, but much more elemental carbon comes from the softwood. The remaining material is assumed to be a combination of organically-bound oxygen and hydrogen as well as water and ash. This sizable difference in the fractions of elemental and organic carbon may be related to the composition of woods [11]. In general, hardwoods contain 78% holocellulose (cellulose and hemicellulose) and 22% lignin. Softwoods contain only 71% holocellulose and 29% lignin. Results by Tillman indicate that holocellulose promotes the release of volatiles during combustion, while lignin promotes elemental carbon formation [11]. This is consistent with the results here in which softwoods, which have a higher lignin content, emit larger amounts of elemental carbon.

Carbon Emissions from Furnaces — The filterable particulate emissions from furnaces were extremely low, averaging 0.09 ng/J when methane was the fuel and 0.28 ng/J when propane was the fuel. The laboratory results indicate that the filterable particulate was only 8% organic carbon and 4% elemental carbon. When the furnace was operated in on-off cycles, as in a residential setting, the same amount of particulate was emitted as during continuous operation. However, the cyclic mode of operation did increase the elemental carbon content. One series of tests showed 0.4% elemental carbon when the furnace ran continuously and 7% when the furnace was cycled. The organic carbon was about 17% under both operating conditions. Most of the remaining filterable particulate was nitrate or sulfate.

In the laboratory experiments, diluted exhaust air was sampled 2 m downstream of the furnace. For comparison, two samples were also collected from residential chimneys. The filterable emissions from these samples were within the range measured in the laboratory. However, the two samples showed organic carbon contents of 33% and 40%, respectively, values considerably higher than the average of 8% measured in the laboratory. There may be an effect of time, temperature, or dilution which leads to an underestimate of organic carbon in the laboratory tests. On the other hand, the elemental carbon emissions were within the range measured in the laboratory.

When the furnace was operated with the air shutters completely closed leading to a highly fuel-rich condition, the mass emissions increased by a factor of thirty. Emissions were 2.7 ng/J when methane was the fuel and 9.2 ng/J when propane was the fuel. Under this rich condition, the organic carbon content was about 5% and the elemental carbon content was 89%. We have not estimated the fraction of furnaces that are run in a fuel-rich manner. Average carbon results from all tests are shown in Table 3.

TABLE 3
Carbon Content of Filterable Furnace Particulate.

Operating Conditions	Sampling Site	Number of Tests	% Organic Carbon	% Elemental Carbon
Normal	Laboratory	10	8	4
Normal	Residential Chimney	1	40	8
Fuel-Rich	Laboratory	6	5	89
Fuel-Rich	Residential Chimney	1	33	46

Carbon Emissions from Automobiles — *Sea Level* – A wide variety of automobiles have been tested in our laboratory to determine the particulate emission rates of elemental and organic carbon. All cars emitted measurable amounts of both organic and elemental carbon. The average emission results by category are shown in Table 4 for testing in Warren, Michigan. For areas where freeway driving is predominant, the data from the HFET cycle would be more appropriate than the lower speed FTP cycle. In all cases, both the HFET cycle and cruise operation result in lower per mile emission rates of particulate carbon than does the FTP.

TABLE 4
Summary of Carbon Emissions from Automobiles (tested at 180 m MSL).

Car Type	Number of Cars	Driving Mode	Total Particulate	mg/mile Organic Carbon	Elemental Carbon
Diesel	4[a]	FTP	680	137	520
		88 km/h	424	157	258
Precatalyst	5[b]	FTP	119	10	3.6
		HFET	196	9.3	1.9
Oxidation	4[c]	FTP	14	3.0	2.5
catalyst		HFET	25	4.2	1.5
Computer-Command	5[d]	FTP	12	1.9	1.1
Control		88 km/h	35	1.3	0.1

[a] *'78 Oldsmobile Delta; '79 Oldsmobile Cutlass; '78 Opel Rekord; '80 Oldsmobile Delta.*
[b] *'70 Chevrolet Malibu; '70 Chevrolet Impala; '74 Chevrolet Caprice; '78 Chevrolet Caprice with catalyst removed; '78 Pontiac Sunbird with catalyst removed.*
[c] *'78 Chevrolet Caprice; '79 Buick Century; '79 Oldsmobile Cutlass; '79 Pontiac Catalina.*
[d] *'78 Pontiac Sunbird, California system; '79 Buick Regal, California system; 4.3 L prototype; 5.0 L prototype; 5.1 L prototype.*

High Altitude – Ten privately-owned cars in Denver were obtained by a contractor (Automotive Test Laboratories) and were tested by GM personnel at the GM Denver Vehicle Emissions Laboratory. Five 1972-1974 model-year cars running on a single fuel, Amoco leaded gasoline from the Denver area, were used to represent precatalyst cars. Similarly, five 1975-1978 model-year cars running on Chevron unleaded gasoline, also purchased in Denver, were used to represent catalyst-equipped cars. The emission results for the ten Denver cars are shown in Table 5.

Particulate carbon emissions from the precatalyst cars averaged 0.044 g/mile organics and 0.024 g/mile elemental carbon. The particulate carbon HFET emission rates were indistinguishable from the corresponding FTP rates. Carbon emissions from the catalyst-equipped cars averaged 0.005 g/mile organic and 0.006 g/mile elemental carbon for the FTP cycle. In the HFET cycle, carbon particulate emissions from catalyst cars averaged 0.009 g/mile organic carbon and 0.003 g/mile elemental carbon. It should be noted that the differences between the catalyst and precatalyst car emissions were not solely due to the catalyst, but most likely included effects of other emission control factors, fuel composition, and mileage as well.

The Denver testing included three cars from which regulated emissions and particulate emissions had also been measured at Warren, Michigan. The cars were transported to Denver without modification and were tested in Denver on the same fuel that had been used in Warren. The results of the altitude comparison tests are

TABLE 5
Particulate Emissions from Denver Vehicles.

Vehicle	Driving Mode	Total Particulate	mg/mile Organic Carbon	Elemental Carbon
'72 Chevrolet	FTP	97.3	19.1	7.4
Monte Carlo	HFET	263	10.5	2.4
'73 Oldsmobile	FTP	120	13.3	8.9
Delta	HFET	330	5.6	4.0
'73 Pontiac	FTP	153	11.9	13.9
Catalina	HFET	301	9.6	3.9
'74 Chevrolet	FTP	288	126	79.8
Vega	HFET	460	182	89.8
'74 Ford	FTP	99.5	48.8	10.5
Pinto	HFET	144	23.2	4.2
Precatalyst	FTP	152	43.8	24.1
Average	HFET	300	46.2	20.9
'75 Chevrolet	FTP	21.4	6.5	10.3
Vega	HFET	13.3	8.1	3.0
'77 Oldsmobile	FTP	11.4	1.8	1.6
Cutlass	HFET	13.9	4.3	2.2
'77 Ford	FTP	21.3	3.9	5.9
Granada	HFET	52.5	19.4	9.0
'78 Buick	FTP	26.4	8.7	7.5
Regal	HFET	49.8	9.2	0.7
'78 Chevrolet	FTP	12.4	3.4	2.9
Monte Carlo	HFET	17.9	3.4	0.6
Catalyst	FTP	18.6	4.9	5.6
Average	HFET	29.5	8.9	3.1

summarized in Fig. 2. The reduced air pressure at Denver (1550 m MSL) has a direct effect on the air/fuel ratio which, in turn, affects the amounts of gaseous and particulate emissions. The HC and CO were higher at the higher altitude, while NO_x was lower. Elemental carbon emissions increased 290% for the precatalyst car, increased 600% for the catalyst car, but decreased slightly for the diesel car. The particulate organic carbon emission rates changed less. These results for cars tested at both altitudes are consistent with the emission results we have reported above for cars native to the two different altitudes.

Benzo(a)pyrene Results — The benzo(a)pyrene content of emissions from wood burning was determined using several filters as shown in Table 6. Obviously, the results cover a wide range, and the two tests which used willow suggest source repeatability problems. The median value from the wood-burning tests was 58 micrograms BaP/kg wood. Two aliquots of impinger water were also analyzed. One sample had no detectable BaP and the second sample yielded 0.08 micrograms BaP/

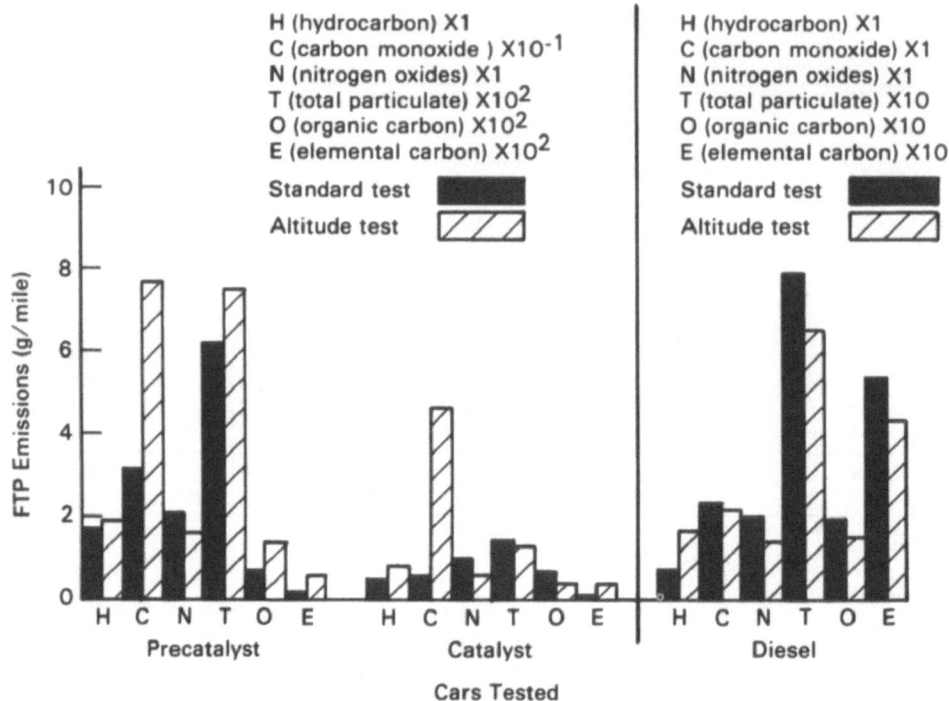

Fig. 2. A comparison of emissions from three cars tested under standard (180 m MSL) and high altitude (1550 m MSL) conditions.

kg for the impinger water, which is insignificant compared to the BaP in the filter catch. This suggests that the BaP is almost totally associated with the particulate emissions. For comparison, BaP emissions per kilogram of fuel are listed in Table 6 for other fuels. The only BaP levels that are higher than those from wood-burning appear to be those from very inefficient combustion processes such as residential coal burning. In contrast, efficient coal burning, as in power plants, produces much lower BaP levels. As shown in Table 6, older model gasoline-powered automobiles without catalysts emit about 16 micrograms BaP/kg gasoline, but BaP emissions are reduced more than 97% by the catalytic converters on later model automobiles. Diesel-powered automobiles also have intermediate emission rates averaging 16 micrograms BaP/kg diesel fuel.

DISCUSSION

The ratio of elemental carbon to total carbon (C_e/C_T) has been calculated for each source and is shown in Fig. 3. The elemental carbon fraction varied from 7.7%

TABLE 6
Benzo(a)pyrene Content of Filterable Particulate Emissions.

Source	μg BaP/g particulate	μg BaP/kg fuel
Wood — Fireplaces		
Ponderosa Pine	24	50
Willow	105	700
Willow	141	1900
White Ash	3	5
White Ash	7	17
Sugar Maple	11	45
Hickory	120	130
Peanut 'Log'	18	58
Sawdust 'Log'	40	400
Coal — Power Plant [12]		2
Residential [12]		25 000
Gasoline — Noncatalyst Automobile [13]		16
Catalyst Automobile [13]		0.4
Diesel Fuel — Automobile [13]		16

from hardwoods burned in fireplaces to 94 % from furnaces running rich. The C_e/C_T ratio of ambient particulate will therefore depend on the local mix of sources, the ratio from each source and the amount of secondary particulate carbon formation. Based on the data in Fig. 3, it does not seem likely that the amount of secondary particulate carbon can be deduced from the ambient C_e/C_T ratio alone.

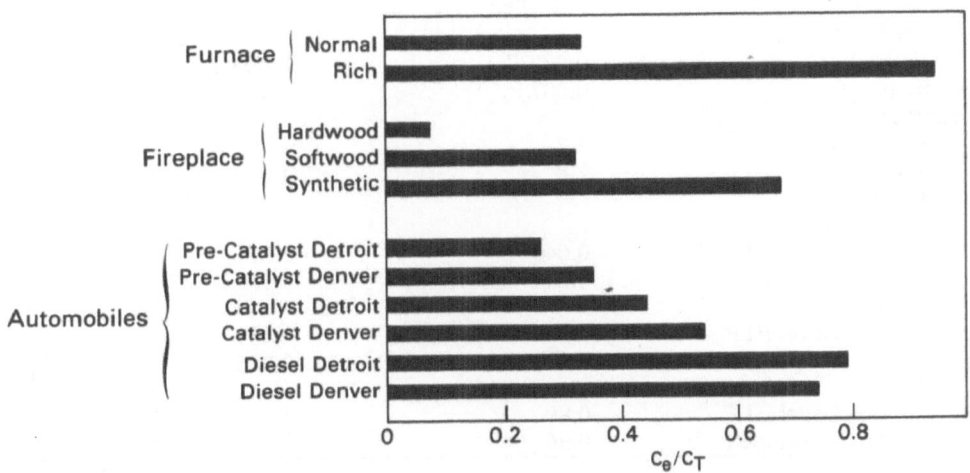

Fig. 3. The ratio of elemental carbon to total carbon from selected sources.

It is also possible to compare the different sources on the basis of total carbon emissions. To facilitate such a comparison, it is necessary to express fuel-specific emission rates in common units. For this purpose, the emission rates in grams per kilogram of fuel have been calculated and are shown in Table 7. Based on these emission rates and on national fuel consumption information, we have constructed an inventory of annual nationwide emissions from the sources we have studied, as shown in Table 8. Fireplaces turn out to be the largest source of both organic and elemental carbon, while natural gas furnaces are the smallest. Heavy-duty diesel emissions were not included in Table 8 because we have not made emission measurements from heavy-duty diesels. However, based on emission rates found in the literature [14], it is possible to estimate the nationwide emissions for heavy-duty diesels. The fuel consumption of heavy-duty diesel trucks and buses in 1975 was 230 x 10^6 barrels [14]. Approximately 25% of these were 2-stroke engines with average particulate emissions of 4.74 g/kg fuel [14]. The other 75% were 4-stroke engines with average particulate emissions of 2.64 g/kg. Based on the diesel automobile emissions, we estimate that the particulate is 20% organic carbon and 80% elemental carbon. Thus, the annual nationwide emissions for heavy-duty diesels are estimated to be 19 x 10^9 g organic carbon particulate and 76 x 10^9 g elemental carbon particulate.

The inventory we have constructed deals with annual, nationwide emissions. However, if a local or short-term inventory is required, it is necessary to consider a number of factors, such as the season of the year, the meteorological conditions and

TABLE 7
Fuel-Specific Emission Rates (g carbon/kg fuel).

Source	Organic C	Elemental C
Furnace[a]		
Normal	0.00037	0.00018
Rich	0.0070	0.12
Fireplace		
Hardwood	4.7	0.39
Softwood	2.8	1.3
Automobiles[b]		
Precatalyst		
Sea Level FTP	0.040	0.014
High Altitude FTP	0.240	0.130
Catalyst		
Sea Level FTP	0.014	0.011
High Altitude FTP	0.028	0.033
Diesel		
Sea Level FTP	0.89	3.4
High Altitude FTP	0.96	2.8

[a]5.18 x 10^7 joules/kg natural gas.
[b]Based on the actual mileage of the individual cars tested.

TABLE 8

Estimated Annual Nationwide Emissions of Particulate Carbon
from Selected Sources (in units of 10^9g).

Source	Organic C	Elemental C
Residential Gas Furnaces[a]	0.037	0.018
Fireplaces[b]	86	11
Automobiles		
Precatalyst[c]	5.6	2.0
Catalyst[c]	2.0	1.5
Diesel[d]	1.1	4.2

[a]*Residential usage of natural gas was 1.4×10^{11} m^3 in 1975 [15] or 1.0×10^{11} kg. The estimate is based on emissions from a properly set furnace.*
[b]*Based on several estimates of firewood usage, we chose a value of 2.0×10^{10} kg burned in a year [11, 16, 17]. This estimate is based on a national mix of 80% hardwood and 20% softwood burned [16].*
[c]*The gasoline usage for highway vehicles in 1978 was 110×10^9 gallons [18]. The 1980 fuel mix is approximately 50% leaded and 50% unleaded. The emissions are calculated based on sea level emissions.*
[d]*We estimate that there are 800,000 diesel automobiles on the road, travelling 10,000 miles per year each, with an average fuel economy of 20 miles/gallon.*

the geographic location. Fireplace and furnace use is seasonal whereas vehicular use is not. The peak usage for fireplaces and furnaces occurs during winter evenings when dispersion is very poor. The effect of such emissions on local ground-level concentrations may be disproportionately large compared to the size of the source. The geographic location plays a role "in comparing urban areas with rural areas." Most of the population, and therefore most of the energy use is concentrated in the cities. However, many of the power plants are located outside urban centers. Likewise over 80% of heavy-duty diesel travel is in non-urban areas. Therefore, the effect of suburban power plants and diesel trucks on urban ambient concentrations is less than suggested by the national inventory.

Obviously, many factors affect the real emission values, such as the fuel type and the stoichiometry in each combustion system. We have considered these complicating factors to some extent in our study by including automobile emission results at high altitude and furnace results under rich combustion. There are, of course, other complicating factors which have not been considered, such as the effect of low temperatures on vehicle emissions or the effect of altitude on furnace and fireplace emissions. In addition, there are a multitude of sources which have not been considered in this study. In the final analysis, the real emissions from a large number of independent sources cannot be determined accurately. Therefore, while we have used the best emissions data available, the emission estimates in Table 8 should be considered as approximations.

References p. 198

ACKNOWLEDGMENTS

The following people have all made valuable contributions to this report: S. G. Anderson, J. R. Collins, K. G. Kennedy, A. I. Ricci, D. P. Stroup, T. L. Gibson, S. H. Cadle, P. J. Groblicki, R. L. Klimisch, D. L. Eggemeyer, J. I. Gutting, and W. K. Langhorst of General Motors and M. Koelling of Michigan State University.

REFERENCES

1. A. D. A. Hansen, H. Rosen, R. L. Dod, W. H. Benner, and T. Novakov, "Optical Characterization of Ambient and Source Particulates," in Atmospheric Aerosol Research Annual Report, LBL-8696, 1977-1978.
2. S. G. Chang, R. Brodzinsky, R. Toossi, S. S. Markowitz, and T. Novakov, "Catalytic Oxidation of SO_2 on Carbon in Aqueous Suspensions," in Proceedings of the Carbonaceous Particles in the Atmosphere Conference, T. Novakov, Ed., LBL-9037, June 1979.
3. R. A. Reck and J. R. Hummel, "Influence of Aerosol Optical Properties on the Atmospheric Temperature at the Earth's Surface," General Motors Research Laboratories, Publication GMR-3146, November 1979.
4. R. L. Johnson and J. J. Huntzicker, "Analysis of Volatilizable and Elemental Carbon in Ambient Aerosols," in Proceedings of the Carbonaceous Particles in the Atmosphere Conference, T. Novakov, Ed., LBL-9037, June 1979.
5. S. H. Cadle, P. J. Groblicki, and D. P. Stroup, Anal. Chem., Vol. 52, 2201 (1980).
6. H. B. Palmer and C. F. Cullis, "The Formation of Carbon from Gases," in The Chemistry and Physics of Carbon, Vol. 1, Marcel Dekker, Inc., New York, NY, 1965.
7. "Standards of Performance for New Stationary Sources," Federal Register, Vol. 42, August 1977.
8. W. D. Snowden, D. A. Alguard, G. A. Swanson, and W. E. Stolberg, "Source Sampling Residential Fireplaces for Emission Factor Development," U.S EPA-450/3-76-010, November 1975.
9. S. H. Cadle and P. J. Groblicki, "An Evaluation of Methods for the Determination of Organic and Elemental Carbon in Particulate Samples," These Proceedings.
10. S. J. Swarin and R. L. Williams, "Liquid Chromatographic Determination of Benzo(a)pyrene in Diesel Exhaust Particulate: Verification of the Collection and Analytical Methods," General Motors Research Laboratories, Publication GMR-3127, October 1979.
11. D. A. Tillman, Wood as an Energy Resource, Academic Press, New York, NY, 1978.
12. N. L. Nagda, D. J. Pelton, and J. L. Swift, "Emission Factors and Emission Inventories for Carcinogenic Substances," presented at 72nd Annual Meeting of the Air Pollut. Control Assoc., Cincinnati, OH, June 1979.
13. R. L. Williams and S. J. Swarin, "Benzo(a)pyrene Emissions from Gasoline and Diesel Automobiles," General Motors Research Laboratories, Publication GMR-2881, February 1979.
14. T. M. Baines, J. H. Somers, and C. A. Harvey, J. Air Pollut. Control Assoc., Vol. 29, 616 (1979).
15. Mineral Yearbook 1975, Vol. 1, U.S. Dept. of Interior, Bureau of Mines, Washington, DC, 1977.
16. Yearbook of Forest Products, 1966-1977, Food and Agriculture Organization of the United Nations, Rome, Italy, 1979.
17. America's Demand for Wood 1929-1975, Stanford Research Institute, Tacoma, WA, 1954.
18. MVMA Motor Vehicle Facts and Figures, '80, Motor Vehicle Manufacturers Association, Detroit, MI, 1980.

DISCUSSION

R. Stevens, *(US. Environmental Protection Agency)*

That was a very excellent paper. I am fascinated at the amount of data that was collected. I have two comments. We have been collecting samples for two years in Denver every twelve hours at two sites. As a result of our studies we observed a vastly increased concentration of potassium in the fine particle component which correlated with an increase in the blackness of the filters. The periods when this took place were on weekends, holidays, and at night as you pointed out during very cold days. In your emissions studies, did you make any trace metal measurements so that we can look at tracer elements?

J. Muhlbaier

All the samples we have collected so far were on glass fiber filters so the blanks were too high. The samples that I am collecting now are on cleaner filters and we will make potassium and iron measurements.

J. Huntzicker, *(Oregon Graduate Center)*

In the northwest, and I presume in other parts of the country, most serious wood heating is done with wood stoves rather than fireplaces, and people tend to cut the air flow way down so that they have very slow burning. I suspect this would affect the organic to elemental carbon ratio dramatically. Have you looked at wood stoves at all?

J. Muhlbaier

No, we have not. There are some papers by people who have done that and they did find higher particulate emissions and higher carbon monoxide emissions which would suggest that there would be more elemental carbon.

J. Heicklen, *(Pennsylvania State University)*

What are the molecular weights of your organic carbon particulates?

R. Williams, *(General Motors Research Laboratories)*

The diesel goes up to very high molecular weights which are on the order of five thousand. We have not looked at the full distribution on the fireplace particulate.

J. Heicklen

And, tell me, when this organic carbon gets into the atmosphere, is some of it going to vaporize?

R. Williams

I think that the best answer to that is that nobody really knows. We do have a very high concentration of absorbing particles. We do not know how tightly bound the organics are to the particles or how rapidly they might attach to some other particle.

R. Bradow, *(U.S. Environmental Protection Agency)*

I have a comment relative to that matter. Clearly, the ability of organics associated with the particles to enter the gas phase depends on the nature of the condensed phase. There are measurements recently made of the heat absorption of various organic compounds present in diesel particles. These suggest numbers on the order of 2 to 5 kcal per mole surface energy. It appears that the diesel organics are probably not very volatile and it takes quite a bit to drive them off. For example, to completely displace the organics requires heating in vacuum to about 300°C. So, apparently, the material is not very volatile. But in the case of fireplace emissions there is a great deal of organic matter and no one knows for sure if it is going to be bound on carbon exclusively or exist as liquid droplets.

J. Heicklen

Have you done Ames mutagenicity tests or measured benzo-a-pyrene in these samples?

J. Muhlbaier

We decided that the thrust of this conference was not health effects and so I did not speak about it. The benzo-a-pyrene results however are in the paper. We thought the Ames test results were not appropriate for this symposium. We have done Ames tests on diesel exhaust and on fireplace particulate. The diesel exhaust is similar to what is in the literature. The fireplace particulate all gave positive results in the Ames test primarily with S-9 activation.

R. Draftz, *(I.I.T. Research)*

Did you do any size distribution measurements on the fireplace emissions to find out whether the absorptive carbon was principally in the submicron size range?

J. Muhlbaier

We had the distribution on the front catch and you saw the proportion of material there which should be greater than ten microns. I ran an in-stack cascade impactor test a few days ago and the results don't quite make sense because over 90% of the material was on the back up filter . . . which would catch only particles less than five-tenths of a micron. So, I was not sure where the large particle material went.

According to what is collected in the cyclone there is a large fraction of large particle material but it did not show up in the impactor. So, I am not sure if I am getting bounce-off.

J.Huntzicker

I would like to comment on Dr. Heicklen's question concerning the volatility of some of these things. If you collect a sample on a hi-vol in an area where there is considerable residential wood combustion, the filter carries with it a very pungent aroma that stays for days and days. I am sure that there would be continuous mass transfer between the particulate and vapor phases.

T. Hansen, *(Lawrence Berkeley Laboratory)*

Do you intend to expand your vehicle testing to be more representative of the actual vehicle fleet? The reason I mention this is that coming here from California, I see that here there are not very many old cars. In California, there are many ten and twenty year old cars on the roads. Some very crude testing which we did, indicated that some of these older cars may be contributing orders of magnitude more than the newer ones.

R. Williams

I think testing older cars quickly becomes an area of diminishing returns. Once you go back to leaded vehicles, you start dealing with a big variety of mileages and treatment that the customers have given their vehicles. We have looked as far back as 1970 and up to 80,000 miles. I think that is probably enough of a data base.

H. Gerber, *(Naval Research Laboratories)*

What was the fraction of volatile organics between the cold trap and the filter?

J. Muhlbaier

I recall that it was something like 50-50 organic to elemental on the filter. When we added in all the condensed organic, the total catch went down to about 20% elemental. But it depends on the type of wood.

E. Macias, *(Washington University)*

Did those numbers which you showed include cars as old as 1970 with the mileage as high as 80,000?

J. Muhlbaier

Yes.

S. Arnold, *(Colorado Department of Health)*

On the pictures of the fireplace filters that you showed, you indicated that there was no elemental carbon on it. Later, however, you indicated that there was about 20% elemental carbon and it appeared that when you fired the filter at 350°C, all the coloration disappeared. Would that indicate that there are some samples which contain no elemental carbon?

J. Muhlbaier

The filter on which all of the coloration disappeared upon heating to 350°C had less than 5% elemental carbon.

A. Waggoner, *(University of Washington)*

Are the measurements that you made different enough from the source apportionment that was used in Denver to significantly change those results? My inspection of it looks like you are now finding more organic carbon from fireplaces and less from diesel vehicles and furnaces.

J. Muhlbaier

Our fireplace numbers were used in the Denver inventory and in fact our softwood was sent here from Denver. Therefore, it should be representative. The vehicle numbers used were from Ron's work. The natural gas numbers were taken from a different source. This work wasn't done at that time.

G. Wolff, *(General Motors Research Laboratories)*

I would like to respond to that also. We did not use the diesel numbers from this study because there were essentially no light-duty diesels in Denver during our study in Nov.-Dec., 1978. Most of the diesels were heavy-duty diesels or diesel trucks. And those emission rates were taken from the literature. The only source that would change significantly, if we use Jean's numbers, is the natural gas contribution to elemental carbon which will drop the estimate substantially.

R. Draftz

Did I understand you to say that the noncatalyst car emissions were over 90% lead?

R. Williams

I think that the question you are struggling with concerns the fact that what Jean has reported as lead is actually lead *and* the associated anions. The anions make up about half of the mass of the non-carbonaceous fraction which you saw. Those numbers are very much in agreement with the numbers we and other people have published in the past both in terms of the composition and in terms of the emission rate of lead.

L. Newman, *(Brookhaven National Laboratory)*

In your last figure I recall that fireplaces emit 10 times more carbon than cars on an annual basis. I assume it is as cars and fireplaces now exist in the United States. Is that correct? It sounds like a very high ratio for a fireplace.

J. Muhlbaier

As I said, the main problem there is getting a good estimate of the amount of wood burned. And I have three different estimates.

L. Newman

Do they vary by a factor of ten.

J. Muhlbaier

The three estimates of the amount of wood burned varied by a factor of 4, so, I used the average. I recognize the result is surprisingly high.

L. Gundel, *(Lawrence Berkeley Laboratory)*

You did not mention emissions from fuel of oil-fired furnaces. I mention this because it is important on the east coast.

J. Muhlbaier

We just have not measured them yet.

L. Gundel

Do you know of any other work related to carbon emissions from oil fired furnaces?

J. Muhlbaier

I think George Wolff used some in his Denver inventory.

G. Wolff

In the Denver inventory we used fuel oil emission rates that were measured in the Portland study. John Watson's thesis* has a considerable amount of data from oil fired boilers.

S. Dattner, *(Texas Air Control Board)*

Did you measure any older cars that had ring damage so that they emitted a characteristic blue smoke? I am wondering if those are rather large emitters of carbon.

R. Williams

We did not have what I would classify as a smoker. We did, however, have two high mileage 1970 vehicles which emitted about 10 times the average of the other three vehicles. In a sense we have taken a case . . . a realistic case, I think, where perhaps the amount of oil consumption has already gone up to some significantly inflated value. So, we tried to stay in what we would call representative high mileage normal use vehicles but we have not measured the elemental carbon from a legitimate smoker.

R. Bradow, *(U.S. Environmental Protection Agency)*

I think all of our measurements of particulate emission from cars burning leaded gasoline are in general agreement with the averages that you report. One thing I do not recall is whether you showed the differences between the Highway Fuel Economy Test and the Cold Start Federal Test Procedure (FTP). In our labs those tests give quite different emission factors.

J. Muhlbaier

All of our data was from FTP's.

R. Bradow

The lead emission factors, particularly the FTP emissions, are usually much higher than steady state or even warmed up emissions. Frequently in the literature, you can find numbers that are characteristic of cold starts in which organics are higher than in your tests. However, generally speaking, our FTP data are very similar to yours for automobiles going back to 1967. We have one 1963 car which had a higher percentage of organic compounds present in the particles than in any other car we ever tested and that was because it had a particularly high lubricant consumption. It is possible to find cars that run quite a lot richer than other cars in the FTP. And those will have higher organic emissions.

---

*J. G. Watson, Ph.D. Thesis, Environmental Science Dept., Oregon Graduate Center, Portland, OR, 1979.

R. Draftz

On the national inventory, how did you deal with residential versus commercial uses of natural gas?

J. Muhlbaier

The results shown were for residential usage only. However, other gas appliances would be included in addition to furnaces.

EMISSIONS AND AIR QUALITY RELATIONSHIPS FOR ATMOSPHERIC CARBON PARTICLES IN LOS ANGELES

G. R. CASS

California Institute of Technology
Pasadena, California

P.M. BOONE and E.S. MACIAS

Washington University
St. Louis, Missouri

ABSTRACT

An emission inventory for fine particle aerosol carbon has been constructed for the Metropolitan Los Angeles area and compared to the results of ambient sampling. Information on the level of source activity, total aerosol emission rate, particle size and chemical composition was used to estimate the mass emission rate of particulate organic and elemental carbon in this urban area. It was found that carbonaceous particle emissions to the atmosphere arise from more than fifty classes of mobile and stationary air pollution sources. Emission data for elemental carbon at present contain significant uncertainties. Methods for verifying the consistency of carbonaceous aerosol emission inventories thus are important. Verification tests based on comparison of the ratio of organic carbon to elemental carbon and lead in source emissions versus that measured in the atmosphere were developed in this paper. Using these tests, excellent agreement was found between the emission inventory and results of ambient sampling in Los Angeles during January and February 1980.

INTRODUCTION

Aerosol samples taken in urban areas almost invariably show that large amounts of fine carbonaceous particulate matter are present in the atmosphere of cities [1-12]. These airborne carbon particles consist principally of organic material accompanied by black nonvolatile soot components which have a chemical structure similar to impure graphite [13, 14]. Graphitic or ''elemental'' carbon particles are thought to be the most abundant light absorbing aerosol species in the atmosphere

References pp. 222-225.

[15], and have been associated with visibility deterioration in both Denver and Los Angeles [16, 5, 8, 9, 10]. Polycyclic aromatic hydrocarbons which are adsorbed onto soot particles show mutagenic activity in the Ames Test, and hence are of public health concern [17, 18].

Primary aerosol carbon emissions to the atmosphere arise from more than fifty classes of mobile and stationary sources. Development of control strategies for aerosol carbon abatement requires that the relative contribution of each source type be identified. Assessment of the contribution of different classes of emission sources to observed air quality has been attempted by three methods; (a) chemical element balance receptor models [19,20,21]; (b) emission inventories [22, 23] and (c) hybrid combinations of (a) and (b) [9, 10]. Applications of these methods to date occasionally have led to conflicting conclusions concerning the amount of secondary organic aerosol in the atmosphere [22]. Carbon particles emitted from many source types are chemically indistinguishable. Hence, it is difficult to obtain unambiguous separation between sources contributing to the carbon content of an ambient sample by the multivariate statistical methods employed in chemical element balance studies. Emission inventories, on the other hand, suffer from the possibility that some important sources will be overlooked. Hybrid combinations of the emission inventory approach combined with chemical element tracers have a better chance of overcoming these potential problems.

The objective of this paper is to establish a formal framework for addressing emission estimation problems posed by carbonaceous aerosols. An approach is taken in which an energy balance on an air basin's fuel supply, combined with data on industrial process and fugitive source activities, is used to compute total suspended particulate emissions. Next the fine particle fraction of those emissions is identified along with the fraction of that fine particle mass emitted as aerosol carbon. Finally, aerosol carbon present is subdivided into organic carbon and elemental carbon fractions based on source test data and engineering estimates.

At present, emission estimation procedures for elemental carbon contain significant uncertainties. However, one seldom discovers which uncertainties are really critical until an attempt is made to work a difficult problem quantitatively. Therefore the approach to aerosol carbon emission estimation developed in this study will be tested by application to the greater Los Angeles metropolitan area. Methods for verifying the consistency of that emission inventory will be developed based on comparison of the ratio of organic carbon to elemental carbon and lead in source emissions versus that measured in the Los Angeles atmosphere by Conklin *et al.* [5].

APPROACH

Fuel Use — The geographic area of interest is the South Coast Air Basin which surrounds Los Angeles, as shown in Fig. 1. An energy balance performed on commerce data for fuels entering and leaving that region during the year 1973 has been reported by Cass, *et al.* [24]. Fuel with a total energy content totaling 6×10^{12} BTUs/day was consumed within the air basin during 1973. This fuel consumption was subdivided between thirty-four classes of mobile and stationary combustion sources, as itemized in Tables A.1 and A.2 in the appendix of this paper.

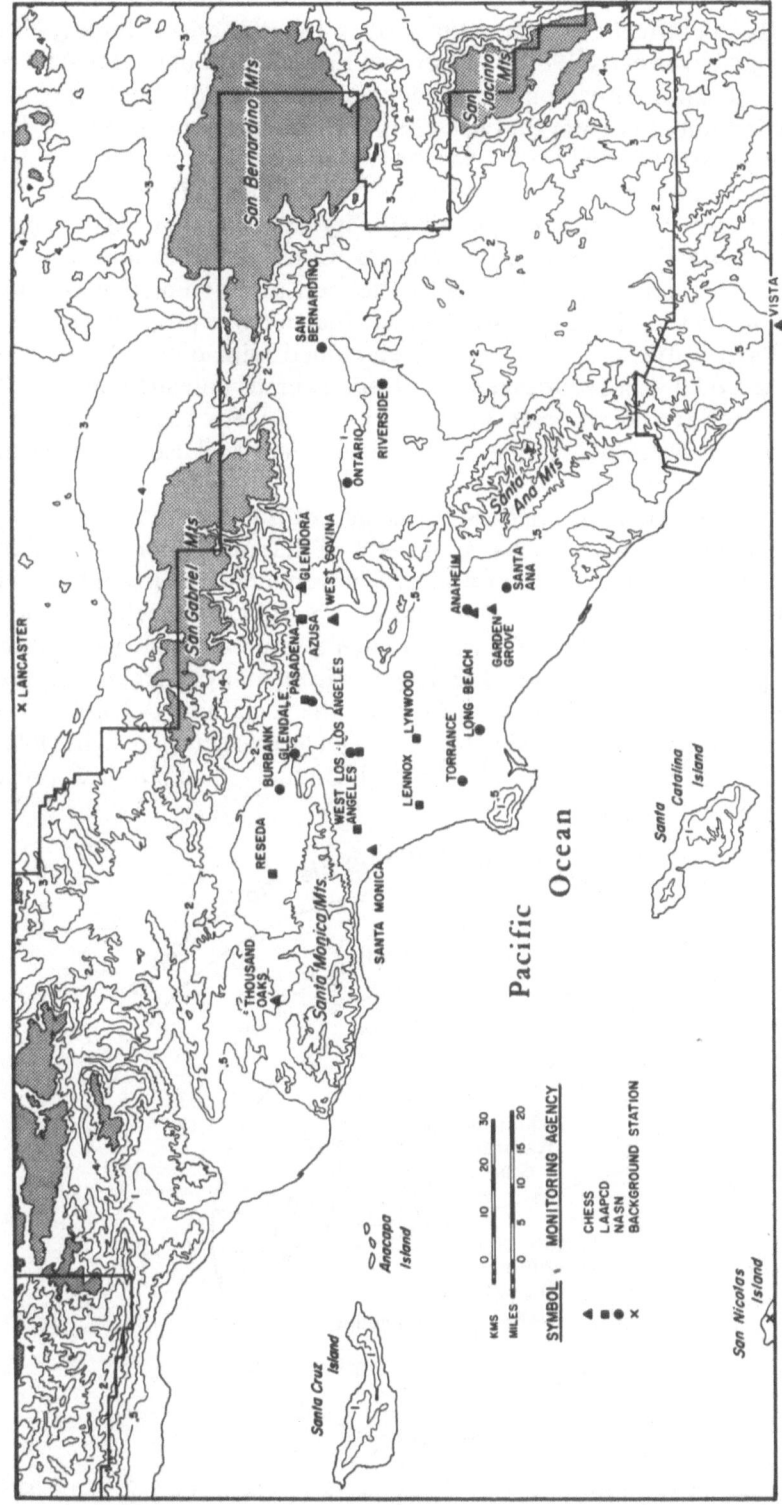

Fig. 1. The South Coast Air Basin which surrounds the Los Angeles area.

In order to better reflect current circumstances, the 1973 energy budget for the air basin was projected to early 1980 (January and February). Commerce data needed to determine current fuel use only become available several years after the fact. Thus a combination of current fuel use reports plus emissions inventory *forecast* techniques must be used to arrive at a winter 1980 fuel use estimate for Los Angeles. Stationary source fuel combustion data at electric utilities and petroleum refineries were obtained from the South Coast Air Quality Management District based on actual January and February fuel use reports. Combustion estimates at smaller sources were based on projections of 1980 fuel consumption made by natural gas utilities [25] and by Cass [26]. Highway vehicle fuel use was projected to 1980 from 1973 using local traffic growth data [26] combined with data on fuel economy standards and model year requirements for leaded versus unleaded fuel (see Tables A.3 - A.5). Aircraft and shipping activity forecasts were based on a survey of transport mode growth [26]. Military and miscellaneous fuel uses were assumed to persist from 1973 to the present.

Estimated 1980 fuel use within the South Coast Air Basin is summarized in Fig. 2. On an equivalent heat content basis, about 41 % of that fuel supply is consumed by industrial firms, including electric utilities. The remainder represents residential and commercial use. Close to half of the energy input to Los Angeles is supplied by natural gas, process gas and LPG. Use of such a large amount of clean burning gaseous fuel probably distinguishes Los Angeles from many other cities. One quarter of the fuel supply is consumed as gasoline. The remaining one fourth consists mostly of fuel oils. Diesel powered vehicles and jet aircraft burning light and middle distillate fuel oils account for only 5 % of total local fuel consumption at present.

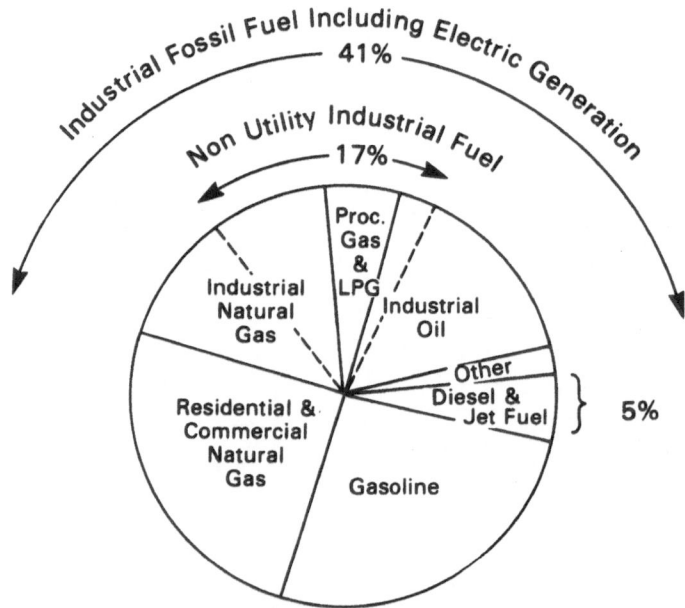

Fig. 2. Greater Los Angeles area fossil fuel combustion estimate for Jan. and Feb. 1980.

Emission Estimates — Fuel combustion data combined with information on the level of industrial process and fugitive source activity were used to estimate fine aerosol organic and elemental carbon emissions in the Los Angeles area. Results are summarized in Tables 1 through 4. Detailed calculations are outlined in the appendix to this paper. Data in these tables should be read to no more than two significant figures.

TABLE 1
Emissions Estimates for Mobile Sources

	Estimated 1980 Fuel Use (10^9 BTU/day)	Fine Non-volatile Carbon (kg/day)	Total Aerosol Carbon (kg/day)
MOBILE SOURCES			
Highway Vehicles			
Catalyst Autos & Lt. Trucks	963.78	369	803
Non-catalyst Autos & Lt. Trucks	566.77	563	7042
Medium & Heavy Gasoline Vehicles	249.26	286	3571
Diesel Vehicles	137.16	4319	5917
LPG use for Carburetion	6.07		
Civil Aviation			
Jet Aircraft	45.50	547	749
Aviation Gasoline	1.32	1	16
Commercial Shipping			
Residual Oil-Fired Ships Boilers	32.05	67	333
Diesel Ships	22.43	540	740
Railroad			
Diesel Oil	38.11	1530	2096
Military			
Gasoline	6.03	7	86
Diesel Oil	17.81	686	940
Jet Fuel	16.71	462	633
Residual Oil (Bunker Fuel)	0.27	1	3
Miscellaneous			
Off-Highway Vehicles (diesel)	39.73	<u>1531</u>	<u>2097</u>
TOTAL		10909	25026

Fine aerosol mass as used in those tables refers to particles with aerodynamic diameter less than 10 μm. Fine particle emissions are emphasized for two reasons: fine aerosols are respirable while coarse material is not [27], and particles larger than 10 μm settle out of the atmosphere rapidly enough that most of the coarse material does not reach the regional air monitoring sites used for comparison to emission data [28].

Nonvolatile carbon data shown in Tables 1 through 4 provide an estimate of elemental carbon emissions. The term "nonvolatile carbon" is used to alert the reader to the fact that most current procedures for analysis of the graphitic or

TABLE 2
Emissions Estimates for Stationary Combustion Sources

	Estimated 1980 Fuel Use (10^9 BTU/day)	Fine Non-volatile Carbon (kg/day)	Total Aerosol Carbon (kg/day)
STATIONARY SOURCES			
Fuel Combustion			
Electric Utilities			
Natural Gas	666.25	—	48
Residual Oil (0.25 % S)	885.72	312	1560
Landfill & Digester Gas	0.85	—	0.06
Refinery Fuel			
Natural Gas	148.54	—	90
Refinery Gas	347.63	—	210
Residual Oil	5.57	3	15
Non-Refinery Industrial Fuel			
Natural Gas	421.64	—	212
LPG	2.74	—	1
Residual Oil	53.42	28	141
Distillate Oil	42.74	85	147
Digester Gas (IC Engines)	6.30	6	27
Coke Oven Gas	37.53	—	19
Residential/Commercial			
Natural Gas	1592.26	483	1465
LPG	18.08	6	17
Residual Oil	22.19	12	59
Distillate Oil	22.19	41	70
Coal	0.55	27	124
TOTAL		1003	4205

"elemental carbon" content of source samples are based on quantification of carbon present which resists volatilization or oxidation in a manner similar to graphite. These three terms will be used interchangeably in this paper. Advances in measurement techniques may at some point in the future permit more careful distinction between high temperature behavior and graphitic structure.

Emission inventory procedures for elemental carbon aerosols are in their formative stages at present. Emission factors derived from the average of a large number of well-defined source tests are rare. In the present study, emission factors had to be constructed from composites of a number of different types of tests, often performed on similar but separate pieces of equipment. For example, one data set might be available which gives the total aerosol mass emission rate from oil-fired boilers, while a separate reference describing tests on a different boiler might provide chemical composition information without providing total mass. A composite of these two tests would be used to estimate the mass emission rate of a particular chemical species, like elemental carbon. The emission estimates were based on source tests taken in Los Angeles whenever possible.

TABLE 3
Emission Estimates for Industrial Process Sources

	Fine Nonvolatile Carbon (kg/day)		Total Aerosol Carbon (kg/day)
STATIONARY SOURCES			
Industrial Process Point Sources			
Petroleum Industry			
Production	0	≤	103
Refining	7		30
Marketing	0	<	207
Organic Solvent Use			
Surface Coating	0		1511
Degreasing	0	≤	21
Other	0	≤	10
Chemical			399
Metallurgical	111		3686
Mineral	—		—
Waste Burning at Point Sources	37	≤	56
Wood Processing	0		79
Food and Agriculture	0		275
Miscellaneous Industrial	0		164
TOTAL	155	≤	6541

TABLE 4
Emission Estimates for Fugitive Sources

	Fine Nonvolatile Carbon (kg/day)	Total Aerosol Carbon (kg/day)
FUGITIVE SOURCES		
Road and Building Construction	—	—
Agricultural Tilling	—	—
Refuse Disposal Sites	—	—
Livestock Feedlots	0	65
Unpaved Road Travel	—	—
Paved Road Travel	—	—
Forest Fires (seasonal)	226	3761
Structural Fires	73	149
Fireplaces	183	373
Cigarettes	17	1692
Agricultural Burning	58	647
Tire Attrition	1729	5240
Brake Lining Attrition	388	2159
Sea Salt		
TOTAL — Annual Average Day	2674	14086
TOTAL — Winter Day with no Forest Fires	2448	10325

References pp. 222-225.

In constructing composite emission estimates, personal judgement must enter into the process. This is especially true when two conflicting sources of information are available with no third data set present to resolve the dispute. As a result, the emission inventory which follows should be viewed as a series of estimates based on presently available data and assumptions, and should be revised as better data are published.

An outline of the emission estimation approach for all source types is given in the appendix to this paper. A few critical source classes which dominate the emission inventory merit further discussion.

Mobile Sources — Approximately 25 metric tons per day of aerosol carbon are emitted from motor vehicles in the South Coast Air Basin, as shown in Table 1. About half of that total carbon burden arises from gasoline-powered engines while the other half is emitted largely from diesel engines. It is important to note that diesel engines presently are used in a wide variety of applications such as ships, railroad locomotives, and off-road vehicles in addition to highway trucks and busses. When all of these diesel applications are considered, diesel engines dominate elemental carbon emissions from motor vehicle exhaust.

Aerosol carbon emissions from mobile sources were computed as shown in Table A.6 of the appendix to this paper. Emissions from catalyst-equipped autos and light trucks are based on experiments by Muhlbaier and Williams [29], which show low total carbon emissions of about 5.5 mg/mile. Characterization of total aerosol carbon emissions from older high-mileage pre-catalyst cars was based on work by Ter Haar *et al.* [30], Habibi [31], and Huntzicker *et al.* [28], plus qualitative discussions with Robert Gorse of Ford Motor Company. A fine carbonaceous aerosol emission rate of 115 mg/mile was obtained for older cars burning leaded fuel. That emission rate is much higher than reported by Muhlbaier and Williams [29] for pre-catalyst cars at low altitude, but is in closer agreement with data reported to date for pre-catalyst autos at high altitude (101.2 mg C/mile) [32]. Emission factors for older pre-catalyst cars based on Habibi's [31] summary have worked well in field studies and modeling applications in Los Angeles [28, 33], and were adopted for use in this study. Separation of carbon emitted by vehicles burning leaded gasoline into elemental and organic carbon was based on Johnson *et al.* [34] plus qualitative discussions with Robert Gorse of Ford Motor Company [35].

Diesel engines emit much greater amounts of fine black carbonaceous material per unit of fuel burned than do comparable gasoline engines. Hence accurate characterization of these sources is crucial to an inventory of elemental carbon emissions. Total mass emission rates for diesel highway trucks and busses were estimated at 0.83 g/km for a typical Los Angeles driving cycle from Baines, *et al.* [36]. This is a lower emission rate than expected in more congested traffic like that in New York City [36], and is lower than that used in recent studies in Denver [10]. The prospect that diesel engine emission estimates if anything are biased *low* in the present study is important, as will become apparent shortly. The carbon content and composition of diesel exhaust aerosol was estimated based on the summary by Pierson and Brachaczek [37] and the data of Johnson *et al.* [34].

Total solid particulate emissions from jet aircraft were computed using U.S. Environmental Protection Agency [38] procedures. These estimates reflect landing and takeoff operations below 1100 meters elevation from both civilian and military air traffic. The fraction carbon in jet engine aerosol is given by Heywood, Fay, and Linden [39]. In the absence of any elemental carbon test data for jet aircraft, it was assumed that jet engine exhaust aerosol has an elemental carbon to organic carbon ratio similar to diesel engine exhaust aerosol [40]. Additional source test data on jet aircraft should be sought in order to improve these emissions estimates.

Stationary Sources — Particulate carbon emissions from stationary source fuel combustion totals only about four metric tons per day as shown in Table 2. The important source types are industrial residual fuel oil combustion due to its relatively high emission rate per unit of fuel burned, and commercial/domestic natural gas combustion due to the very large amount of fuel consumed.

Key references on stationary source total particulate emissions include the work by Danielson *et al.* [41] based on Los Angeles Air Pollution Control District and South Coast Air Quality Management District source tests, and the source test program performed for the California Air Resources Board by Taback *et al.* [23]. Separation of emissions into fine and coarse particle mass and computation of the fraction of fine particle mass present as carbon was based on Taback *et al.* [23] in most cases. Partition of carbon present into elemental carbon and organics followed Johnson *et al.* [34] for stationary residual and distillate oil burning sources. *Large* industrial and utility natural gas-fired boilers and heaters were estimated to emit little or no elemental carbon under normal operating conditions [42, 43, 44, 23]. The split between elemental and organic carbon emitted from domestic natural gas combustion was based on Mulhbaier and Williams [29], and source tests performed as part of our study.

Industrial Processes — Carbonaceous aerosols are emitted from a wide variety of industrial processes, as shown in Table 3. The industrial process emission inventory used here is based largely on Taback *et al.* [23] plus assumptions about which source types dominate the industrial classifications shown. These emissions estimates should be considered less reliable than for the fuel burning sources. It is clear however that there are many potentially large noncombustion sources of carbonaceous aerosols. Several of these processes (e.g., incinerators, wood working, food and agricultural operations) act as industrial sources of contemporary (non-fossil) carbon emissions to the atmosphere.

Fugitive Emissions — Fugitive emissions arise from area-wide activities which often do not emit through stacks and hence are difficult to test in a conventional manner. Fugitive carbon aerosols are inventoried in three broad classes: fugitive dust, fugitive combustion, and dust from vehicle tire plus brake wear. Estimates of total fine aerosol carbon emissions from fugitive dust and fugitive combustion sources are based on Taback *et al.* [23]. The partition of carbon present into elemental and organic fractions is based on the data of Watson [21] and others. Tire dust emissions in suspendible particle sizes are estimated at 0.0063 g/tire mile [45], with chemical composition similar to tire tread rubber [46]. Total fine aerosol emissions from brake lining attrition are as estimated by Taback *et al.* [23], with a chemical composition similar to brake lining material [47].

Lead Emissions — Aerosol lead in ambient samples often is used as a tracer for leaded automotive exhaust [19, 20, 21, 22, 24, 37]. The extent to which ambient samples are enriched in carbon relative to motor vehicle exhaust can be assessed if both aerosol carbon and lead emissions are known within our study area.

An inventory of fine lead emissions in the South Coast Air Basin was constructed. The most recent data available on the composition of gasoline in Southern California [48] show that lead content averaged 1.36 g/gal during winter 1978-79 (average of leaded regular and leaded premium grades). Those data were combined with the fuel use inventory of Table A.1 and used to estimate that almost 9 metric tons per day of lead were consumed in leaded gasolines within the air basin during January and February 1980.

The fate of lead consumed in gasoline in Los Angeles has been explored by Huntzicker *et al.* [28]. Their analysis shows that only 30.3% of the lead originally present in fuel is emitted as a fine suspendible aerosol in particle sizes less than 9 μm aerodynamic diameter. Fine lead aerosol emissions in the South Coast Air Basin thus were estimated at 2.7 metric tons per day. That value may be revised to better reflect January and February 1980, once data on the Winter 1979-80 Southern California gasoline pool become available.

EMISSION INVENTORY SUMMARY

Fine aerosol carbon emissions from all sources combined are summarized in Fig. 3. A total of 46 metric tons per day of particulate carbon was emitted within the South Coast Air Basin during January and February 1980. These emissions are diversified over a large number of source types.

- autos and gasoline trucks — 25%
- highway diesels — 13%
- other diesel engines — 13%
- aircraft — 3%
- ship boilers — 1%
- tire and brake wear — 16%
- fugitive combustion — 6%
- stationary source fuel combustion — 9%
- industrial processes — 14%

In the aggregate, transportation sources including tire and brake wear account for two thirds of total fine aerosol carbon emissions in the greater Los Angeles area.

Elemental carbon emissions in particle sizes less than 10 μm diameter total 15 metric tons per day as shown in Fig. 4. These emissions are dominated by internal combustion engines burning light and middle distillate fuel oils (diesels and jet aircraft). Comparing Figs. 2 and 4 we note that diesel engines which consume less than 5% of the total fuel burned in the air basin are contributing 60% of the estimated elemental carbon emissions present. This finding differs from previous studies in Denver [9, 10] because:

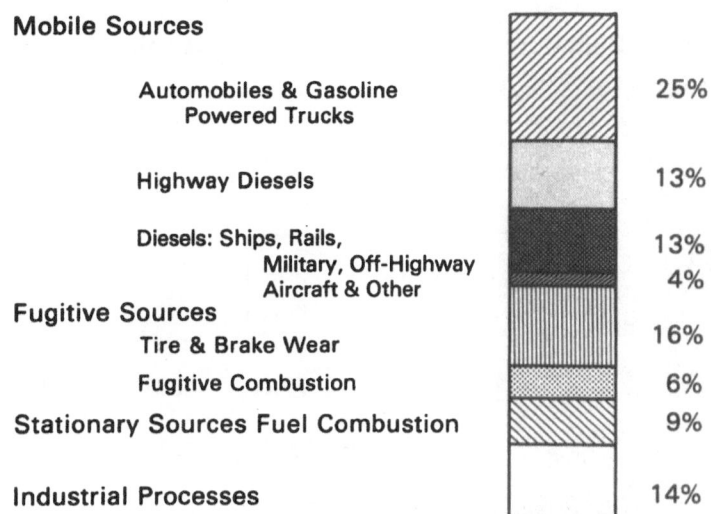

Fig. 3. Total carbonaceous aerosol emissions, greater Los Angeles: Jan.-Feb. 1980.

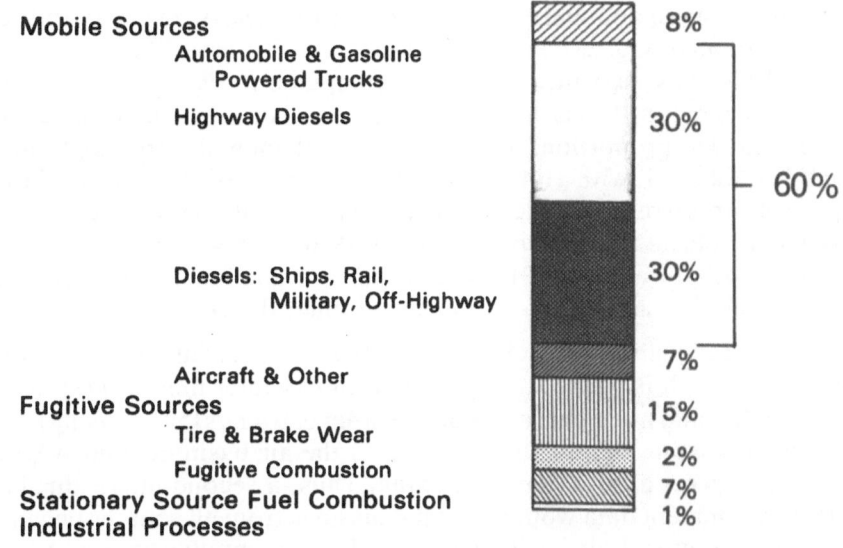

Fig 4. Sources of nonvolatile (elemental) carbon, greater Los Angeles: Jan.-Feb. 1980.

1. diesels other than highway vehicles are present;
2. there is little fireplace combustion of wood in the Los Angeles area; and
3. natural gas combustion is only a modest source of elemental carbon.

To summarize the emission inventory for comparison to ambient data, the abundance of total fine aerosol carbon emissions relative to lead and elemental carbon has been computed as shown in Fig. 5.

References pp. 222-225.

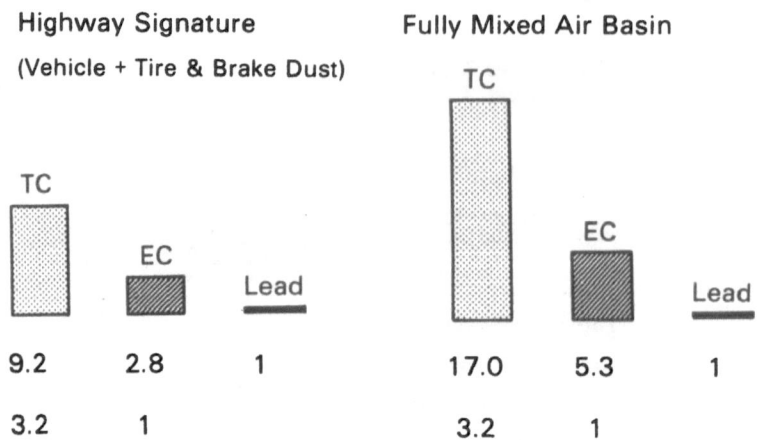

Fig. 5. Normalized fine particle emissions.

The minimum likely carbon to lead ratios in the atmosphere would arise if a sampling station were dominated by emissions from nearby highway traffic. This case is of particular interest in a high traffic density area like downtown Los Angeles. A "highway signature" has been computed in Fig. 5 from the emission inventory by combining the carbon and lead emitted from highway vehicles plus tire and brake wear. The proportions of total carbon to elemental carbon and lead in that case would be 9.2:2.8:1, with a total carbon to elemental carbon ratio of 3.2 to 1. The abundance of carbon relative to lead is higher than that often reported for measurements made in vehicle traffic during the early 1970's. This is due to recent drastic reductions in both the amount of lead per gallon in leaded fuel and the reduction in total leaded fuel sales as catalyst equipped cars enter the vehicle fleet.

Los Angeles air quality is affected by a daily sea breeze/land breeze reversal in wind direction which is particularly pronounced in winter months [49]. Resultant wind speed in January and February is on the order of 0.4 m/s [50], causing hold-over of pollutants from one day to the next within the air basin combined with long retention times needed for extensive mixing. Thus, a second interesting case for comparison to ambient data would arise if emissions from all sources became fully mixed. This fully mixed air basin signature also is computed in Fig. 5 from the emission inventory and shows a total carbon to elemental carbon to lead ratio of 17.0:5.3:1.

Interestingly, the total carbon to elemental carbon ratio for the fully mixed air basin case is 3.2:1, which is the same as found for the highway signature alone. Furthermore, emissions from tire plus brake dust, which are difficult to assess, also are present with a total carbon to elemental carbon ratio of about 3:1. *In the absence of much secondary organic aerosol formation,* it should be difficult to obtain *long term average* total carbon to elemental carbon ratios in the Los Angeles atmosphere which deviate much from 3:1.

AMBIENT SAMPLING PROGRAM

A series of field experiments was performed to characterize existing levels of ambient organic and elemental carbon in the Los Angeles atmosphere [5]. Sampling times and locations were sought which would encounter peak levels of primary combustion products while minimizing exposure to secondary organic aerosols. Examination of historical data on primary combustion products like CO and NO_x show that peak values occur in the early morning hours during winter months [51, 52]. Hence the sampling program was conducted during morning peak traffic periods in January and February 1980.

Sampling sites were established along side of South Coast Air Quality Management District air monitoring stations in downtown Los Angeles and in Pasadena. The downtown Los Angeles station is in an industrial area within a ring of freeway interchanges. Large railroad yards are located to the east of the downtown Los Angeles site. The Pasadena station is located in a mixed residential and commercial district two blocks south of the Interstate 210 freeway.

The sampling protocol for this experiment has been described previously by Conklin, et al. [5]. Three types of low volume filter samples were collected simultaneously at each site. Sampling intervals were 51.5 min in duration with consecutive samples collected each hour from 7 a.m. to 1 p.m. Filter substrates were chosen that were suited to subsequent analysis for total carbon, elemental carbon and trace metals content. The three filter assemblies contained a 47 mm Pallflex Tissuequartz filter (2500 QAO), a 47 mm Nuclepore polycarbonate filter (0.40 μm pore size), and a 13 mm Pallflex Tissuequartz filter (2500 QAO).

The 47 mm Pallflex Tissuequartz filters were chosen for carbon determination. Filters were prefired to 900°C for 1-1/2 hours to reduce their carbon blank. These filters were masked until a 0.32 cm² area remained exposed to the air flow. Ambient air was drawn through each filter at a rate of about 12 ℓ/min. Following sample collection, each filter was sealed between a pair of plastic petri dishes and chilled. Filter deposits were analyzed for total carbon by the Gamma Ray Analysis of Light Elements (GRALE) technique [6]. Elemental carbon determination on the same filters was obtained by reflectance calibrated against heated butane soot standards [6]. Elemental carbon concentrations also can be measured by the GRALE technique after heating the samples to eliminate volatile organics. Elemental carbon determinations by GRALE are in progress at present; all elemental carbon data reported in this paper were obtained by the reflectance technique. Both methods are as described by Delumyea, et al. [6].

Nuclepore filters were used for determination of total aerosol mass, aerosol light absorption coefficient and trace metals content. Following sample collection at approximately 20 ℓ/min, the filters were sealed and chilled. Filters were subsequently quartered, and a quarter filter was analyzed for trace metals, including lead, by X-ray fluorescence [53]. A second wedge of each filter was used for determination of the aerosol light absorption coefficient using the opal glass integrating plate technique [54] as modified by Ouimette [55]. Light absorption by elemental carbon particles was found to be responsible for approximately 17% of total light extinction at downtown Los Angeles, as reported elsewhere [5].

References pp. 222-225.

Average total carbon, elemental carbon and lead concentrations measured at downtown Los Angeles and Pasadena are shown in Fig. 6. Total carbon concentrations present during winter mornings averaged 24 μg m^{-3} at downtown Los Angeles and 17 μg m^{-3} at Pasadena. Elemental carbon concentrations were present at 9 μg m^{-3} and 5 μg m^{-3} at Los Angeles and Pasadena, respectively. Lead concentrations were much lower than originally expected. Annual mean lead concentrations measured at downtown Los Angeles ten years ago averaged between 4.6 and 4.8 μg m^{-3} [33]. Mean values observed during the present study totaled 1.7 μg m^{-3} at downtown Los Angeles and 1 μg m^{-3} at Pasadena. This sharp decline in ambient lead levels is probably due to a reduction in the total amount of leaded gasoline sold plus a reduction in the lead content of each gallon, as mentioned previously.

The relative abundance of total carbon, elemental carbon and lead in atmospheric samples observed at Pasadena and Los Angeles is illustrated in Fig. 7. Pollutant concentrations are present at those two sites in approximately the same proportions. Elemental carbon constitutes about one third of the total carbon burden at both sites. That total carbon to elemental carbon ratio is about the same as calculated from the emissions inventory for both the highway signature and the fully mixed air basin examples (see Fig. 5). The total carbon to elemental carbon to lead ratio in ambient air at downtown Los Angeles lies part way between the highway signature and the fully mixed basin examples pictured in Fig. 5, while the Pasadena samples closely match the fully mixed basin case. Total carbon, elemental carbon and lead proportions computed from emissions data resemble winter morning ambient samples.

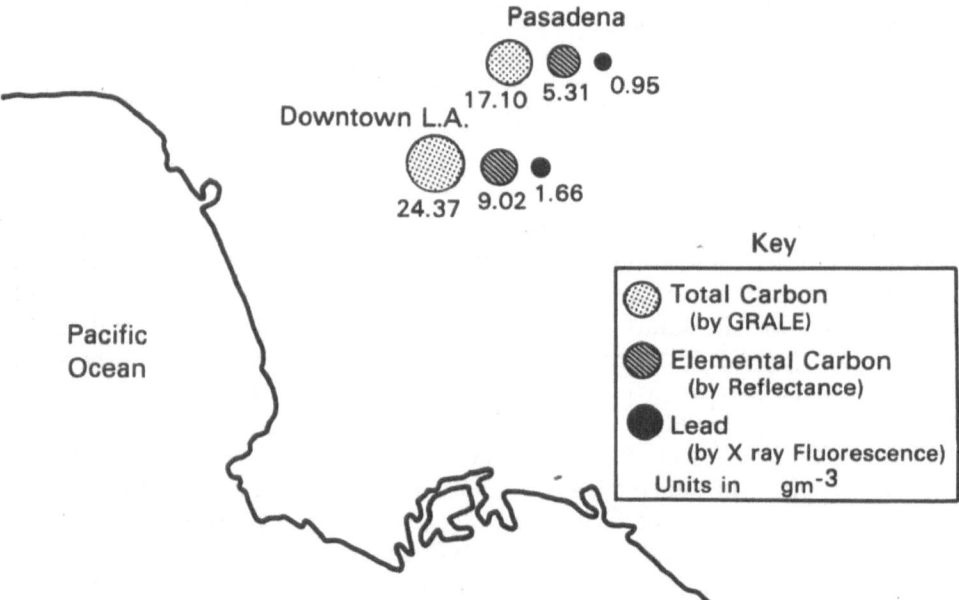

Fig. 6. Results of the ambient sampling program — January and February 1980. Pollutant concentrations shown adjacent to monitoring sites are mean values for the hours 7:00 a.m. to 1:00 p.m.

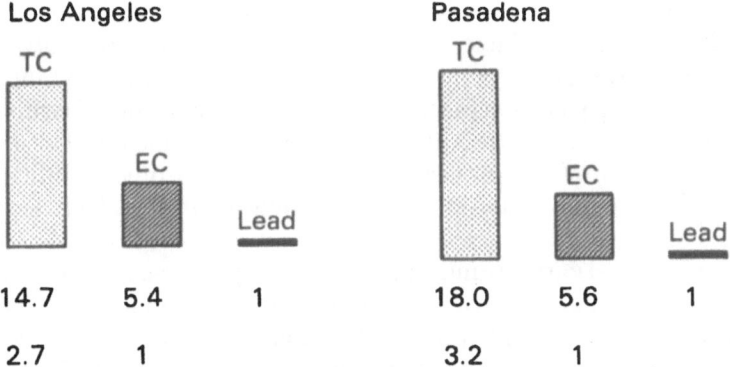

Fig. 7. Normalized ambient measurements.

CONCLUSIONS

Estimates of primary aerosol carbon emissions to the atmosphere are needed before strategies for control of aerosol carbon air quality can be constructed. An inventory of organic and elemental carbon particle emission sources can be assembled in urban areas. A fuel consumption budget for the region of interest is needed, combined with data on industrial process and fugitive source activities, which can be used to estimate total suspended particulate emissions. The fine particle fraction of those emissions then can be estimated on a source by source basis followed by assessment of the organic and elemental carbon content of the fine particles contributed by each source.

This approach to aerosol carbon emissions estimation was applied to Los Angeles during January and February 1980. It was found that 46 metric tons of fine aerosol carbon are emitted daily in the greater Los Angeles area. Thirty-three percent of that material is emitted as nonvolatile ("elemental") carbon. Indeed, it would be difficult to assemble a mixture of emissions sources in Los Angeles at this time which contributes aerosol carbon with a total carbon to elemental carbon ratio greatly different from 3:1. That is because either a 1980 blend of highway vehicle exhaust and tire plus brake dust or a fully mixed air basin source profile both display about a 3:1 ratio of total carbon to elemental carbon.

Presently available references are able to supply only one or two pieces of source test data at many critical points throughout the emission inventory application attempted in Los Angeles. As a result, estimates made for organic and elemental carbon emissions are uncertain and require further confirmation. Methods for verifying the consistency of emission inventories were developed and tested during this research project. These methods are based on comparison of the chemical composition of source exhaust to that observed in atmospheric samples taken under conditions which would be expected to minimize the amount of secondary organic aerosol present. It was found that total carbon, elemental carbon and lead proportions in emissions resemble winter morning ambient samples in Los Angeles and Pasadena.

References pp. 222-225.

Emission estimates for Los Angeles will undoubtedly merit revision as better source test data become available. However, as the inventory presently stands, two features with significance for control of aerosol carbon concentrations are apparent. First, total primary carbon particle emissions are distributed over at least five major source types of roughly equal importance. This means that a basin-wide control strategy which addresses a large number of source types would be needed in order to achieve a major reduction in total primary carbonaceous aerosol in this city.

Secondly, elemental carbon emissions in Los Angeles appear to be dominated by internal combustion engines burning light and middle distillate fuel oil (diesels and jets). Diesel engines at present consume less than 5% of the fuel burned in the air basin but account for perhaps as much as 60% of the estimated elemental carbon emissions.

ACKNOWLEDGEMENT

This work has been supported in part by the U.S. Department of Energy under Grant No. EY-76-G-03-1305, by the U.S. Environmental Protection Agency under Grant No. R806005, and by the Mellon Foundation. Thanks are due to the staff of the South Coast Air Quality Management District for their cooperation throughout the course of this project.

REFERENCES

1. P. Cukor, L. Ciaccio, E. Lanning, R. Rubino, Environ. Sci. Technol., Vol. 6 (1972), p. 633.
2. D. Grosjean, S. K. Friedlander, J. Air Pollut. Control Assoc., Vol. 25 (1975), p. 1038.
3. B. R. Appel, P. Colodny, J. J. Wesolowski, Environ. Sci. Technol., Vol. 10 (1976), p. 359.
4. B. R. Appel, E. M. Hoffer, E. L. Kothny, S. M. Wall, M. Haik, Environ. Sci. Technol., Vol. 13 (1979), p. 98.
5. M. H. Conklin, G. R. Cass, L-C Chu, E. S. Macias, "Wintertime Carbonaceous Aerosols in Los Angeles: An Exploration of the Role of Elemental Carbon," In: E. S. Macias and P. Hopke (eds.) Chemical Composition of Atmospheric Aerosols: Source/Air Quality Relationships, American Chemical Society, Washington, D.C. 1981.
6. R. G. Delumyea, L-C Chu, E. S. Macias, Atmos. Environ., Vol. 14 (1980), p. 647.
7. L. Gundel in T. Novakov (ed.), "Proceedings, Carbonaceous Particles in the Atmosphere," Lawrence Berkeley Laboratory; Berkeley, California, (1978), p. 91.
8. W. R. Pierson, P. A. Russell, Atmos. Environ., Vol. 13 (1979), p. 1623.
9. S. L. Hiesler, R. C. Henry, J. G. Watson, G. M. Hidy, "The 1978 Denver Winter Haze Study, Vol II, Final Report," Environmental Research and Technology Inc., Document P-5417-1, Westlake Village, California, 1980.
10. G. T. Wolff, R. J. Countess, P. J. Groblicki, M. A. Ferman, S. H. Cadle, J. L. Muhlbaier, Atmos. Environ., Vol. 15 (1981), p. 2485.
11. C. W. Lewis, E. S. Macias, Atmos. Environ., Vol. 14 (1980), p. 185.
12. H. Malissa in T. Novakov (ed.), "Proceedings, Carbonaceous Particles in the Atmosphere," Lawrence Berkeley Laboratory, Berkeley, California, (1978), pp. 221-228.
13. H. Rosen, T. Novakov, Atmos. Environ., Vol. 12 (1978), p. 923.

14. *H. Rosen, A. D. A. Hansen, R. L. Dod, T. Novakov, Science, Vol. 208 (1980), p. 741.*

15. *H. Rosen, A. D. A. Hansen, R. L. Dod, T. Novakov, "Application of the Optical Absorption Technique to the Characterization of the Carbonaceous Component of Ambient and Source Particulate Samples," Lawrence Berkeley Laboratory, Berkeley, California, 1977.*

16. *A. P. Waggoner, R. J. Charlson, "Measurements of Aerosol Optical Properties" in P.A. Russell (ed.), "Proc. Symp. on Denver Air Pollution Study – 1973," Vol II, document EPA-600/9-77-001, U.S. Environmental Protection Agency, Research Triangle Park, North Carolina, (1977), pp. 35-55.*

17. *IARC Working Group, Cancer Res., Vol. 40 (1980), p. 1.*

18. *E. T. Wei, Y. Y. Wang, S. M. Rappaport, J. Air Pollut. Control Assoc., Vol. 30 (1980), p. 267.*

19. *M. S. Miller, S. K. Friedlander, G. M. Hidy, J. Colloid Interface Sci., Vol. 39 (1972), p. 165.*

20. *G. Gartrell Jr., S. K. Friedlander, Atmos. Environ., Vol. 9 (1975), pp. 279-299.*

21. *J. G. Watson Jr., "Chemical Element Balance Receptor Model Methodology for Assessing the Sources of Fine and Total Suspended Particulate Matter in Portland, Oregon," Ph.D. Thesis, Oregon Graduate Center, Beaverton, Oregon, 1979.*

22. *A. D. A. Hansen, W. H. Benner, T. Novakov, "A Carbon and Lead Emission Inventory for the Greater San Francisco Bay Area," in "Atmospheric Aerosol Research Annual Report 1977-78," Lawrence Berkeley Laboratory, document LBL-8696, Berkeley, California, 1978.*

23. *H. J. Taback, A. R. Brienza, J. Macko, N. Brunetz, "Fine Particle Emissions from Stationary and Miscellaneous Sources in the South Coast Air Basin," KVB Inc., document No. KVB 5806-783, Tustin, California, 1979. (Includes an important appendix volume).*

24. *G. R. Cass, P. S. McMurry, J. E. Houseworth, "Methods for Sulfate Air Quality Management," Appendix A3. Environmental Quality Laboratory, Report 16, California Institute of Technology, Pasadena, California, 1980.*

25. *"1977 California Gas Report," (A report prepared pursuant to California Public Utilities Commission Decision Number 62260; authorship and publisher are unknown, but copies are obtainable from the Southern California Gas Company).*

26. *G. R. Cass, "Sulfur Oxides Emissions in the Early 1980's under Conditions of Low Natural Gas Supply," working paper, Environmental Quality Laboratory, California Institute of Technology, Pasadena, California, 1980.*

27. *Task Group on Lung Dynamics, Health Phys., Vol. 12 (1966), p. 173.*

28. *J. J. Huntzicker, S. K. Friedlander, C. I. Davidson, Environ. Sci. Technol., Vol. 9 (1975), p. 448.*

29. *J. L. Muhlbaier and R. L. Williams, These Proceedings, p. 185.*

30. *G. L. Ter Haar, D. L. Lenane, J. N. Hu, M. Brandt, J. Air Pollut. Control Assoc., Vol 22 (1972), p. 39.*

31. *K. Habibi, Environ. Sci. Technol., Vol. 7 (1973), p. 223.*

32. *R. A. Gorse and G. T. Wolff, "Effects of altitude on particulate organics and elemental carbon emissions from pre-1970 automobiles," Motor Vehicle Manufacturers' Association, Detroit, Michigan (in preparation).*

33. *G. R. Cass, "Lead as a Tracer for Automotive Particulates: Projecting the Sulfate Air Quality Impact of Oxidation Catalyst Equipped Cars in Los Angeles," Environmental Quality Laboratory, Memorandum 12, California Institute of Technology, Pasadena, California, 1975.*

34. *R. L. Johnson, J. J. Shah, R. A. Cary, J. J. Huntzicker, "An Automated Thermal-Optical Method for the Analysis of Carbonaceous Aerosol," presented at the Second Chemical Congress of the North American Continent, Las Vegas, Nevada (1980) in E. S. Macias and P. K. Hopke (eds.), Chemical Composition of Atmospheric Aerosols: Source/Air Quality Relationships, American Chemical Society, Washington, D.C., 1981. Plus additional data transmitted by letter from J. J. Huntzicker dated September*

10, 1980, giving the composition of source samples. Source data were presented at that conference but are not in the proceedings volume.

35. R. Gorse, personal communication, Ford Motor Company, October 1, 1980. Tests on vehicles at Denver elevation show 89.6 mg/mile fine organics plus 16.1 mg/mile fine elemental C. Low altitude tests are not yet completed, but filter samples examined visually appear to be about 40% lighter in shade, indicating less elemental carbon.

36. T. M. Baines, J. H. Somers, C. A. Harvey, J. Air Pollut. Control Assoc., Vol. 29 (1979), p. 616.

37. W. R. Pierson and W. W. Brachaczek, SAE Trans., Vol. 85 (1976), p. 209.

38. U.S. Environmental Protection Agency, "Compilation of Air Pollutant Emission Factors, Second Edition," including Supplements 1-8, U.S. Environmental Protection Agency, document AP-42, Research Triangle Park, North Carolina, 1976.

39. J. B. Heywood, J. A. Fay, L. H. Linden, AIAA Journal, Vol. 9 (1971), p. 841.

40. R. C. Flagan, personal communication, California Institute of Technology, Pasadena, California, 1980, suggested this approximation.

41. J. A. Danielson, R. W. Graves, "Fuel Use and Emissions from Stationary Combustion Processes," Southern California Air Pollution Control District, El Monte, California, 1976.

42. S. C. Hunter and H. Taback, personal communication, KVB Incorporated, Tustin, California, September 29, 1980. Filter samples taken during source tests of large natural gas combustors are not dark in color.

43. N. Mansour, personal communication, Southern California Edison Co., Rosemead, California. Samples taken during source tests on Los Angeles Department of Water and Power Scattergood Unit 3 while burning natural gas showed mostly iron, silica, sodium and calcium. Samples were not noticeably darkened. See source test referenced to Manfredi and Mansour [44].

44. M. Manfredi and N. Mansour, "Test Conducted at Department of Water and Power City of Los Angeles (Scattergood Station) 12700 Vista Del Mar, Playa Del Rey, California on September 26, 1975 – Report on the Emissions from Steam Generator No. 3 Under Natural Gas Firing," Southern California Air Pollution Control District, Source Test Section Report No. C-2354, El Monte, California, 1975.

45. W. R. Pierson and W. W. Brachaczek, Rubber Chem. Technol., Vol. 47 (1974), p. 1275.

46. M. Morton, (ed.), Rubber Technology, Litton Educational Publishing, New York, New York, 1973.

47. J. R. Lynch, J. Air Pollut. Control Assoc., Vol. 18 (1968), p. 824.

48. E. M. Shelton, "Motor Gasolines, Winter 1978-79," U.S. Department of Energy, document BETC/PPS-79/3, Bartlesville, Oklahoma, 1979.

49. G. A. DeMarrias, G. C. Holzworth and C. R. Hosler, "Meteorological Summaries Pertinent to Atmospheric Transport and Dispersion over Southern California," U.S. Weather Bureau, Technical Paper No. 54, Washington, D.C., 1965.

50. G. R. Cass, "Methods for Sulfate Air Quality Management with Applications to Los Angeles," Ph.D. Thesis, California Institute of Technology, Pasadena, California, 1978.

51. G. C. Tiao, M. S. Phadke, M. Grupe, S. Hillmer, S. T. Liu, W. Fortney, "Los Angeles Aerometric Carbon Monoxide Data 1956-1972," Department of Statistics, Technical Report No. 377, University of Wisconsin, Madison, Wisconsin, 1974.

52. M. S. Phadke, G. C. Tiao, M. Grupe, S. T. Liu, W. Fortney, S. Wu, "Los Angeles Aerometric Data on Oxides of Nitrogen 1957-1972," Department of Statistics, Technical Report No. 395, University of Wisconsin, Madison, Wisconsin, 1974.

53. J. A. Cooper and C. Frazier, personal communications, NEA Laboratories, Beaverton, Oregon. Performed trace metal analysis by X-ray fluorescence.

54. C. Lin, M. Baker, R. J. Charlson, Applied Optics, Vol. 12 (1973), p. 1356.

55. J. R. Ouimette, "Chemical Species Contributions to the Extinction Coefficient," Ph.D. thesis, California Institute of Technology, Pasadena, California, 1980.

56. *South Coast Air Quality Management District, personal communication, 1980. Provided copies of fuel use reports for January and February 1980 filed by all electric utilities and petroleum refiners in Los Angeles, Orange, Riverside and San Bernardino Counties, California.*

57. *K. Rosenthal, personal communication, Southern California Edison Co., 1980. Provided January and February 1980 fuel use by electric utility generating stations located in Ventura County, California.*

58. *Southern California Gas Company, "South Coast Air Basin Sales to Firm and Interruptible (Excluding Steam Electric) Customers – Year 1974," Southern California Gas Company data tables, Los Angeles, California, 1975.*

59. *C. Marks, "The Fuel Quality and Quantity Requirements of Future Automobiles," Engineering Staff, General Motors Corp., Warren, Michigan, presented at 1977 National Petroleum Refiners Association Annual Meeting, San Francisco, March 29, 1977.*

60. *Bureau of National Affairs, Environ. Reporter, Vol. 7 (43), pp. 1638-1639; Vol. 7 (52), p. 1982; Vol. 9 (3), p. 78.*

61. *H. S. Goodman et al., "A Mobile Source Emission Inventory System for Light Duty Vehicles in the South Coast Air Basin," TRW Environmental Engineering Division, Redondo Beach, California, 1977.*

62. *K. W. Arledge and R. L. Tan, "A Heavy Duty Vehicle Emission Inventory System," TRW Environmental Engineering Division, Redondo Beach, California, 1977.*

63. *A. Laresgoiti and G. S. Springer, Environ. Sci. Technol., Vol. 11 (1977), p. 285.*

64. *Federal Aviation Administration, "FAA Air Traffic Activity, Fiscal Year 1973," (and 1974), U.S. Department of Transportation, 1973, 1974.*

65. *Federal Aviation Administration, "Military Air Traffic Activity Report, Calendar Year 1974," U.S. Department of Transportation, 1974.*

66. *South Coast Air Quality Management District, "Emission Inventory – 1976," Air Programs Division, South Coast Air Quality Management District, El Monte, California, revision of May 1979.*

67. *South Coast Air Quality Management District, "Emission Inventory of Sources at Kaiser Steel, 1976," South Coast Air Quality Management District, El Monte, California, 1978.*

68. *D. M. Roessler, F. R. Faxvog, Appl. Opt., Vol. 19 (1980), p. 578.*

APPENDIX — FUEL USE AND EMISSIONS ESTIMATES

TABLE A.1
Estimated Fuel Consumption by Mobile Sources

	1973 Annual Average (10^9BTU/day)[a]	Growth Factor 1973 to 1980[b]	1980 Average Day (10^9BTU/day)
Highway Vehicles			
Catalyst Autos & Lt. Trucks	0.00	(c)	963.78
Non-catalyst Autos & Lt. Trucks	1566.58	(c)	566.77
Medium & Heavy Gasoline Vehicles	213.64	1.167	249.26
Diesel Vehicles	117.53	1.167	137.16
LPG for Carburetion	5.20	1.167	6.07
Civil Aviation			
Jet Aircraft	47.40	0.96	45.50
Aviation Gasoline	1.37	0.96	1.32
Commercial Shipping			
Residual Oil-Fired Ship's Boilers	24.66	1.30	32.05
Diesel Ships	17.26	1.30	22.43
Railroad			
Diesel Oil	29.32	1.30	38.11
Military			
Gasoline	6.03	1.00	6.03
Diesel Oil	17.81	1.00	17.81
Jet Fuel	16.71	1.00	16.71
Residual Oil (Bunker Fuel)	0.27	1.00	0.27
Miscellaneous			
Off-Highway Vehicles	39.73	1.00	39.73

(a) From Cass et al. [24]; Table A3.6.

(b) From survey of growth in transport mode activity [26].

(c) See Table A.5, this paper.

TABLE A.2
Estimated Fuel Consumption at Stationary Sources

	1973 Annual Average (10^9BTU/day)	January and February 1980 (10^9BTU/day)
ELECTRIC UTILITIES		
Natural Gas	440.55 (a)	666.25 (e)
Residual Oil	1058.36 (a)	885.72 (e)
Distillate Oil (Turbines)	7.40 (a)	small (e)
Landfill and Digester Gas	0.82 (a)	0.85 (e)
REFINERY FUEL		
Natural Gas	132.88 (b)	148.54 (e)
Refinery Gas	410.41 (b)	347.63 (e)
Fuel Oil	32.33 (b)	5.57 (e)
NON-REFINERY INDUSTRIAL FUEL		
Natural Gas	421.64 (c)	421.64 (f)
LPG	2.74 (c)	2.74 (f)
Residual Oil	53.42 (c)	53.42 (f)
Distillate Oil	42.74 (c)	42.74 (f)
Digester Gas	6.30 (c)	6.30 (f)
Coke Oven Gas	37.53 (d)	37.53 (f)
Coal	0.00	Included Within Cement Kiln
RESIDENTIAL/COMMERCIAL		
Natural Gas	1181.92 (c)	1592.26 (g)
LPG	18.08 (c)	18.08 (h)
Residual Oil	22.19 (c)	22.19 (h)
Distillate Oil	22.19 (c)	22.19 (h)
Coal	0.55 (c)	0.55 (h)

(a) Cass et al. [24]; Table A3.5.

(b) Cass et al. [24]; Table A3.4.

(c) Cass et al. [24]; Table A3.6.

(d) Danielson et al. [41], p. 51, 2.17 x 10^6 equivalent bbl/yr. A fuel oil equivalent barrel is 6.3 x 10^6 BTU.

(e) South Coast Air Quality Management District [56] fuel burning records plus personal communication, Rosenthal [57].

(f) Examination of 1977 California Gas Report [25] "Requirements" for industrial natural gas shows that 1980 projected industrial fuel use (heat input) is about the same as in 1973. Refinery and electric utility fuel use records cited above show a high level of gas supply to industry during January and February 1980. Survey by Cass [50] shows little seasonal variation in industrial natural gas demand. Hence winter 1980 industrial fuel use will be approximated by 1973 annual average data.

(g) Survey by Cass [26] shows little growth in residential/commercial gas demand since 1974. January and February 1974 natural gas combustion data from Southern California Gas Co. [58] is used to represent winter 1980.

(h) Assumed same as 1973 annual average day.

TABLE A.3
Fuel Economy Calculation for 1980 Auto Fleet

Age (years)	Model Year	Fraction of Total Vehicles In Use (c)	Annual Avg Mileage Driven (c)	Fraction of Annual Travel	Fuel Economy (mpg)	Weighted Average Fuel Economy (mpg)
1	1980	0.083	15,900	0.116	20.0(a)	wt. avg.(e) 17.8
2	1979	0.103	15,000	0.135	19.0(a)	
3	1978	0.102	14,000	0.125	18.0(a)	
4	1977	0.106	13,100	0.122	17.5(b)	
5	1976	0.099	12,200	0.106	16.5(b)	
6	1975	0.087	11,300	0.086	14.7(b)	
7	1974	0.092	10,300	0.083		
8	1973	0.088	9,400	0.072		
9	1972	0.068	8,500	0.051		
10	1971	0.055	7,600	0.037		13.6(c)(d)
11	1970	0.039	6,700	0.023		
12	1969	0.021	6,700	0.012		
≥13	1968(−)	0.057	6,700	0.033		

(a) Energy Policy and Conservation Act Goals (see Marks, [59]).

(b) U.S. Environmental Protection Agency, Fleet Average Fuel Economy Data, reduced to 94% of measured value (see Bureau of National Affairs, [60]).

(c) Environmental Protection Agency [38].

(d) Pre-catalyst auto fleet average fuel economy.

(e) Weighted average fuel economy for catalyst equipped automobiles.

TABLE A.4
Percentage of Vehicle Miles Traveled and Fuel Economy for
Each Vehicle Type in 1980

Vehicle Type	Fraction of Daily Total Vehicle Miles Traveled (a)(b)	Weighted Average Fuel Economy (miles/gallon)
Autombiles		
Catalyst-equipped	53.0%	17.8(c)
Noncatalyst type	23.8%	13.6(d)
Light trucks		
Catalyst-equipped	9.2%	13.1(e)
Noncatalyst type	4.1%	10.0(d)
Medium and heavy duty gasoline trucks	6.5%	6.83(f)
Diesel trucks and buses	3.5%	4.6(d)

(a) *Fraction of vehicle miles traveled by automobiles, light trucks, medium and heavy gasoline trucks and buses and diesel trucks and buses computed from 1975 data reported for the South Coast Air Basin by TRW (Goodman et al. [61]; Arledge and Tan, [62]).*

(b) *Light duty vehicle miles traveled are divided into 69% by catalyst equipped vehicles and 31% by noncatalyst vehicles, as computed from Table A.3.*

(c) *Computed in Table A.3.*

(d) *See Environmental Protection Agency [38].*

(e) *Assuming improvement in newer light truck fuel economy proportional to that observed for newer automobiles.*

(f) *Heavy trucks computed at 6 mpg (Environmental Protection Agency, [38]), medium trucks evaluated at 8 mpg.*

TABLE A.5
Computation of 1980 Fuel Use by Gasoline-Powered Highway Vehicles

	Col 1 Percentage of Total Highway Vehicle Miles Traveled(a)	Col 2 Fraction of Miles Traveled Within Group Shown	Col 3 Total 1973 Gasoline Use By Entire Group (10⁹BTU/day) (b)	Col 4 Growth in Vehicle Miles Traveled: 1973 to 1980(c)	Col 5 Change in Fuel Economy Relative to 1973(d)	Col 6 Estimated 1980 Fuel Consumption (10⁹BTU/day)(e)
VEHICLE TYPE						
Group 1						
Catalyst equipped autos	53.0%	0.69	1264	1.167	0.764	777.6
Non-catalyst autos	23.8%	0.31	1264	1.167	1.00	457.3
Group 2						
Catalyst light trucks	9.2%	0.69	303	1.167	0.764	186.4
Non-catalyst light trucks	4.1%	0.31	303	1.167	1.00	109.6

(a) From Table A.4.
(b) From Cass, McMurry and Houseworth [24]
(c) Computed from survey of growth in vehicle miles traveled; Cass [26]
(d) Catalyst equipped auto fleet fuel economy computed at 17.8 mpg in Table A.4; non-catalyst older cars estimated at 13.6 mpg; relative improvement of catalyst fleet fuel economy is 13.6/17.8 = 0.764.
(e) Col (2) x Col (3) x Col (4) x Col (5)

TABLE A.6
Emissions Estimates for Mobile Sources

	Estimated 1980 Fuel Use (10⁹BTU/day) (a)	Total Particulate Emission Factor	Particulate Emissions (kg/day)	Mass Fraction Less Than 10μm Diameter	Fine Particle Emissions (kg/day)	% Total Carbon in <10μm Fraction	Partition of Carbon Present — Volatile Organics Carbon	Partition of Carbon Present — Non-volatile Carbon	Fine Organic Carbon (kg/day)	Fine Non-volatile Carbon (kg/day)	Total Carbon (kg/day)
MOBILE SOURCES											
Highway Vehicles											
Catalyst Autos & Lt. Trucks	963.78	2.137 kg/10⁹BTU (b)	2059	~100% (o)	2059	39% (o)	54%	46% (o)	434	369	803
Non-catalyst Autos & Lt. Trucks	566.77	37.814 kg/10⁹BTU (c)	21432	53% (p)	11359	62% (p)	92%	8% (p)	6479	563	7042
Medium & Heavy Gasoline Vehicles	249.26	43.596 kg/10⁹BTU (d)	10867	53% (q)	5760	62% (q)	92%	8% (q)	3285	286	3571
Diesel Vehicles	137.16	64.20 kg/10⁹BTU (e)	8806	96% (r)	8453	70% (w)	27%	73% (x)	1598	4319	5917
LPG use for Carburetion	6.07	Assumed Negligible (f)									
Civil Aviation											
Jet Aircraft	45.50	U.S. EPA [38] Procedure (g)	780*	~100% (s)	780	96% (s)	27%	73% (y)	202	547	749
Aviation Gasoline	1.32	9.08 g/LTO cycle (h)	50	53% (t)	26	62% (q)	92%	8% (q)	15	1	16
Commercial Shipping											
Residual Oil-Fired Ships Boilers	32.05	85.386 kg/10⁹BTU (i)	2737	87% (u)	2381	14% (u)	80%	20% (x)	266	67	333
Diesel Ships	22.43	49.102 kg/10⁹BTU (j)	1101	96% (v)	1057	70% (v)	27%	73% (v)	200	540	740
Railroad Diesel Oil	38.11	81.837 kg/10⁹BTU (k)	3119	96% (v)	2994	70% (v)	27%	73% (v)	566	1530	2096
Military Gasoline	6.03	43.596 kg/10⁹BTU (l)	263	53% (t)	139	62% (q)	92%	8% (q)	79	7	86
Diesel Oil	17.81	78.564 kg/10⁹BTU (m)	1399	96% (v)	1343	70% (v)	27%	73% (v)	254	686	940
Jet Fuel	16.71	U.S. EPA [38] Procedure (n)	659*	~100% (s)	659	96% (s)	27%	73% (y)	171	462	633
Residual Oil (Bunker Fuel)	0.27	85.386 kg/10⁹BTU (i)	23	87% (u)	20	14% (u)	80%	20% (x)	2	1	3
Miscellaneous Off-Highway Vehicles	39.73	78.564 kg/10⁹BTU (m)	3121	96% (v)	2996	70% (v)	27%	73% (v)	566	1531	2097
								TOTAL	14117	10909	25026

*Solid particulate matter only

Notes for TABLE A.6

(a) *From Table A.1 of this paper.*

(b) *Laresgoiti and Springer [63] at 0.3 % sulfur in unleaded gasoline obtained 0.016 g/mile for oxidation catalyst car; Muhlbaier and Williams [29].*

(c) *0.35 gm/mile from Habibi [31] Table VIII (leaded fuel cars) adjusted upward to 0.0705 g/mile lead emitted (see note p below); pre-1975 auto fuel economy of 13.6 miles/gallon is assumed; emission factor is 10.5 lb/10³gal or 37.814 kg/10⁹BTU.*

(d) *U.S. Environmental Protection Agency [38] Table 3.1.4-13, 12 lb/1000 gal or 43.596 kg/ 10⁹BTU.*

(e) *Los Angeles diesel usage emission factor of 0.83 g/km plus fuel consumption rate of 0.3536 l/km from Baines, Somers, and Harvey [36] gives 64.20 kg/10⁹BTU.*

(f) *Negligible particulate emissions, as assumed by U.S. Environmental Protection Agency [38] Section 3.1.6-1.*

(g) *Average daily emissions from jet engines at commercial airports were obtained from Federal Aviation Administration [64] air traffic data plus U.S. Environmental Protection Agency ([38], part 3.2.1) calculation procedure.*

(h) *U.S. Environmental Protection Agency ([38] part 3.2.1) calculation procedure plus general aviation traffic data from Federal Aviation Administration [64].*

(i) *Industrial boiler emission factor of 21.619 kg/10⁹BTU at 0.4 % sulfur in fuel was scaled up to 85.386 kg/10⁹BTU at 1.58 % sulfur in bunker fuel based on evidence of Taback et al. [23] that shows that particulate emissions rate from residual oil-fired boilers is roughly proportional to fuel oil sulfur content. The underlying reason for this is that higher sulfur fuels are higher in ash and asphaltenes. The fuel use shown for ships includes shipping lane activity from Pt. Conception to the southern boundary of Orange County.*

(j) *U.S. Environmental Protection Agency [38] Table 3.2.3-2 gives 15 lb/10³gal or 49.102 kg/10⁹BTU.*

(k) *U.S. Environmental Protection Agency [38] Table 3.2.2-1 gives 25 lb/10³gal or 81.837 kg/10⁹BTU.*

(l) *From medium and heavy duty highway vehicle emission factor (see note d).*

(m) *U.S. Environmental Protection Agency [38] Table 3.2.7-1; assumed at 24 lb/10³gal or 78.564 kg/10⁹BTU from mid-range of off-highway diesels listed.*

(n) *Average daily emissions from jet aircraft at military airports calculated from Federal Aviation Administration [65] traffic data plus U.S. Environmental Protection Agency [38] procedure.*

(o) *All catalyst equipped auto particulate emissions are assumed to be in fine particle sizes 10 μm diameter. Muhlbaier and Williams [29] report that catalyst equipped cars at low elevation emit aerosol with the following properties: total mass 14 mg/mile, organic carbon 3.0 mg/mile, elemental carbon 2.5 mg/mile. This is in reasonable agreement with Johnson et al. [34] who report 69 % of carbon is present as organics, 31 % of carbon is present as elemental carbon from cars burning unleaded gasoline. Partition of carbon present was based on proportions given by Muhlbaier and Williams [29].*

(p) *The fine particle emission factor from leaded fuel auto fleet was assembled as follows: average lead content of gasoline in winter 1978-79 was 1.36 g/gal in southern California (1/2 regular, 1/2 premium; data from Shelton, [48]). At 13.6 miles/gal, lead consumption is 0.100 g/mile. From Huntzicker et al. [28], 70.5 % of lead consumed in gasoline in Los Angeles is emitted as aerosol and 43 % of aerosol lead emitted is in sizes less than 9 μm, giving 30 mg/mile fine aerosol lead or 50 mg/mile fine lead salts as 2 PbBrCl · NH₄Cl. Carbon emissions from pre-catalyst cars burning leaded gasoline are 0.071 g/mile for newer cars (0-1000 miles) and 0.115 g/mile for older cars (30,000- 100,000 miles accumulated) from Ter Haar in Habibi [31]. By 1980 nearly all non-catalyst cars will be in the older group, giving average carbon emissions of 0.115 g/mile (as carbon). Carbon aerosols are assumed to be concentrated in the fine particle fraction of the auto exhaust. Fine elemental carbon emissions are estimated as 0.009 mg/mile in two ways: personal communication with Robert Gorse [35] at Ford Motor Co. and by applying elemental/total carbon factor of 8% found by Johnson et al. [34]. Organic carbon remaining is (0.115 g/mi - 0.009 g/mi) = 0.106 g/mile, which becomes about 0.127 g/mile as organic material. Total auto fine particle emissions become 0.186 g/mile in 1980 (0.050 g/mile lead salts, 0.009 g/mile elemental carbon, 0.127 g/mile organic material). This fine aerosol is about 53% of total aerosol emission factor (i.e. 0.186/0.35 = 0.53).*

(q) *Other leaded gasoline fueled vehicles assumed to have relatively the same chemical composition as leaded auto exhaust.*

(r) *Taback et al. [23]. Table pA-5 plus inference from Figure 7 of Pierson and Brachaczek.*

(s) *Heywood, Fay and Linden [39].*

(t) *Assumed similar to leaded auto exhaust.*

(u) *Assumed similar to industrial boilers burning residual oil, Taback et al. [23] Table pA-3.*

(v) *Assumed similar to heavy duty diesel highway vehicles.*

(w) *Based on data summarized by Pierson and Brachaczek [37], Table 13.*

(x) *From Johnson et al. [34].*

(y) *Jet aircraft emissions measured in this manner are all "soot" (Heywood, Fay and Linden, [39]); assume chemical composition looks like diesel soot (Flagan, [40]).*

TABLE A.7
Emission Estimates for Stationary Combustion Sources

	Estimated 1980 Fuel Use (10⁹BTU/day)	Total Particulate Emission Factor (kg/10⁹BTU)	Particulate Emissions (kg/day)	Mass Fraction Less Than 10μm Diameter	Fine Particle Emissions (kg/day)	% Total Carbon in ≤10μm Fraction	Partition of Carbon Present — Volatile Organics	Non-volatile Carbon	Fine Organic Carbon (kg/day)	Fine Non-volatile Carbon (kg/day)	Total Carbon (kg/day)
STATIONARY SOURCES											
Fuel Combustion											
Electric Utilities											
Natural Gas	666.25 (a)	1.081 kg/10⁹BTU (e)	720.2	95% (r)	684.2	7% (cc)	100%	SMALL (rr)	48	—	48
Residual Oil (0.25% S)	885.72 (a)	9.080 kg/10⁹BTU (f)	8042.3	97% (s)	7801.1	20% (dd)	80%	20% (ss)	1248	312	1560
Landfill & Digester Gass	0.85 (a)	1.081 kg/10⁹BTU (g)	0.9	95% (t)	0.9	7% (ff)	100%	SMALL (tt)	0.06	—	0.06
Refinery Fuel											
Natural Gas	148.54 (a)	9.080 kg/10⁹BTU (h)	1348.7	95% (v)	1281.3	7% (gg)	100%	SMALL (uu)	90	—	90
Refinery Gas	347.63 (a)	9.080 kg/10⁹BTU (h)	3156.5	95% (w)	2998.7	7% (hh)	100%	SMALL (vv)	210	—	210
Residual Oil	5.57 (a)	21.619 kg/10⁹BTU (i)	120.4	87% (x)	104.8	14% (ii)	80%	20% (ss)	12	3	15
Non-Refinery Industrial Fuel											
Natural Gas	421.64 (b)	7.567 kg/10⁹BTU (j)	3190.6	95% (r)	3031.1	7% (jj)	100%	SMALL (ww)	212	—	212
LPG	2.74 (b)	7.567 kg/10⁹BTU (j)	20.7	95% (r)	19.7	7% (kk)	100%	SMALL (xx)	1	—	1
Residual Oil	53.42 (b)	21.619 kg/10⁹BTU (i)	1154.9	87% (x)	1004.8	14% (ii)	80%	20% (ss)	113	28	141
Distillate Oil	42.74 (b)	23.52 kg/10⁹BTU (k)	1005.2	97.6% (y)	981.1	15% (ll)	42%	58% (ss)	62	85	147
Digester Gas (IC Engines)	6.30 (b)	20.430 kg/10⁹BTU (l)	128.7	99% (u)	127.4	21% (mm)	77%	23% (yy)	21	6	27
Coke Oven Gas	37.53 (b)	7.567 kg/10⁹BTU (m)	284.0	95% (z)	269.8	7% (nn)	100%	SMALL (zz)	19	—	19
Residential/Commercial											
Natural Gas	1592.26 (c)	8.071 kg/10⁹BTU (n)	12851.1	95% (r)	12208.6	12% (oo)	67%	33% (aaa)	982	483	1465
LPG	18.08 (d)	8.071 kg/10⁹BTU (o)	145.9	95% (r)	138.6	12% (pp)	67%	33% (bbb)	11	6	17
Residual Oil	22.19 (d)	21.619 kg/10⁹BTU (p)	479.7	87% (x)	417.3	14% (ii)	80%	20% (ss)	47	12	59
Distillate Oil	22.19 (d)	21.619 kg/10⁹BTU (p)	479.7	97.6% (aa)	468.2	15% (ll)	42%	58% (ss)	29	41	70
Coal	0.55 (d)	378.333 kg/10⁹BTU (q)	208.1	62% (bb)	129.0	96% (qq)	78%	22% (qq)	97	27	124
								TOTALS	3202	1003	4205

Notes for TABLE A.7

(a) *South Coast Air Quality Management District [56] fuel burning records.*

(b) *See note (f) Table A.2, this paper.*

(c) *See note (g) Table A.2, this paper.*

(d) *Assumed same as 1973 annual average day from Cass, McMurry and Houseworth [24].*

(e) *Danielson et al. [41], p. 77; 0.015 lb/equivalent barrel (1.081 kg/10⁹BTU) from Los Angeles tests.*

(f) *Taback et al. [23], p. 2-9; 3lb/1000 gal (9.080 kg/10⁹BTU) used for 0.5% sulfur oil.*

(g) *Electric utility boiler natural gas combustion emission factor used.*

(h) *Danielson et al. [41], p. 77; 0.126 lb/equivalent bbl (9.080 kg/10⁹BTU).*

(i) *Danielson et al. [41], p. 77; 0.30 lb/bbl (21.619 kg/10⁹BTU) for combustion of industrial residual fuel oil.*

(j) *Danielson et al. [41], p. 77; 0.105 lb/equivalent bbl (7.567 kg/10⁹BTU); LPG combustion is assumed to emit at the same rate as industrial natural gas combustion on an equivalent heat input basis. This is the same assumption as made by the U.S. Environmental Protection Agency [38], Table 1.5-1, except that our natural gas emission factor is lower.*

(k) *Taback et al. [23], Table 2-1, KVB data, 7.2 lb/10³gal (23.52 kg/10⁹BTU).*

(l) *Average of 2 tests by Taback et al. ([23], pp. 4-95, 4-101) performed on digester gas-fired IC engine; 0.045 lb/10⁶ BTU (20.43 kg/10⁹ BTU).*

(m) *Assumed similar to industrial natural gas combustion on an equivalent heat input basis.*

(n) *Danielson et al. [41], p. 77; 0.112 lb/equivalent bbl (8.071 kg/10⁹BTU).*

(o) *Assumed same as residential/commercial natural gas on an equivalent heat input basis. This is the same assumption as made by U.S. Environmental Protection Agency [38], Table 1.5-1, except that our natural gas emission factor is lower.*

(p) *Danielson et al. [41], p. 77; 0.300 lb/equivalent bbl (21.619 kg/10⁹BTU).*

(q) *U.S. Environmental Protection Agency [38], Table 1.1-2, assuming hand-fired stove use, 20 lb/ton coal (378.33 kg/10⁹BTU).*

(r) *Assumed similar to size distribution from large refinery heater burning natural gas, see note (v).*

(s) *Taback et al. [23]; table pA-8.*

(t) *Assumed similar to size distribution from burning natural gas, see notes (r) and (v).*

(u) *Taback et al. [23]; table p4-99.*

(v) *Taback et al. [23]; table pA-29.*

(w) *Assumed similar to refinery heater natural gas size distribution, see note (r).*

(x) *Assumed similar to industrial boiler burning residual fuel oil, Taback et al. [23] table pA-3.*

(y) *Taback et al. [23], table pA-4.*

(z) *Assumed similar to size distribution used for industrial natural gas combustion.*

(aa) *Assumed similar to industrial boiler distillate oil, Taback et al. [23], table pA-4.*

(bb) *Assumed similar to fireplace wood combustion profile from Watson [21]. Possibly a poor assumption, but data on fireplace coal combustion are lacking.*

(cc) *Assumed based on refinery heater test by Taback et al. ([23], table pA-29); electric utility source test by Manfriedi and Mansour [44] showed that particulate matter emitted from LADWP Scattergood Unit 3 when burning natural gas consisted mostly of Fe, Na, Si, and Ca compounds. Mansour [43] confirms that power plant samples when burning gas during that test were not dark in color.*

(dd) *Taback et al. [23], table pA-8.*

(ee) *note deleted.*

(ff) *Assumed similar to natural gas combustion at power plants (see note cc).*

(gg) *Taback et al. [23], pp 4-243 to 4-249.*

(hh) *Assumed similar to natural gas combustion at refinery heaters, see note (gg).*

  (ii) *Taback et al. [23], table pA-3.*
  (jj) *Assumed similar to natural gas combustion at refinery heaters, see note (gg).*
  (kk) *Assumed similar to natural gas combustion at refinery heaters, see note (gg).*
  (ll) *Taback et al. [23] table pA-4.*
 (mm) *Taback et al. [23] table 4-36, average of impinger catches.*
  (nn) *Assumed similar to natural gas combustion at refinery heaters, see note (gg).*
  (oo) *Muhlbaier and Williams [29] report that 12% of aerosol mass is carbon from 10 samples taken downstream of small boiler; two thirds of that carbon was present as organics. Their data have been used in this table. As part of our study, we source tested a small domestic floor furnace burning natural gas. That test showed 5.7% of aerosol emitted was carbon (filters also showed signs of rust from inside of furnace). Note that our overall emission factor for carbonaceous aerosol from domestic gas combustion becomes (8.071 kg/10⁹BTU x 0.12) = 0.97 kg C/10⁹BTU = Q97 µg C/BTU. Source tests by Hansen, Benner and Novakov [22] show that domestic natural gas combustion sources emit carbon at a rate of between 0.2 to 2.5 µg C/BTU, in good agreement with the above discussion.*
  (pp) *Assumed similar to natural gas combustion (see note oo).*
  (qq) *No data are available on carbon mass as a fraction of fine particle mass emissions from coal combustion in fireplaces, therefore an extreme upper limit has been used. Chemical composition of carbon present is based on volatile organic to non-volatile carbon ratio for fine particles from fireplace combustion of wood given by Watson [21]. Assumptions made for this source class are poor, but source class is very small.*
  (rr) *Assumed similar to large refinery heaters burning natural gas (see note uu). Personnel who have tested utility boilers while burning natural gas report that samples are not dark in color (see note cc).*
  (ss) *Johnson et al. [34] for residual oil combustion with pyrolysis correction.*
  (tt) *Assumed similar to natural gas combustion (see notes rr and uu).*
  (uu) *Taback et al. [23] Table pA-29. All carbon collected during source test was volatile.*
  (vv) *Assumed similar to refinery heater burning natural gas (see note uu).*
 (ww) *Assumed similar to refinery heater burning natural gas (see note uu).*
  (xx) *Assumed similar to industrial natural gas combustion (see notes uu and ww).*
  (yy) *Taback et al. [23] Table 4-36, average percentage from tests 7S-IC and 7J-IC, impinger catch only.*
  (zz) *Assumed similar to industrial natural gas combustion (see notes uu and ww).*
 (aaa) *Speciation based on Muhlbaier and Williams [29], see note (oo) for their data. Our two source tests of a domestic space heater gave conflicting results: one test produced 74% organic C, 26% non-volatile C; the second test produced 5% organic C, 95% non-volatile C (by test procedures of Johnson et al. [34] with pyrolysis correction).*
 (bbb) *Assumed similar to residential/commercial natural gas combustion (see note aaa).*

TABLE A.8
Emission Estimates for Industrial Processes

	Particulate Emissions (kg/day) (a)	Mass Fraction Less than 10μm Diameter	Fine Particle Emissions (kg/day)	% Total Carbon in < 10μm Fraction	Partition of Carbon Present — Organic Carbon	Partition of Carbon Present — Non-volatile Carbon	Fine Organic Carbon (kg/day)	Fine Non-volatile Carbon (kg/day)	Total Carbon (kg/day)
STATIONARY SOURCES									
Industrial Process Point Sources									
Petroleum Industry									
Production	124	(b)	≤124	≤83% (h)	100%	0% (h)	≤103	0	≤103
Refining	1493	61% (c)	911	3.3% (c)	75%	25% (c)	23	7	30
Marketing	249	(b)	≤249	≤83% (h)	100%	0% (h)	≤207	0	≤207
Organic Solvent Use									
Surface Coating	2861	96% (d)	2747	55% (d)	100%	0% (d)	1511	0	1511
Degreasing	25	(b)	≤25	≤83% (h)	100%	0% (h)	≤21	0	≤21
Other	12	(b)	≤12	≤83% (h)	100%	0% (h)	≤10	0	≤10
Chemical	1343	90% (e)	1209	33% (i)	100%	0% (j)	399	0	399
Metallurgical	10448	98% (f)	10239	36% (f)	97%	3% (f)	3575	111	3686
Mineral	313447	20% (e)	62689	~0% (k)	—	—	—	—	—
Waste Burning at Point Sources	187	~100% (g)	≤187	30% (g)	33%	67% (g)	19	37	≤56
Wood Processing	323	60% (e)	194	41% (l)	100%	0% (l)	79	0	79
Food and Agriculture	1144	80% (e)	915	30% (m)	100%	0% (m)	275	0	275
Miscellaneous Industrial	1095	50% (e)	548	30% (n)	100%	0% (n)	164	0	164
						TOTAL	≤6386	155	≤6541

(a) *From Taback et al. [23] Table 1-1.*

(b) *unknown.*

(c) *Based on FCC unit CO boiler profile, Taback et al. [23] pA-30.*

(d) *Based on paint spray booth profile (oil based paint), Taback et al. [23] pA-24.*

(e) *From Taback et al. [23] Table 2-20.*

(f) *Metallurgical source emissions partitioned 62% primary ferrous metals, 38% secondary metals in proportions given by South Coast Air Quality Management District [66]. Primary metals emissions represented as a composite of 21% basic oxygen furnace profile, 19% sinter machine profile, 10% open hearth furnace profile from Taback et al. [23] plus 50% coke oven volatiles which were assumed to be <10μm organics (proportions taken approximately from Kaiser Steel inventory, South Coast Air Quality Management District [67]). Secondary metals represented by aluminum foundry profile from Taback et al. [23] table pA-13.*

(g) *Based on Taback et al's [23] wood waste boiler profile, pA-7. Actually, these emissions are probably from special purpose permitted incinerators for which we lack local source test data on chemical composition.*

(h) *Assumed to be a typical organic liquid: organic mass ≈1.2 times carbon mass present.*

(i) *Based on Urea manufacturing profile as a typical organic chemical product; Taback et al. [23] pA-26.*

(j) *Assuming that few chemical products contain graphitic carbon.*

(k) *Assumed to be entirely mineral matter.*

(l) *Based on an average of wood resawing and wood sanding operation profiles from Taback et al. [23], ppA-27, A-28.*

(m) *Based on carbon content of emissions from feed and grain opeations from Taback et al. [23] pA-29. Source category actually includes wool and cotton fabrication emissions, meat packing, rendering, cooking, etc. All carbon present is assumed to be organic carbon.*

(n) *No data, value assumed in the middle of the range found for other industries listed above.*

TABLE A.9
Emissions Estimates for Fugitive Sources

	Fine Particle Emissions <10 μm Fraction (kg/day)	% Total Carbon in <10μm Fraction	Partition of Carbon Present		Fine Organic Carbon (kg/day)	Fine Non-volatile Carbon (kg/day)	Total Carbon (kg/day)
			Organic Carbon	Non-volatile Carbon			
FUGITIVE SOURCES							
Road and Building Construction	176625 (a)	— (d)	—	—	—	—	—
Agricultural Tilling	23633 (a)	— (e)	—	—	—	—	—
Refuse Disposal Sites	746 (a)	— (f)	—	—	—	—	—
Livestock Feedlots	3234 (a)	2% (g)	100%	0%	65	0	65
Unpaved Road Travel	57216 (a)	— (h)	—	—	—	—	—
Paved Road Travel	(b)				(b)	(b)	(b)
Forest Fires (seasonal)	5970 (a)	63% (i)	94%	6% (i)	3535	226	3761
Structural Fires	498 (a)	30% (j)	51%	49% (j)	76	73	149
Fireplaces	1244 (a)	30% (j)	51%	49% (j)	190	183	373
Cigarettes	1990 (a)	85% (k)	99%	~1% (l)	1675	17	1692
Agricultural Burning	1244 (a)	52% (m)	91%	9% (m)	589	58	647
Tire Attrition	6023 (c)	87% (n)	67%	33% (n)	3511	1729	5240
Brake Lining Attrition	7712 (a)	28% (o)	82%	18% (o)	1771	388	2159
Sea Salt	49753 (a)	— (p)	—	—	—	—	—
TOTAL — ANNUAL AVERAGE DAY					11412	2674	14086
TOTAL — WINTER DAY WITH NO FOREST FIRES					7877	2448	10325

(a) *Taback et al.* [23] *Table 2-4.*
(b) *Potentially a very large source of resuspended vehicle exhaust particles.*
(c) *Daily tire-miles traveled estimated from Taback et al.* [23] *Table 2-15; suspended aerosol emissions rate of 0.0063 g/tire mile based on average of results by Pierson and Brachaczek* [45]. *Some of these particles may be larger than 10 μm.*
(d) *Taback et al.* [23], *table pA-46.*
(e) *Taback et al.* [23], *table pA-50.*
(f) *Taback et al.* [23], *table pA-49.*
(g) *Taback et al.* [23], *table pA-44, assumed to be organic carbon.*
(h) *Taback et al.* [23], *table pA-45.*
(i) *Based on Watson* [21]; *Slash burning fine particles.*
(j) *Taback et al.* [23]; *tables ppA-37 and A-39 give 30% carbon in particulate emissions from wood waste boiler. Muhlbaier as cited by Wolff et al.* [10] *would partition that carbon as 51% organic C, 49% elemental C for soft wood fires.*
(k) *Taback et al.* [23]; *table pA-41.*
(l) *Estimate based on Roessler and Faxvog's* [68] *observation that cigarette smoke is very ineffective at absorbing light.*
(m) *Based on Watson* [21]; *fine particles from simulated field burning.*
(n) *87% carbon computed from formula for oil extended synthetic tire rubber given by Morton* [46]; *29% of tire tread batch is ISAF carbon black. Carbon present thus is found as about 67% organics, 33% elemental carbon.*
(o) *Based on automobile brake lining chemical composition given by Lynch* [47], *plus assumption that resins and polymers are 83% carbon by weight, and assuming that carbon present is emitted as aerosol without combustion or pyrolysis.*
(p) *Taback et al.* [23], *table pA-42.*

TABLE A.10
Emission Inventory Summary

Summary	Fine Organic Carbon (kg/day)	Fine Non-Volatile Carbon (kg/day)	Fine Total Carbon (kg/day)	Fine Lead (kg/day)
MOBILE SOURCES				
Highway Vehicles	11796	5537	17333	2691
Other Mobile Sources	2321	5372	7693	24
STATIONARY SOURCES				
Fuel Combustion	3202	1003	4205	
Industrial Processes	≤6386	155	≤6541	
FUGITIVE SOURCES (WINTER)				
Non-vehicular	2595	331	2926	
Tire and Brake Wear	5282	2117	7399	
TOTAL	31582	14515	46097	2715

DISCUSSION

L. Gundel *(Lawrence Berkeley Laboratory)*

Did you make any measurements to get the same data for the summertime or for an entire annual period?

G. Cass

No, but there are a number of summer measurements available from other studies in Los Angeles including the ones from your laboratory. I would invite comparison of our emission inventory to summertime results obtained by other people.

E. Macias *(Washington University)*

We picked wintertime so that there would be the least influence of secondary aerosol carbon formation.

G. Cass

Right. The intention was not to get representative air samples but to try to get air samples that would be useful for comparison to the emission inventory.

T. Hansen *(Lawrence Berkeley Laboratory)*

I believe you said that you were taking one hour samples starting at ten in the morning. Have you looked at them in order to determine if there is a diurnal pattern or something in the time structure?

G. Cass

Yes. I did not mention that another objective in our study was to develop a 25-year historical hourly elemental carbon air quality data base for the Los Angeles area (see Conklin *et al**). This is done by calibrating historical tape sampler records against elemental carbon conccentrations. The elemental carbon concentrations were highest during our early morning samples. That is, the 7 am or 8 am samples were always the high ones and declined until noon on most of our sampling events. If one takes the reflectance data from the LA district tape samplers, one can look at the 24-hour diurnal variation of light absorbing black material in the Los Angeles atmosphere. I do not have those data in front of me, but my memory tells me that there are two peaks during a 24-hour cycle; one during the traffic peak in the morning and the other one around midnight.

---

* M. H. Conklin, G. R. Cass, L-C, Chu, E. S. Macias, *Wintertime Carbonaceous Aerosols in Los Angeles; An Exploration of the Role of Elemental Carbon*, to appear in E. S. Macias and P. Hopke (eds.), "Chemical Composition of Atmospheric Aerosols: Source/Air Quality Relationships," American Chemical Society, Washington, D. C., 1981.

J. Heicklen *(Pennsylvania State University)*

Could you comment on secondary carbon formation compared to primary emissions?

G. Cass

We deliberately did not study that. I think that the existing references on that subject including the ACHEX study* and the data by Grosjean and Friedlander** (1975) are probably the best place to turn for an answer to your question. I can tell from the samples that we took that there appears to be relatively little secondary material present during wintertime mornings because the emission inventory and the ambient samples closely resemble each other.

J. Ogren *(University of Washington)*

You lumped the jet fuel and diesel fuel together. Can you give us an idea what those two would be separately and did you burn all the jet fuel in the basin or was it jet fuel sold there?

G. Cass

The emissions shown are mostly from diesels. I grouped the two together in the slides which accompanied my presentation because the fuels are similar. The aircraft emission estimate, I think, is one of the weakest in our inventory. There are two competitive procedures for estimating TSP emissions from jet aircraft and the Los Angeles district procedure apparently disagrees with EPA's procedure. We used the EPA procedure which involves a landing and takeoff cycle up to around 3,000 feet in altitude. Our fuel burning estimates for aircraft reflect only landing and take-off cycles and not the cross country in-flight use of jet fuel.

J. Ogren

Do you have that emission factor in terms of grams per kilogram fuel?

G. Cass

We calculated emissions on the basis of mass per landing and take-off cycle. We could probably take the fuel use estimate and divide it into the total mass emitted to calculate emissions per gram of fuel. The EPA calculation procedure is based on the number of engines per airplane and number of landings and take-offs by aircraft of different sizes. Aircraft provide the only case where we did not use fuel burning data directly. I think that the jet engine emission estimation procedures could be improved a great deal. I was not in a position to stand behind a jet engine and sample for the purposes of this paper. But, I wish that somebody would.

---

* G. M. Hidy et al., "Characterization of Aerosols in California (ACHEX)," "Science Center, Rockwell International, revision of April 1975
**D. Grosjean, S. K. Friedlander, J. Air Pollut. Control Assoc., Vol. 25, (1975), p. 1038.

W. Fagley (Chrysler Corporation)

Your cigarette data showed 1083 kilograms of elemental carbon and 1692 kg for total carbon which is a ratio of about 1.6. You have about a 3 to 1 ratio for other things. Do you have any explanation for that?

G. Cass

Total mass emissions from cigarettes came from the KVB inventory. I did not calculate that. The partition between organic and elemental carbon was based on data for plant material smoldering within Jim Huntzicker's apparatus. I think that is a very rough approximation of a cigarette. The outcome for cigarettes is not important to the overall totals. But, I think it should be recognized that 10 million people smoke a lot of cigarettes in a day and that the resulting emissions should be estimated. I want to stress that in the case of many of the minor source classes in the inventory, the data are weak but they should be sufficient to gauge the neighborhood of the total mass emitted and place an upper limit on the amount of carbon that could be present.

[*postscript: The emission estimates for cigarettes mentioned in this question have since been revised and they appear correctly in the paper.*]

CONTEMPORARY PARTICULATE CARBON*

L. A. CURRIE

National Bureau of Standards
Washington, D.C.

ABSTRACT

Advances in natural radiocarbon measurement techniques have made it feasible, for the first time, to assess the contribution of biogenic (contemporary) carbonaceous sources to individual chemical fractions in milligram quantities of atmospheric particles. Isotopic measurements for source reconciliation are doubly important when dealing with pure species, such as methane, carbon monoxide or elemental carbon, because they represent the only compositional information obtainable. Elemental carbon is of special interest in this regard because of changing energy patterns associated with both contemporary (wood-burning) and fossil (diesel fuel and unleaded gasoline) carbon. Following a review of the assumptions underlying the use of radiocarbon as a biogenic tracer and the status of minicounter and accelerator techniques for the assay of milligram and microgram samples, a survey is presented of recent observations on urban and rural carbonaceous particles. Results for these particles, which have been fractionated according to size or volatility, have exhibited the full range from fossil to biogenic source dominance.

INTRODUCTION

One of the foremost questions involved in understanding the atmospheric cycle of particulate carbon is the quantitative assessment of perturbations introduced through human activities. Although natural emissions of carbonaceous gases and particles clearly predominate on a global scale [1-3], man's use of coal, petroleum, natural gas and wood as fuel is known to produce severe consequences from particulate pollution ranging from local (especially urban) episodes to long-range effects [3-6]. A major objective of our research program is therefore the deconvolution of

---

References pp. 259-260.

sources of atmospheric carbon through the analysis of ambient samples, with special emphasis on fine particle ($\lesssim 2$ μm) elemental carbon because of its effects on health, climate and visibility [7-9].

Naturally-occurring isotopes of carbon — ^{12}C, ^{13}C, ^{14}C (radiocarbon) — provide important tools for the deconvolution process, because the isotopic ratios of ^{13}C/^{12}C and ^{14}C/^{12}C differ significantly among different classes of potential source materials. As indicated in Table 1, for example, measurement of δ^{13}C (the deviation of the ^{13}C/^{12}C ratio from that of the PDB standard [10]) allows one to distinguish between carbon originating from species such as carbonate minerals, marine plants, petroleum and natural gas, and carbon from C_3 and C_4 terrestrial vegetation [11, 12]. Of greater importance, radiocarbon serves as a unique discriminator be-

TABLE 1
Carbon Isotope Concentrations Characterizing
Different Sources (a), (b)

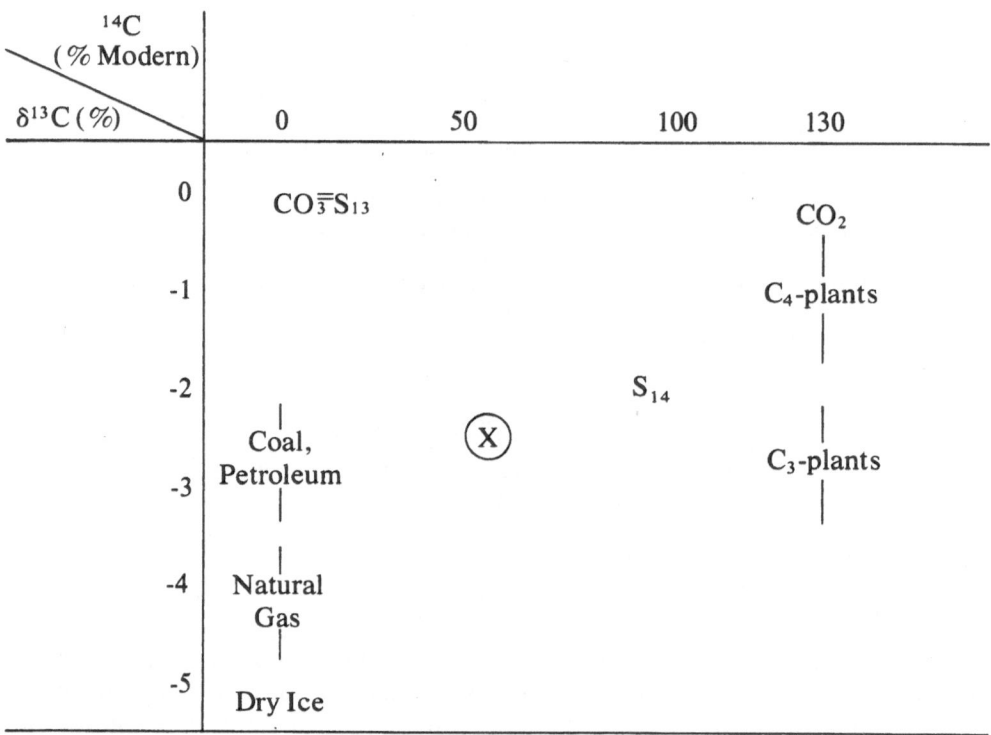

a. ^{14}C/^{12}C relative to the radiocarbon standard, S_{14} (100% $\equiv$ 0.95 x NBS Oxalic Acid), [23]; and ^{13}C/^{12}C deviations (%) from the PDB standard, S_{13}, [10]. Note that the activity of contemporary ($\gtrsim$1979) C-14 is about 30% higher than "modern standard" (pre-nuclear era) C-14.

b. C_4, C_3 plants represent the two major photosynthetic classes, [12]. Point X represents a Los Angeles aerosol sample whose isotopic composition is consistent with 54% temperate (C_3) vegetative emissions and 46% fossil fuel [30].

tween contemporary biogenic (e.g., vegetation) and fossil organic matter [13]. The measurement of radiocarbon thus serves as a very important complement to indirect methods (chemical element balance, factor analysis, dispersion modeling) for inferring sources of ambient carbonaceous particles. Its special characteristics are that it is a direct and unique indicator of biogenic carbon, and that, along with $\delta^{13}C$, it is robust with respect to changes during transport. Also, for single chemical substances such as elemental carbon, isotopic measurements provide the only compositional information attainable.

Since the preceding conference on atmospheric particulate carbon, at which we reported the first radiocarbon results for milligram quantities of carbonaceous particles [14], there have been some important societal and technical developments. In the first category, the need for a reliable, direct tracer for carbonaceous emissions has been magnified because of increases in wood-burning and in the use of diesel fuel and unleaded gasoline. (The relative decrease in the mix of pre-1975 automobiles is, in effect, causing the "loss of a tracer," i.e., lead, upon which the chemical element balance method heavily depended.) Technical progress includes improved control and demonstrated reliability of our mini-gas proportional radiocarbon counting system, and most importantly, the introduction of the accelerator method of atom counting [15, 16], which is likely to extend the range of sample sizes downward by another three orders of magnitude, to about 10 μg of carbon [17]. This development is of special consequence for elemental carbon, because it appears to be the optimal for carbon to be placed in the tandem accelerator sputter ion source [18], and the (sampling) blank for elemental carbon appears to be far lower than that for organic carbon [19].

Following a brief review of recent developments in small sample radiocarbon measurement and critical assumptions involved in interpreting the resulting data, we shall present a summary of several recent applications to urban and rural ambient aerosols which have helped to establish the validity of the method and which have provided a direct measure of the importance of biogenic sources on individual chemical and physical (size) fractions of carbonaceous particulate matter.

BIOGENIC CARBON TRACING — METHODS AND CAUTIONS

The phenonemon which makes possible the use of radiocarbon as a direct tracer for sources of biogenic carbon is the same as that upon which radiocarbon dating is based [20]. That is, all living matter is in (approximate) isotopic equilibrium with atmospheric CO_2 which continually acquires radiocarbon following cosmic ray neutron capture on ^{14}N in the atmosphere. Unlike radiocarbon dating, in which ^{14}C concentrations are related to age, biogenic carbon tracing is based on isotope dilution — the ^{14}C concentration in an ambient sample reflecting directly the mixing ratio of contemporary ("living") carbonaceous material and fossil material. Carbonaceous substances of intermediate age — e.g., 500 to 20,000 years — are largely absent from atmospheric particulate matter.

Progress in the measurement of natural radiocarbon is indicated in Table 2. Following its discovery by W. F. Libby in the late forties, radiocarbon dating was recognized with a Nobel prize in 1960, and its use has continued to increase since

TABLE 2

Radiocarbon Measurement — Evolution, Sensitivity

(See text for references and explanation of symbols)

	Mass (form)	Precision	Notes
Radiocarbon Dating (1960); Small Sample Decay Counting (1976):	10 g (gas, liquid)	0.5% ($\approx$40 yr)	Minimal Chemistry
Liquid Scintillation	100 mg (C_6H_6)	2%	Individual Chemical, Size Fractions
Gas Proportional	10 mg (CO_2)*	6%	
Accelerator Atom Counting (1977)	10 μg (C)**	3%	Exploratory; 1st Dedicated Machine (1981)

Number of Counts =

$$AE\Delta t = (0.18\ min^{-1})\ (0.8)\ (2880\ min) = 415^*$$

$$AE'\tau = (95\ yr^{-1})\ (0.001)\ (8270\ yr) = 786^{**}$$

that time. The method, as generally practiced, is characterized by excellent precision, but required sample sizes are extremely large (by air pollution standards) and chemical or physical separations are rare (with the exception of simple "pretreatment", and the extraction of cellulose for tree ring dating). Rapid progress in measurement techniques for small atmospheric samples was marked by (a) the introduction of mini-counter liquid scintillation and gas proportional counting in 1976 [21] for 100 mg and 10 mg size samples, respectively, and (b) the development in 1977 of direct ^{14}C atom (rather than β-decay) counting by means of high energy nuclear accelerators [15, 16]. Although the first dedicated radiocarbon accelerator has not yet been installed, it has been demonstrated that natural radiocarbon may be measured in samples containing only 10-20 μg of carbon [17].

During the past three years, the mini-gas proportional counting method has been applied to a number of interesting problems, ranging from gas and particulate matter in the atmosphere to organic species in sediment [22], and it may be considered validated and almost routine in operation. Atom counting (with accelerators) is still very much in an exploratory stage, but its enormous sensitivity (with consequential sample size reduction) *will* lead to very broad applications. As shown at the bottom of Table 2, decay counting of 10 mg of contemporary carbon yields a Poisson imprecision of about 6 percent (415 sample counts, S/B $\approx$ 2), whereas atom counting of 10 μg yields a value of 3 to 4 percent (786 counts, S/B $\geqslant$ 1). The latter method requires an overall measurement time of hours rather than days. Many other factors which affect the time, cost and reliability of the alternative methods are beyond the scope of this paper. What the two methods share, however, is the exciting opportunity of determining the isotopic composition of individual chemical species and size fractions in atmospheric samples selected according to time and location.

A number of assumptions and cautions involved in the measurement and interpretation of radiocarbon data are highlighted in Table 3. With respect to the (two-source, isotope dilution) model, the only significant assumption-limitation is connected with the ^{14}C variations in the biosphere. The three principal causes of these variations are (a) solar modulation of the cosmic rays, (b) dilution of atmospheric radiocarbon with fossil fuel generated CO_2 (Suess effect) and (c) artificial radiocarbon production (primarily associated with atmospheric nuclear tests) [23]. As shown in Fig. 1, the perturbation due to nuclear testing predominates, causing about a 30 percent radiocarbon excess in currently-living matter. Because of this, radiocarbon concentrations relative to the "modern standard" (0.95 x NBS Oxalic Acid SRM # 4990-B) must be divided by 1.3 to estimate the fraction of contemporary (biogenic) carbon. When the biogenic source spans a number of years, the situation is less straightforward. If a 60 year-old log is burned in a fireplace, for example, its average radiocarbon contribution is enhanced by an additional 9 percent due to the wood formed during the period of major nuclear testing [24]. (One must integrate the input function, Fig. 1, over the life of the vegetation taking into account also the rather small correction for radioactive decay.) Unlike most chemical species, however, radiocarbon directly tracks the biospheric source emissions, and the isotope ratio is extremely resistant to changes during transport, from source to receptor site.

References pp. 259-260.

TABLE 3

Cautions and Assumptions

Model Validity
  Radiocarbon:
    * two-source assumption (soil-carbon, nuclear energy, gasohol)
    * ^{14}C-variations (solar, bomb, Suess effects)
    * vegetation age (firewood)
  Receptor:
    * all sources identified
    * accurate, reproducible profiles at receptor site
Sample and Sample Preparation Limitations
    * the blank
    * large particle contamination (spores, insect parts, . . .)
    * chemical stability, reliable separations (recovery, charring)
Counting Difficulties
    * isotopic heterogeneity (non-quantitative recovery)
    * background fluctuations (sunspots, barometric pressure)
    * radon contamination
    * electronegative contamination (efficiency)
    * spurious pulses

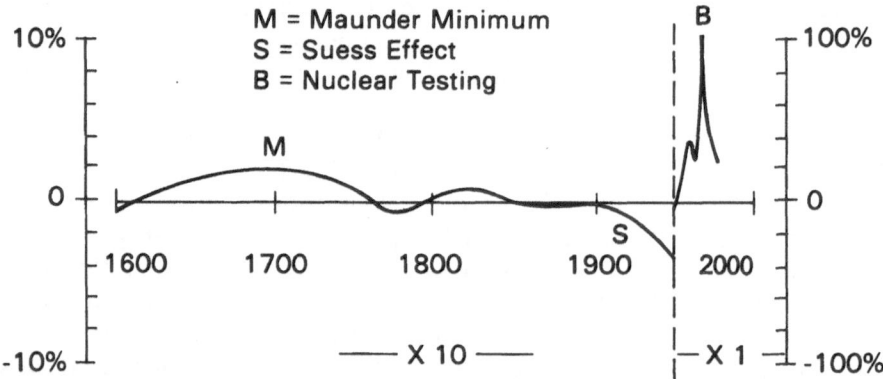

Fig. 1. Natural Radiocarbon Variations. Percentage deviations from steady-state radiocarbon concentration in the biosphere. Scale is expanded by a factor of ten before 1950. Major excursions marked "M", "S", and "B" correspond to: the Maunder minimum in solar activity (and the Little Ice Age), the Suess Effect (fossil CO_2 dilution) and the Bomb effect (atmospheric nuclear testing). See reference [23] for further discussion.

Most of the subtle, yet critical sources of error in the low-level counting of radiocarbon have been discussed previously [14, 21]. To assure reliability we now routinely monitor six parameters during counting periods; and our measurements generally encompass five (hierarchical) levels of quality control.

New information concerning two other factors, developed within the past year, bears special mention. These are the blank and isotopic heterogeneity. Until recently, sufficient material and information concerning the blank was lacking. Blank quartz-fiber filters obtained from R. K. Stevens of the U.S. EPA, however, gave us sufficient filter-blank carbon to carry out a radiocarbon determination. The result, that the blank was indistinguishable from contemporary carbon, must be considered when measuring samples having less than a few milligrams of carbon, as will be noted in the following section.

Isotopic heterogeneity, variation of the isotopic ratio among different chemical and/or size fractions of the ambient sample, is exactly analogous to the problem of non-contemporaneity in radiocarbon dating [25]. That is, although the overall radiocarbon concentration may be meaningful in terms of the average biogenic source contribution (or age), significant information loss takes place unless measurements are made on the individual fractions. As will be discussed subsequently, we have observed particle size and volatility-related isotopic heterogeneity in carbonaceous particulate samples; therefore, the relative mix of biogenic and fossil source contributions must differ among the different fractions, and conclusions based upon just the average isotopic composition may be relatively weak or ambiguous. A corollary to these observations is that quantitative chemical recoveries (from particles to pure CO_2 counting gas) are mandatory; otherwise even the estimated average isotopic composition may be biased due to selective chemical losses.

EXPERIMENTAL RESULTS — CONTEMPORARY CARBON IN URBAN AND RURAL AEROSOLS

The Isotopic Dimension; Chemical and Physical Selectivity — Our initial applications of mini-radiocarbon measurements dealt with total ambient aerosol samples and the carbonaceous material in urban and rural particles was examined in a simple dichotomous manner to determine whether it was primarily "anthropogenic" (fossil) or "natural" (biogenic, vegetative) in origin. The results were not surprising [14]. More recently, it has become important to apply the technique to samples from specific episodes which implied serious health consequences (especially involving fine carbonaceous particles) or where significant contemporary carbon emissions, from wood burning or vegetative emissions, were presumed present. Under these circumstances it has become necessary not only to assay ^{14}C in the total suspended particulate material, but also to examine individual chemical or size fractions. In addition to this "serial selectivity," "parallel" measurements of other chemical or physical characterisitcs of the samples in question have served two important purposes:

a. validation of the isotopic conclusions (or the converse, validation of the Chemical Element Balance conclusions), and

b. complementarity needed for extracting more source information than either approach alone could provide.

The first or quality control application has been and will continue to be of major importance in the development and validation of new methods; its value rests especially on the uniqueness of the radiocarbon tracer, and on the complete independence (with respect to underlying assumptions and methods) of the alternative

References pp. 259-260.

(isotopic, chemical) techniques. The second aspect, which implies complete integration of the isotopic dimensions with the chemical characterization, provides the most powerful use of the isotopic data and should lead to one of the more reliable and precise approaches to future efforts in receptor modeling [26, 27]. The summary in Table 4 of one rural and three urban ambient aerosol investigations, all completed within the last two years, illustrates both of these aspects.

TABLE 4
Recent Particulate Studies

Locale	Carbon Sources	Percent Contemporary	Supporting Data	Reference
Portland, OR	Field, slash and wood burning	57. to 107.	Known impact, particle size, inorganic composition (Chemical Mass Balance)	24
Los Angeles	Urban fossil fuel and vegetative emissions	~54.	Particle size, dispersion data, $\delta^{13}C$, organic composition	30
Rural Utah	Vegetative emissions and oil shale dust	~64.	Pyrolysis — mass spectrometry (Chemical Pattern Recognition)	31
Denver	Urban fossil fuel and wood burning	10. to 55.	Organic/elemental fractions, inorganic composition, emission data and modeling, coarse/fine fractions.	28

It is noteworthy that the percent of contemporary carbon observed in these investigations covered the full range from 10 ± 3 percent to 107 ± 15 percent. The extremes represented periods of severe particulate pollution, but from very different sources! A general observation from these studies is that contemporary sources do contribute significant amounts of particulate carbon to the ambient aerosol from vegetative emissions during the summer (Los Angeles, rural Utah), and from wood-burning in the winter (Portland, Denver). Continuing severe pollution episodes in Los Angeles and Denver ("Denver brown cloud") were the geneses for two of the investigations, and the remaining two were connected with special situations involving contemporary (Portland, field and slash burning) and fossil (Utah, oil shale tracts) carbon. Of particular importance is the nature of the supporting data. As shown in the fourth column of Table 4, serial or parallel information on particle size, emission/dispersion modeling, and inorganic or organic composition were intimately involved in the overall efforts at source reconciliation.

Patterns of Results and Conclusions — The isotopic data served as a truly independent dimension in the urban studies, in that they were more or less formally combined with chemical, meteorological and emission data to infer sources of the carbonaceous particles. Emission, chemical and radiocarbon data collected for Denver samples, for example, were consistent with the conclusion (for the sampling period in the winter of 1978) that wood-burning was responsible for about one-third of the carbonaceous aerosol and two-fifths of the visibility-reducing elemental carbon [28, 29]. Qualitative support for the 54 percent contemporary carbon in the

Los Angeles study was given by observations of significant amounts of primary or secondary organic species characteristic of fossil sources (a large hump of branched cyclic compounds in the GC trace of the hydrocarbon fraction indicated the presence of polycyclic aromatic hydrocarbons, and dicarboxylic acids) on the one hand, and species reflecting vegetative emissions ($\delta^{13}C$, odd carbon-numbered high molecular weight n-alkanes, predominantly even carbon-numbered fatty acids, 2-ketones, camphor) on the other. It should be noted that individual size fractions were examined in the Los Angeles study, and that wind patterns at the location of the receptor site (City of Hope hospital at Duarte in the foothills of the San Gabriel Mountains) favored impacts from both intense urban smog and forest emissions (carried by night-time winds) [30].

The investigations in Portland and Utah provided important opportunities for semi-quantitative tests of consistency between isotopic and chemical data. Table 5 lists results (and Poisson standard deviations) for ^{14}C-inferred contemporary carbon derived from vegetative burning in Portland and vegetative emissions in Utah. Paralleling the radiocarbon results are independent estimates based on inorganic chemical element balance (CEB) in Portland [24] and organic chemical "pattern recognition" (CPR) in Utah [31]. The general pattern of results is quite satisfactory, but two special comments are in order. That is, the CEB estimates (of biogenic carbon) were rather indirect, involving several assumptions, so that the assignment of uncertainties was extremely difficult. Overall consistency between the ^{14}C and CEB estimates indicates an average CEB standard deviation of about 9 percent. The CPR uncertainties reflected measurement error only; sampling and model errors would, of course, increase the uncertainty. It is interesting to note that K/Fe ratios in Portland were qualitatively consistent with the radiocarbon data, but were subject to large fluctuations.

TABLE 5
Quantitative Consistency
(Percent contemporary carbon)

Portland[a]		Utah[b]	
$\underline{^{14}C}$	CEB Emission Ratios (inorganic)	$\underline{^{14}C}$	CPR, Py-MS (organic)
107 ± 15 (field)	99	(sage)	64 ± 2
111 ± 19 (slash)	85	64 ± 4 (pine)	11 ± 2
83 ± 16 (slash)	78	(juniper)	—
79 ± 12 (slash)	97	fossil (shale)	24 ± 2
57 ± 12 (wood)	79		

a. *Fine particle fraction (≤2.5 μm), corrected for wood age (ref. [24]). (CEB = Chemical Element Balance).*

b. *Ref. [31]. Chemical Pattern Recognition (CPR) figures represent the derived percent contributions from the indicated sources.*

Finally, let us briefly examine patterns associated with the Utah and Denver results. Figs. 2 and 3 show the pyrolysis-mass spectra for the fossil and three vegetative components contributing to the ambient sample in Utah, and the resulting separation of these components as shown in non-linear two-space using pattern recognition techniques. An overview of the Denver study is given in Table 6, and a three dimensional projection of the pattern of selected results (TSP) appears in Fig. 4. This figure shows contemporary carbon as a function of three variables: degree of pollution, percent of volatile ("organic") carbon, and chemical fraction (total carbon, "elemental" carbon)*. General conclusions which may be drawn from this overall pattern of results are:

 a. that the radiocarbon blank must not be ignored,
 b. contemporary carbon, comprising up to about 40 percent of the carbonaceous material, is a significant component — presumably due to residential wood burning in the winter-time Denver "brown cloud", and
 c. the fossil component is more pronounced: in the elemental carbon fraction, with increasing pollution, and with decreasing volatility.

Three of the major assumptions/limitations in the isotopic inferences mentioned in the previous section deserve special comment with respect to the Denver results. First, the average age of the wood burned was not taken into account in deriving the results as shown in Fig. 4. The maximum likely effect would be a relative shift of about 9 percent toward fossil carbon.

Second, the effect of the blank, which corresponded to about 0.7 mg-C, has been taken into account. Fortunately, the corrections were small compared to the Poisson counting errors, since the correction involves assumptions concerning the chemical and isotopic compositions and variability in the total mass of the carbon blank [19]. For the (Fig. 4) samples having the smallest and largest blank contributions — sample "P" with 23.9 mg-C and sample "U" with 9.8 mg-C — the adjustment to the percent of contemporary carbon was just -1.5 percent and -3.3 percent, respectively.

The third limitation relates to chemical yield. If the recovery is not quantitative, isotopic heterogeneity can lead to bias for the estimated *average* percent contemporary carbon; and low yields greatly increase the error-propagation uncertainties when the volatile carbon component is deduced from the two (total, elemental) measured components. For example, the pair of samples, "WD", had estimated recoveries of 87 percent (total) and 101 percent (non-volatile), and contemporary carbon compositions of 26 $\pm$ 5 percent and 14 $\pm$ 5 percent, respectively. The derived contemporary carbon estimate for the volatile component was 56 $\pm$ 26 percent. Obviously, quantitative recovery and direct measurement of ^{14}C in the volatile component are desirable for future studies.

---

*The terms organic and elemental are used here in an operational sense. The "elemental" fraction, defined as being non-volatile, may include high molecular weight compounds as well as amorphous or graphitic carbon [28, 29].

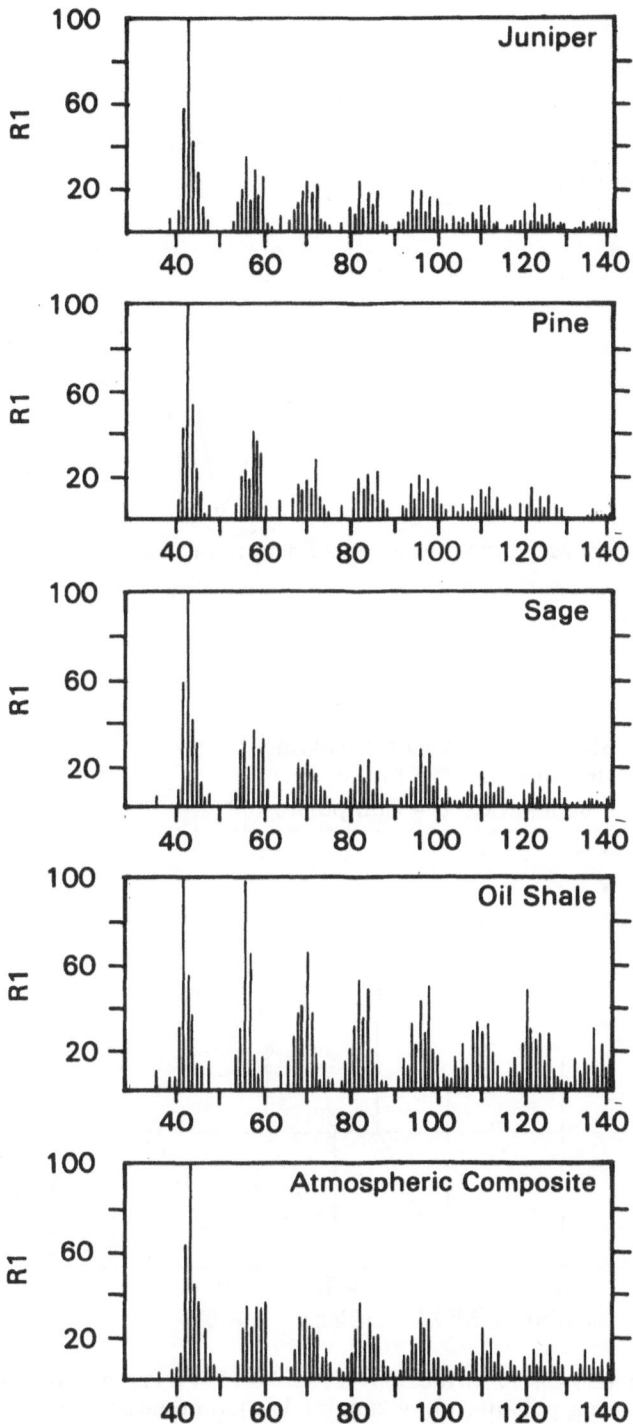

Fig. 2. Summary of Pyrolysis-Mass Spectra (Py-MS) for Source Samples and Ambient Atmospheric Particulate Sample Collected in Utah Oil Shale Tract [31].

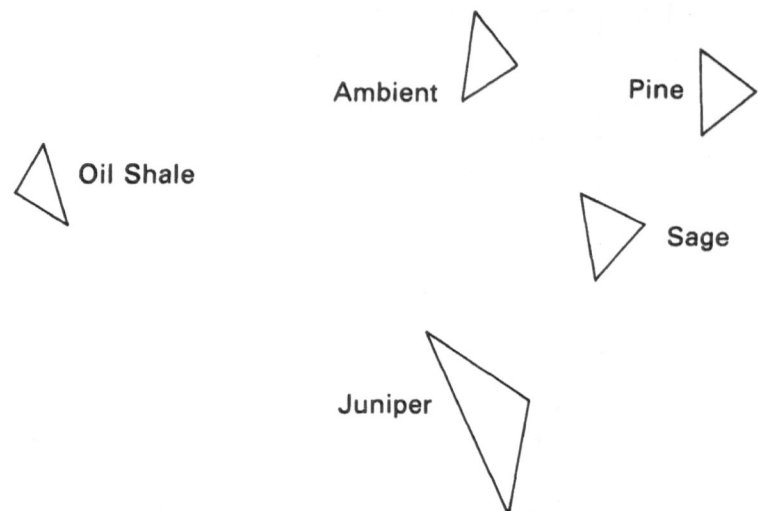

Fig. 3. Non-Linear Map of Pyrolysis-Mass Spectra Data Summarized in Fig. 2 [31].

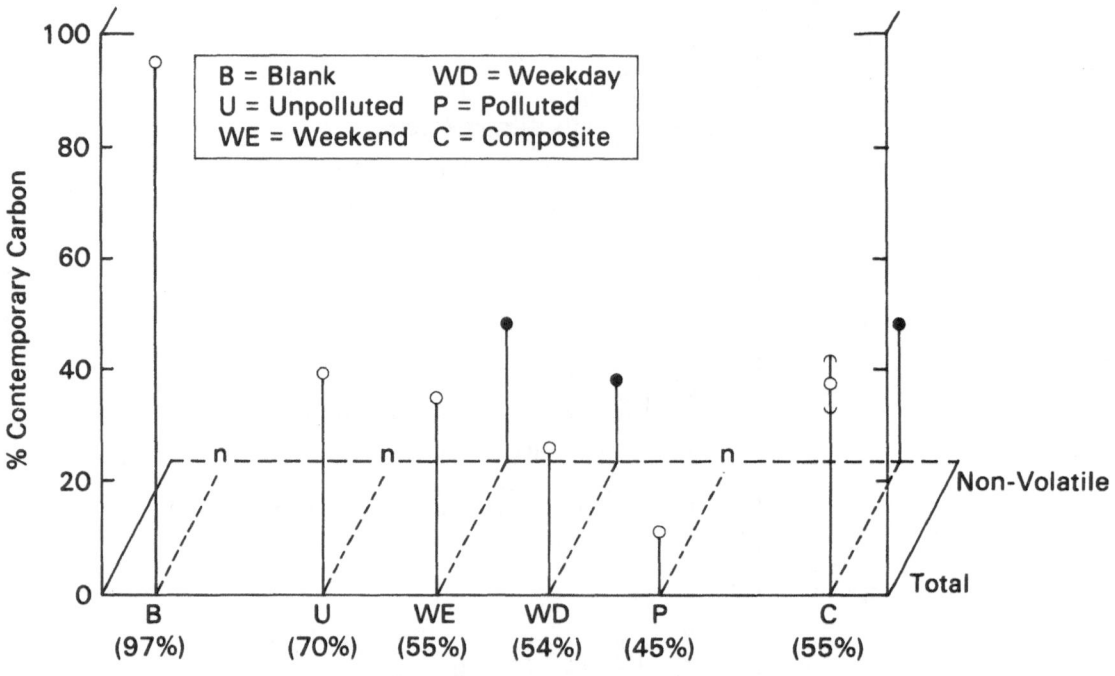

Fig. 4. Contemporary Carbon in the Denver Aerosol (TSP). Percent contemporary is given as function of increasing pollution (for 4 samples, U-P), for total and "non-volatile" carbon. An independent, composite Denver sample (C) also is shown, as well as a quartz filter blank (non-Denver) supplied by R. K. Stevens. The Poisson error is indicated for sample-C, and percent of volatile carbon is shown at the bottom of the figure.

TABLE 6
Denver Samples (Overview)

Conditions:
  Sunny, cloudy, -20 °C to +10 °C
  Clear, weekend, weekday, polluted
  [7. to 43. μg-C/m³]
Fractions:
  Total, non-volatile, fine
Percent Volatile:
  25. to 70.
Percent Recovery (CO_2):
  70 to 108
Mass (carbon):
  5.6 mg to 23.9 mg
  [Blank = 0.7 mg; ~8. μg/cm²]
Percent Contemporary:
  10 ± 3 to 55 ± 13

SUMMARY

Radiocarbon has been demonstrated to be a unique and robust tracer for contemporary carbon in atmospheric particles. Utilization of this isotope, together with selective sampling plus measurement of $\delta^{13}C$ and inorganic and organic chemical composition has indicated the importance of a number of specific vegetative sources contributing to urban and rural ambient particles. Semi-quantitative confirmation of the radiocarbon conclusions has been derived from chemical mass balance (inorganic species, emission ratios) and chemical pattern recognition (organic species).

Rapid progress has occurred in the measurement of radiocarbon, in that (a) small (5-10 mg-C) samples may be measured using mini-gas proportional counters with a method which has been developed with a high degree of internal control and is nearly routine in applicability, and (b) the eventual capability for micro (10-20 μg-C) samples has been demonstrated using direct atom counting with a tandem accelerator.

Assay of contemporary carbon, which has been carried out for the first time on a complete blank and on different *chemical* fractions of small samples of atmospheric particles, has indicated isotopic heterogeneity — *viz.*, the organic and elemental carbon fractions of ambient particles have been shown to depend to differing extents on biogenic carbon sources. These observations, obtained from measurements of "Denver Brown Cloud" particles, are analogous to the contemporary carbon (isotopic) heterogeneity which we observed in different *size* fractions of particles collected both in Denver and in Portland, Oregon. For the particles sampled in these studies, the contemporary carbon fraction — presumably from vegetative burning — tended to be concentrated in the organic carbon fraction and in the fine particle fraction. The overall contemporary carbon fraction depended greatly on locale and conditions, ranging from 10 ± 3 percent (Denver) to 107 ± 15 percent (Portland).

Elemental carbon, already important because of its impact on climate, health and visibility, is of special interest for radiocarbon measurements. This is because the composition of such a single chemical species can only be characterized using isotopes, because it has a very low blank, and because it is a chemically robust tracer for combustion sources. Finally, elemental carbon represents the ideal chemical form for the technique having greatest sensitivity, i.e., accelerator atom counting.

As isotopic measurements of greater sensitivity (smaller sample size) or improved precision become practicable, it will be increasingly important to pay attention to three critical factors: the blank, recovery (and isotopic heterogeneity), and assumed vegetative age. Uncertainties connected with the blank are negligible for elemental carbon and slightly smaller than current Poisson counting errors for total carbon. Isotopic heterogeneity means that low chemical yields will result in biased contemporary carbon estimates, and important information loss will occur if individual chemical and size fractions are not separately analyzed. Vegetation age, if inadequately corrected for, can introduce uncertainties up to about 10 percent due to the injection of nuclear bomb radiocarbon during the last two decades.

The next steps to be taken in the application of carbon isotopes to the investigation of the life cycle of carbonaceous particles will include sampling at remote sites and advances in measurement methodology. The latter will include isotopic enrichment and multiple counter arrays plus exploration of precision and sensitivity improvements using a dedicated radiocarbon accelerator. Also, further utilization of chemical and size selectivity are planned, together with full integration of the isotopic and chemical data in the source deconvolution process. For these purposes, efforts will be made to further characterize a particulate standard reference material, and to provide a standard data set for receptor modeling.

ACKNOWLEDGMENT

The urban and rural studies reviewed in this manuscript, and to be published fully elsewhere, took place in cooperation with the following scientists: Portland — J.A. Cooper (Oregon Graduate Center), Los Angeles — I. R. Kaplan (Global Geochemistry, Inc.), Utah — K. J. Voorhees (Colorado School of Mines), and Denver — R. J. Countess, G. T. Wolff and D. P. Stroup (General Motors Research Laboratory). R. K. Stevens (USEPA) supplied the blank ("B", Fig. 4). G. A. Klouda and R. E. Continetti of NBS were largely responsible for the experimental radiocarbon measurements. Grateful acknowledgment is expressed to all of these colleagues. Partial support for this research was provided by the Office of Environmental Measurements, U.S. National Bureau of Standards, and the Energy-Environment Program (EPA-IAG-D6-E684), U.S. Environmental Protection Agency.

REFERENCES

1. *D. A. Covert, R. J. Charlson, R. Rasmussen and H. Harrison, Review of Geophysics and Space Physics, Vol. 13 (1975), p. 765.*
2. *"Aerosols: anthropogenic and natural sources and transport." Annals N.Y. Acad. Sci., Vol. 338 (1980).*
3. *B. Bolin, Ann. Rev. Energy, Vol. 2 (1977), p. 197.*
4. *J. A. Cooper, J. G. Watson and J. J. Huntzicker, "Summary of the Portland aerosol characterization study," Paper No. 79-24.4, 72nd Air Pollution Control Association Meeting, Cincinnati, Ohio, June, 1979.*
5. *W. J. Courtney, J. W. Tesch, G. M. Russwurm, R. K. Stevens, T. G. Dzubay and C. W. Lewis, "Characterization of the Denver aerosol between December, 1978 and December 1979," Paper No. 80.58-1, 73rd Air Pollution Control Association Meeting, Montreal, Canada, June, 1980.*
6. *K. A. Rahn, C. Brosset, B. Ottar and E. M. Patterson, These Proceedings.*
7. *National Research Council, Controlling Airborne Particles, National Academy of Sciences: Washington, D.C., 1980.*
8. *Geophysics Study Committee, Energy and Climate, NRC Geophysics Research Board, National Academy of Sciences, Washington, D.C., 1977.*
9. *T. Novakov, Ed., Proceedings Conference on Carbonaceous Particles in the Atmosphere, LBL-9037 (Lawrence Berkeley Laboratory) 1978.*
10. *H. Craig, Geochim. Cosmochim. Acta, Vol. 3 (1953), p. 53.*
11. *J. A. Calder and P. L. Parker, Environ. Sci. Technol., Vol. 7 (1968), p. 535.*
12. *J. H. Troughton, "Carbon Isotope Fractionation by Plants," Proceedings of the Eighth International Radiocarbon Dating Conference, Lower Hutt, New Zealand, Vol. 2 (1972), p. 421.*
13. *L. A. Currie and R. B. Murphy, "Origin and residence times of atmospheric pollutants: Application of ^{14}C," in Methods and Standards for Environmental Measurement, W. H. Kirchoff, Ed., NBS Spec. Pub. 464, National Bureau of Standards, Washington, D.C., Nov., (1977), p. 439.*
14. *L. A. Currie, S. M. Kunen, K. J. Voorhees, R. B. Murphy and W. F. Koch, "Analysis of Carbonaceous Particulates and Characterization of Their Sources by Low-Level Radiocarbon Counting and Pyrolysis/Gas Chromatography/Mass Spectrometry," Conference on Carbonaceous Particles in the Atmosphere, University of California, Berkeley, 1978.*
15. *R. A. Muller, Science, Vol. 196 (1977), p. 489.*
16. *H. Gove, Ed., Proceedings of the First Conference on Radiocarbon Dating with Accelerators, University of Rochester, 1978.*
17. *L. A. Currie, G. A. Klouda, D. Elmore, R. Ferraro and H. Gove, "Accelerator Mass Spectrometry and Electromagnetic Isotope Separation for the Determination of Natural Radiocarbon at the Microgram Level," (in preparation).*
18. *M. Rubin, "Sample Preparation for Van de Graaff Accelerator Dating," in L. A. Currie, Ed., Nuclear and Chemical Dating Techniques, American Chemical Society Symposium Series, 1981.*
19. *R. K. Stevens, W. A. McClenny, T. G. Dzubay, M. A. Mason and W. J. Courtney, These Proceedings.*
20. *W. F. Libby, Radiocarbon Dating, University of Chicago Press: Chicago, 1952.*
21. *L. A. Currie, J. Noakes and D. Breiter, "Measurement of Small Radiocarbon Samples: Power of Alternative Methods for Tracing Atmospheric Hydrocarbons," Ninth International Radiocarbon Conference, University of California, Los Angeles and San Diego, 1976.*
22. *J. Swanson, A. Fairhall and L. A. Currie, "Carbon Isotope Analysis of Sedimentary Polycyclic Aromatic Hydrocarbons," (in preparation).*
23. *I. U. Olsson, Ed., Radiocarbon Variations and Absolute Chronology, Proceedings of the 12th Nobel Symposium held at the Institute of Physics at Uppsala University, Wiley-*

Interscience, New York, 1970; and P. E. Damon, J. C. Lerman and A. Long, "Temporal Fluctuations of Atmospheric C-14: Causal Factors and Implications," Annual Review of Earth and Planetary Science, Vol. 6 (1978), p. 457.

24. J. A. Cooper, L. A. Currie and G. A. Klouda, "Assessment of Contemporary Carbon Combustion Source Contributions to Urban Air Particulate Levels Using C-14 Measurements," (to be published in Env. Sci. & Tech.).
25. H. Schultz, L. A. Currie, F. R. Matson and W. W. Miller, Radiocarbon, Vol. 5 (1963), p. 342.
26. J. G. Watson, (Ed.), Proceedings Receptor Modeling Workshop, Quail Roost, N.C., Feb., 1980.
27. G. E. Gordon, Env. Sci. & Tech., Vol. 14 (1980), p. 792.
28. G. T. Wolff, R. J. Countess, P. J. Groblicki, M. A. Ferman, S. H. Cadle and J. L. Muhlbaier, Atmos. Environ. Vol. 15 (1981), p. 2485.
29. S. L. Heisler, R. C. Henry, J. G. Watson and G. M. Hidy, "The 1978 Denver Winter Haze Study," Motor Vehicle Manufacturers Association, Detroit, Michigan, 1980.
30. I. R. Kaplan, L. A. Currie and G. A. Klouda, "Isotopic and Chemical Tracers for Organic Pollutants in the Southern California Air Basin," (to be published).
31. K. J. Voorhees, S. M. Kunen, S. L. Durfee, L. A. Currie and G. A. Klouda, "The Determination of Source Contribution of Organic Matter in Atmospheric Particulates by Pyrolysis/Mass Spectrometry and ^{14}C Analysis," (to be published).

DISCUSSION

E. Macias, *(Washington University)*

Could you comment on how you removed the blank which is 100% contemporary carbon? Is that a problem?

L. Currie

It is an important problem and will become more important as one deals with smaller amounts of carbon and starts using the accelerator. The blank samples I mentioned were obtained from the Soviet Union by Bob Stevens. It is the only collection of filter material where I was able to get enough blank carbon to measure by decay counting, and it amounted to about 7.5 milligrams total. The correction, itself, is based on the relative amounts of carbon in the gross sample and the blank, and the observed isotopic compositions. Second-order corrections take into account sample isotopic heterogeneity and chemical yield.

E. Macias

But are these results corrected for the amount of contemporary carbon in the blank?

L. Currie

Yes, they are. For the samples at hand where the blank carbon did not exceed 7% of the total, the maximum correction was -3.3 percent contemporary carbon.

DISTINGUISHING CARBON AEROSOLS BY MICROSCOPY

R. G. DRAFTZ

IIT Research Institute
Chicago, Illinois

INTRODUCTION

A number of atmospheric aerosols such as quartz, calcite, feldspars, pollens, rubber tire fragments, sulfates, and nitrates hold rank as ubiquitous aerosols. The attention given to elemental carbon at a 1978 conference [1] and this symposium finally elevates elemental carbon to its rightful place among the ubiquitous aerosols. The widespread presence of these aerosols necessitates that they be included in the analytical protocol for any comprehensive air pollution study, especially those related to total suspended particulates non-attainment and visibility reduction.

With any recently recognized ubiquitous aerosol, there is a trial and error period for evaluating analytical methods that will hopefully provide the sensitivity and specificity needed to evaluate the environmental impact of this aerosol specie. The specificity requirement for atmospheric aerosols has two aspects:

a. distinction of sample components, and
b. distinction of sources. It appears that a number of methods exist for analyzing carbon that offer adequate sensitivity, but lack complete specificity for source distinction.

The focus of this paper is a summary of atmospheric particles that contribute to the carbon content of aerosol samples. This summary will hopefully aid those in pursuit of analytical methods to enhance source specificity, and perhaps provide a basis for judging the value of existing data on atmospheric carbon.

References p. 271.

APPROACHES TO DISTINGUISHING COMMON
CARBON-CONTAINING AEROSOLS

Most of the current methods for quantitating the carbon content of atmospheric aerosol samples involve aerosol collection on filters, combustion of captured aerosols and quantitation by non-dispersive infrared spectroscopy [2] or gas chromatography [3]. Several alternate carbon analysis methods utilize optical attenuation [4], gamma ray spectroscopy [5], and carbon isotope analysis [6] or a combination of optical attenuation with temperature programmed oxidation [7]. All of these approaches are bulk analysis techniques though the carbon isotope analysis and optical attenuation with controlled thermal oxidation provide some distinction of carbon aerosol types and their generic sources.

Optical and electron microscopy provide a single particle approach that utilizes particle morphology to distinguish carbon-containing aerosols. This is an indirect approach to carbon analysis since carbon is not detected or determined directly. However, morphological identification does permit distinction of the particles or compounds known to contain carbon, and allows source distinction in many instances.

Morphological analysis has been employed at IIT Research Institute on more than 5000 aerosol filter samples from over 50 cities in the United States. Based on these analyses, carbon-containing aerosols can be segregated by their optical properties and by size. Table 1 lists a number of carbon containing aerosols that have

TABLE 1
Carbon Containing Aerosols in the U.S.

Translucent (white)	Opaque (black)
*Calcite	*Raw Coal
*Dolomite	*Pyrolyzed Coal
Calcium Oxalate	Coke
*Trichomes	Silicon Carbide
*Starch	*Oil Soot
*Pollens	*Spores & Fungal Conidia
Spray Paint Polymers	Insect Parts
Paper Fibers	Wood Ash
*Condensed Organics	*Elemental Carbon
Leather Fibrils	gasoline combustion
	diesel fuel combustion
	gas combustion
	*Rubber Tire Fragments
	Asphalt Fragments
	Kish
	*Humus

*Frequently present in aerosol samples

been segregated into two optical attenuation categories: translucent aerosols that absorb minimal light, and opaque aerosols that strongly absorb light. Figs. 1-13 show the common appearance of some of the aerosols in Table 1. This separation clearly shows that bulk analysis for carbon may provide little correlation with optical absorption measurements due to the presence of appreciable quantities of non-absorbing carbon compounds. For example, a TSP study of an industrial area in Baltimore, Maryland [8] showed that cornstarch accounted for 63 $\mu g/m^3$ of the TSP concentration.

While microscopy may be the only, currently available technique for distinguishing absorbing and non-absorbing carbon aerosols, a practical distinction can be achieved with bulk analysis methods by utilizing size selective sampling. Tables 2-4

TABLE 2
Coarse (> 15 μm) Carbon-Containing Aerosols

Translucent	Opaque
Calcite	Raw Coal
Dolomite	Pyrolyzed Coal
Calcium Oxalate	Oil Soot
Trichomes	Spores & Fungal Conidia
Starch	Wood Ash
Pollens	Rubber Tire Fragments
Spray Paint Droplets	Asphalt Fragments
Paper Fibers	Kish
Leather Fibrils	Insect Parts
	Silicon Carbide
	Humus

TABLE 3
Inhalable (15 μm ≤ 2 μm) Carbon-Containing Aerosols

Calcite	Raw Coal
Dolomite	Pyrolyzed Coal
Starch	Coke
Pollens	Oil Soot
Spray Paint Droplets	Spores
Paper Fibers	Wood Ash
	Rubber Tire Fragments
	Asphalt Fragments
	Kish
	Insect Parts
	Silicon Carbide
	Humus

TABLE 4
Fine (≤ 2 μm) Carbon-Containing Aerosols

Translucent	Opaque

Calcite
Dolomite
Condensed Organics
	Raw Coal
	Pyrolyzed Coal
	Cokes
	Oil Soot
	Elemental Carbon
	   gasoline combustion
	   diesel combustion
	   gas combustion

show a reclassification of the carbon aerosols in Table 1 into three size categories: coarse (> 15 μm), inhalable (15 < 2 μm) and respirable (≥ 2 μm). The size range of greatest potential impact on visibility and health is the respirable range. The number of carbon-containing constituents in this size range is greatly reduced from those that may be present in a total aerosol sample commonly collected by high volume sampling. Moreover, this size range is frequently dominated by the highly absorbing elemental carbon emitted from gasoline, diesel and gas combustion in large urban areas.

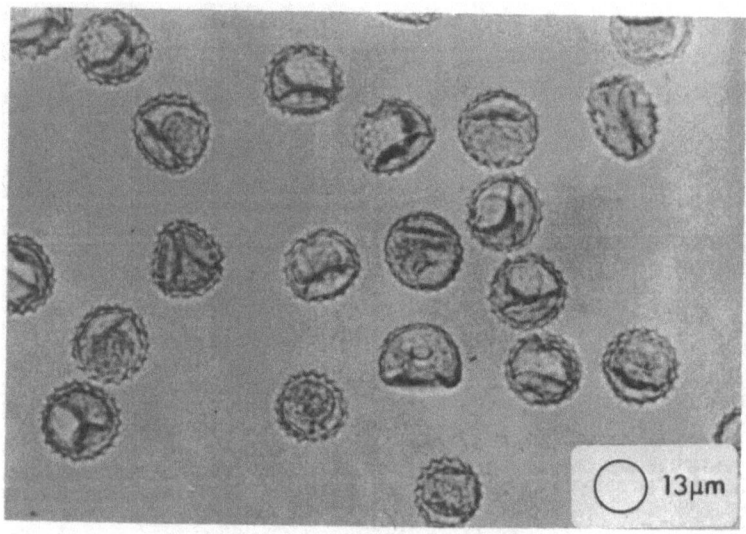

Fig. 1. Ragweed pollen.

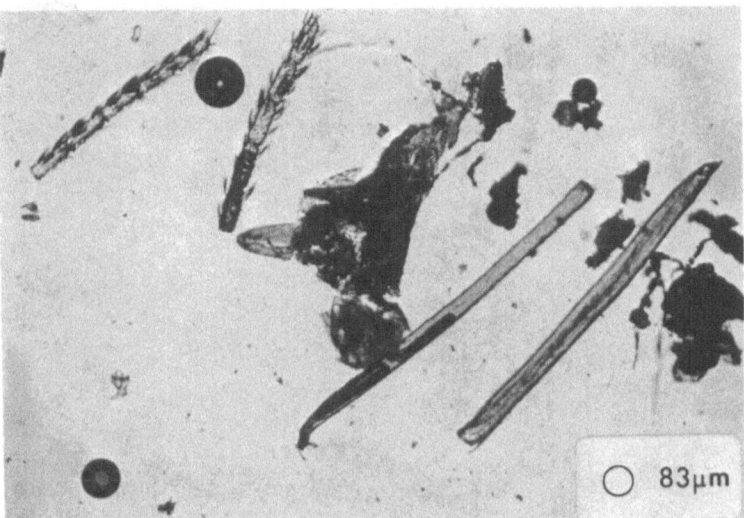

Fig. 2. Insect fragments; the black circular objects are air bubbles entrapped in the immersion media.

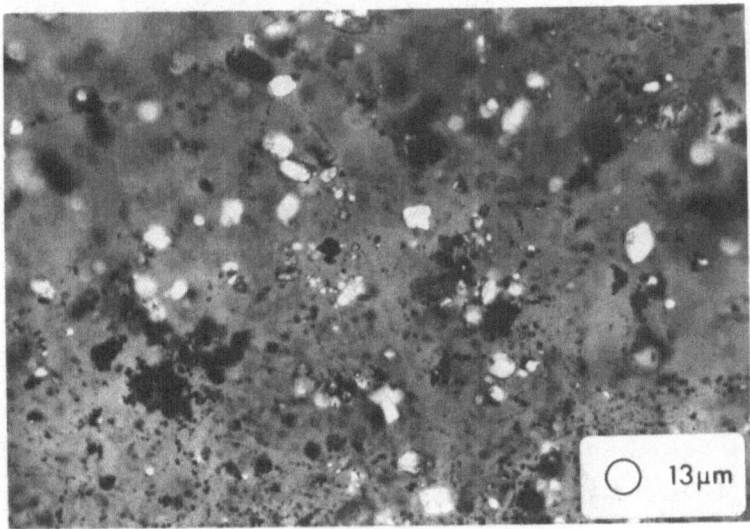

Fig. 3. The bright angular fragments are calcite ($CaCO_3$) from asphalt pavement.

References p. 271.

Fig. 4. Black, cigar-shaped particles of rubber tire tread; the black circular particles are oil soot and most of the bright angular particles are calcite.

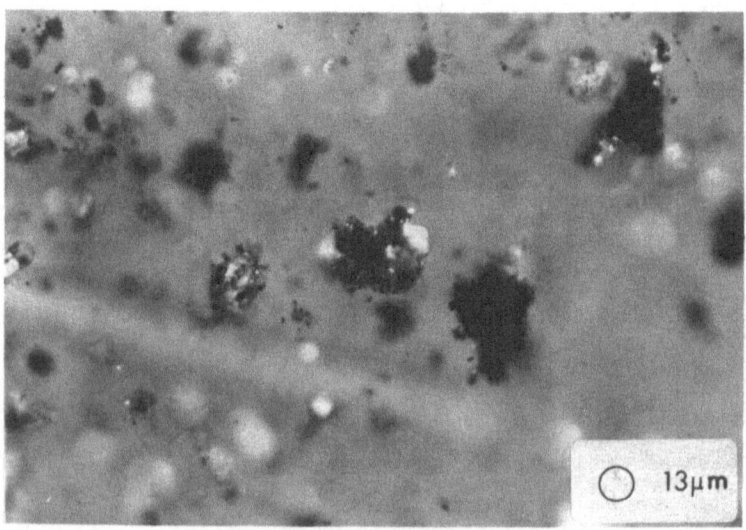

Fig. 5. Calcite coated with asphalt.

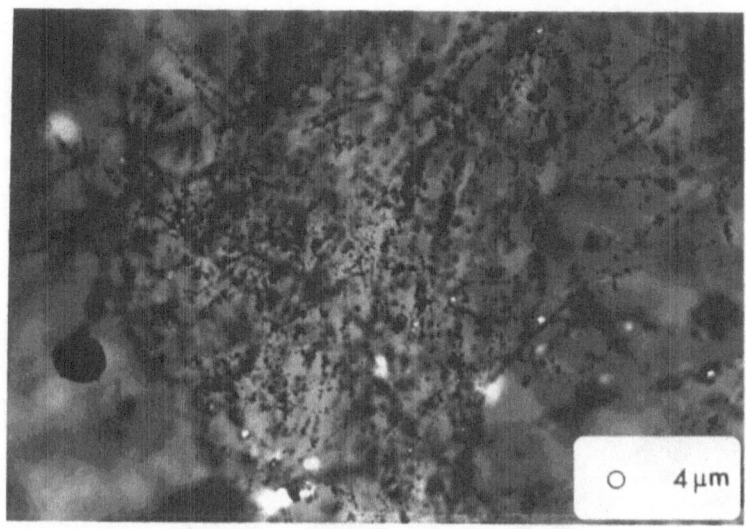

Fig. 6. Submicrometer clusters of vehicle exhaust carbon, collected within the matrix of a glass fiber filter.

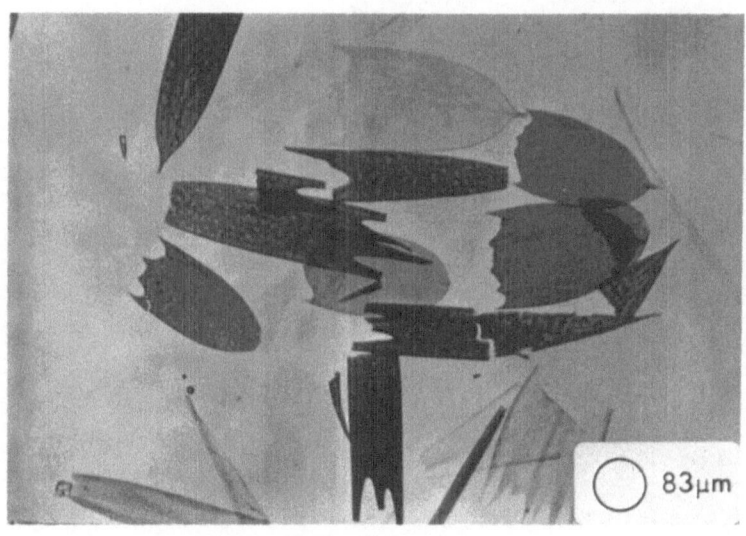

Fig. 7. Moth scales.

Fig. 8. Silicon carbide (abrasive).

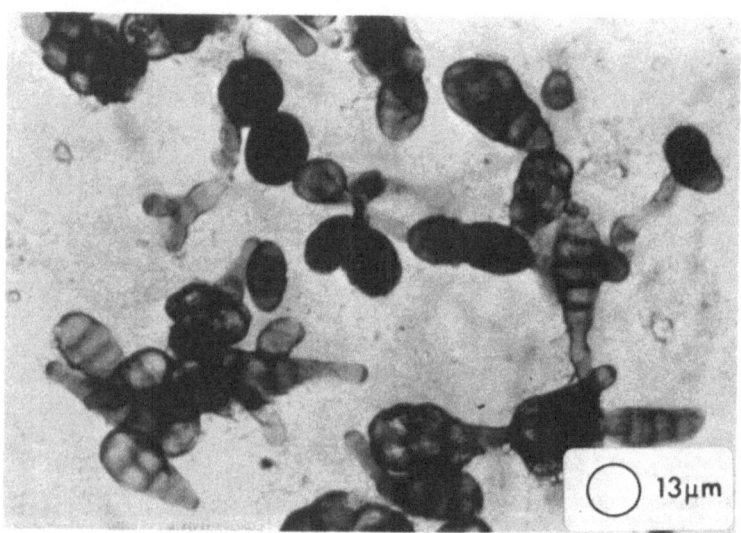

Fig. 9. Fungal conidia frequently found on hivol samples though usually in low concentrations.

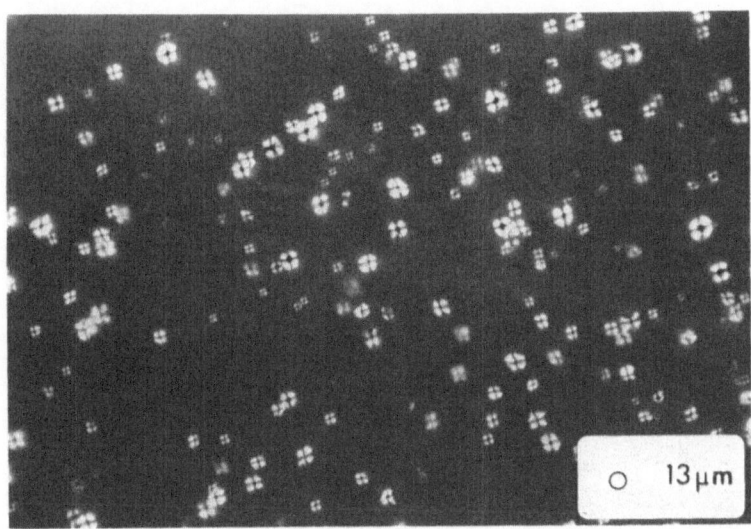

Fig. 10. Cornstarch showing a maltese cross centered on the hilum, characteristic of starches.

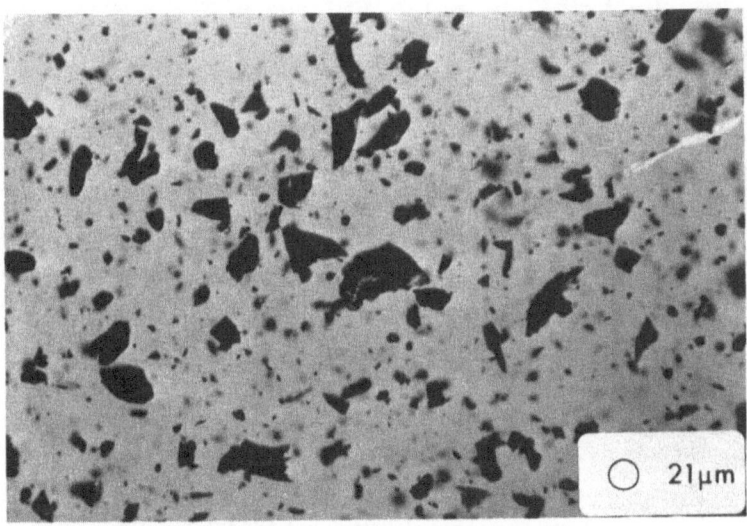

Fig. 11. Bituminous raw coal.

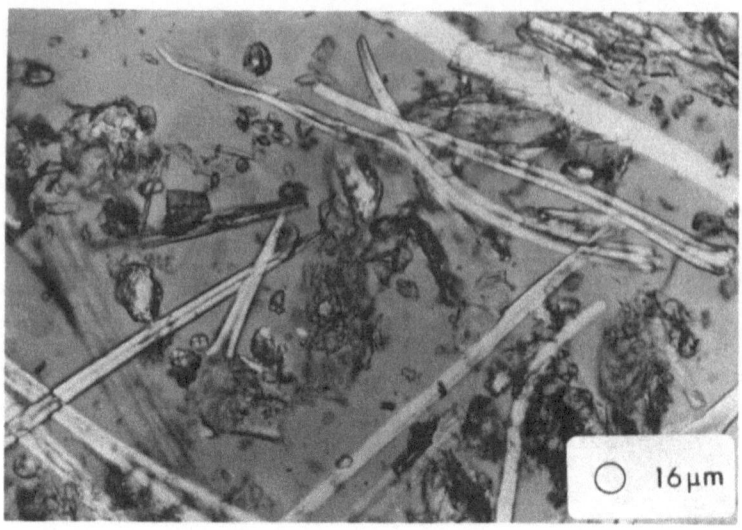

Fig. 12. The long, white fibrous particles are trichomes (plant hairs).

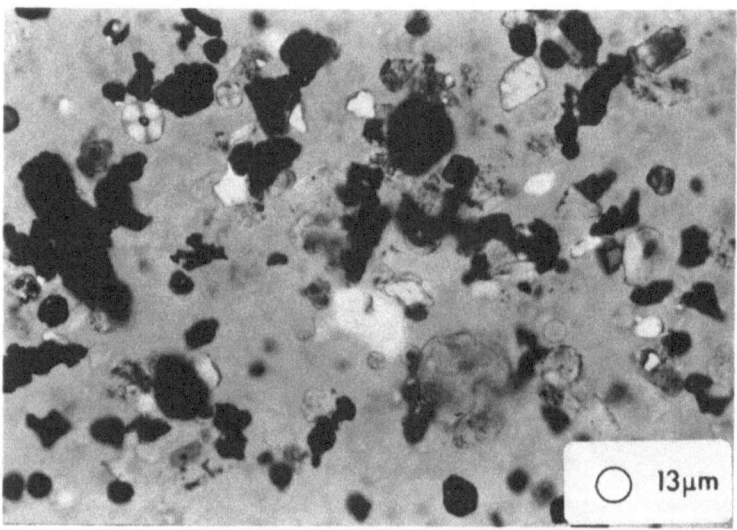

Fig. 13. Aerosol sample from commercial district in Philadelphia, PA; contains a variety of carbon-containing aerosols — calcite, cornstarch, rubber tire dust, oil soot, humus and spores.

CONSIDERATIONS IN THE SAMPLING AND ANALYSIS OF CARBON AEROSOLS

In many ambient, atmospheric aerosol studies, it has become quite common and necessary to use a number of simultaneous sampling techniques such as the high volume sampler and size segregating inertial impactors to obtain samples for various analytical methods. The high volume samples are normally collected on glass or quartz fiber filters while the impactor samples are often collected on polymeric membrane filters. It is also quite common to use the high volume samples for carbon analysis by combustion techniques since the polymeric filters obviously interfere with carbon determinations. Unfortunately, the carbon determinations from the high volume samplers are often attributed to the less than 2 μm size class without independent verification that larger particle size carbon aerosols are absent.

Now that carbon is on every atmospheric scientist's analytical menu, the obvious solution is to collect size segregated samples on substrates suitable for direct carbon analysis. The added time and cost for collecting and analyzing these additional samples should produce substantial benefits in allowing data comparisons of independent studies that can clarify the impact of elemental carbon on health and visibility.

REFERENCES

1. T. Novakov (ed), Conference on Carbonaceous Particles in the Atmosphere, Lawrence Berkeley Laboratory, University of California, 20-22 March 1978.
2. S. H. Cadle, P. J. Groblicki and D. P. Stroup, Analytical Chemistry, Vol. 52 (1980), p. 2201.
3. R. L. Johnson and J. J. Huntzicker, "Analysis of volatilizable and elemental carbon in ambient aerosols." Proceedings of the Conference on Carbonaceous Particles in the Atmosphere, T. Novakov, editor, Lawrence Berkeley Laboratory, University of California (June 1979).
4. H. Rosen, A. D. A. Hansen, L. Gundel and T. Novakov, "Identification of the graphitic carbon component of source and ambient particulates by Raman spectroscopy and an optical attenuation technique."
5. E. S. Macias, C. D. Radcliffe, C. W. Lewis and C. R. Sawicki, Analytical Chemistry, Vol. 50 (1978), p. 1120.
6. L. A. Currie, S. M. Kuren, K. J. Voorhees and W. F. Koch, "Low-level radiocarbon counting and pyrolysis/gas chromatography/mass spectrometry: methods for analysis of carbonaceous particles and characterization of their sources."
7. R. L. Dod, H. Rosen and T. Novakov, "Optical-thermal analysis of the carbonaceous fraction of aerosol particles." Atmospheric Aerosol Research, 1977-78 Annual Report, LBL-8696, Lawrence Berkeley Laboratory, University of California.
8. R. Koch, J. Schakenback and K. G. Severin, "Identifying sources and quantities of fugitive particulate emissions in Baltimore." Third Symposium on Fugitive Emissions: Measurement and Control, San Francisco, California, 23-25 October 1978.

GRAPHITIC CARBON IN URBAN ENVIRONMENTS AND THE ARCTIC

H. ROSEN, A.D.A. HANSEN, R.L. DOD, L.A. GUNDEL and T. NOVAKOV

University of California
Berkeley, California

ABSTRACT

The application of the Raman scattering, photoacoustic, optical attenuation, opticothermal analysis, and solvent extraction techniques to the characterization of the graphitic component of urban and remote aerosols will be described. These analyses allow unambiguous identification and quantification of the optically absorbing component of the aerosol. Recent observations in the Arctic show graphitic carbon concentrations which are comparable to those found in urban environments. The possible effects of these highly absorbing species on the heat balance will be discussed.

INTRODUCTION

Recent measurements of aerosol particles indicate the presence of an optically absorbing component which can cause visibility degradation and climatic effects. The nature of this absorbing species has been investigated by a variety of modern methods of analysis, which has led to its identification as graphitic (black) carbon [1]. In our paper this methodology will be described and extended to the quantification of graphitic carbon and its absorption coefficient. We will present new results which show that graphitic carbon concentrations in the Arctic are comparable to those found in urban environments. In the first section of this paper, the Lawrence Berkeley Laboratory (LBL) laser transmission method will be described and compared to both photoacoustic measurements [2] and measurements using the integrating plate method [3]. A theoretical model will be presented which gives physical insight into why these transmission methods selectively measure the absorbing properties of the aerosol and are insensitive to its scattering properties. In the second section, the use of Raman spectroscopy to identify the optically absorbing

References p. 292

species in urban particulates will be described. In the third section, preliminary results using a thermal analysis technique [4] for quantifying the graphitic carbon content of the aerosol will be presented. Finally, in the fourth section the methods of analysis described above will be applied to the characterization of the carbonaceous aerosol in the Arctic.

DETERMINATION OF THE ABSORPTION COEFFICIENT OF AEROSOL PARTICLES BY THE LASER TRANSMISSION METHOD

The LBL laser transmission method is an extension of the integrating plate technique [5] to certain other filter media (Millipore, quartz fiber, Teflon), which act as efficient collection substrates as well as play the role of the opal glass diffuse scatterer. The laser transmission apparatus compares the transmission of a 633-nm He-Ne laser beam through a loaded filter relative to the transmission through a blank filter (Fig. 1). The loaded filters are placed in the beam with the loaded side toward the laser. After multiple scattering through the filter substrate, the light is collected by an f/1 lens and focused on a photo-multiplier tube. The absorption coefficient, σ_{ab}, is determined from the Beer-Lambert Law in a fashion similar to that outlined for the integrating plate method. This method measures the absorbing component of aerosol particles and is apparently insensitive to its scattering properties, as was demonstrated by a photoacoustic study [2].

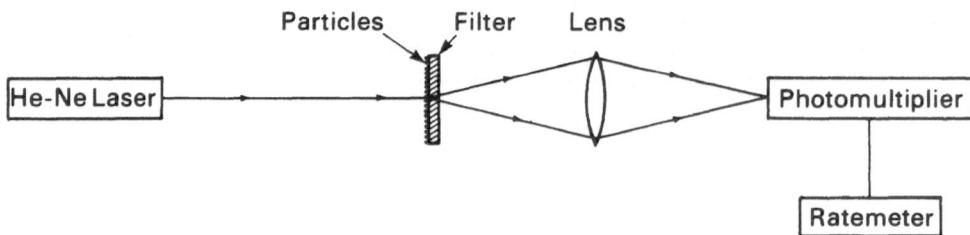

Fig. 1. Schematic of LBL laser transmission apparatus.

Unlike conventional optical absorption techniques, photoacoustic spectroscopy measures the energy deposited in a sample due to absorption. Since questions have been raised as to whether the LBL laser transmission method exclusively measures the absorbing rather than the scattering component of the aerosol, a comparison between photoacoustic and optical attenuation measurements on the same aerosol sample should help resolve this ambiguity. This work was a collaborative effort between the Aerosol Research Group & the Nabil Amer Laser Spectroscopy Group headed by LBL.

The photoacoustic measurements were made in an acoustically nonresonant detector with cylindrical geometry (Fig. 2). A Knowles microphone (Model BT-1759) was used, and the cell dimensions were 2.1 cm in diameter and 0.3 cm in length. The gas in the detector cell was air at atmospheric pressure. An He-Ne laser operating at 632.8 nm with 0.5 mW of power was used as the light source, and the

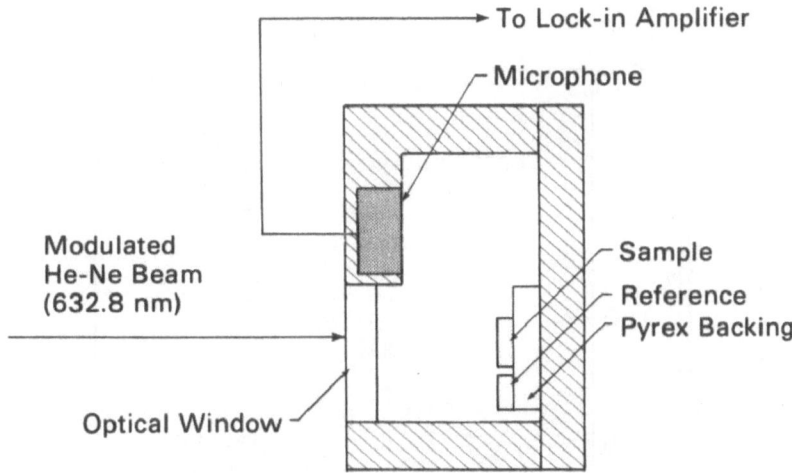

Fig. 2. Photoacoustic experimental arrrangement.

Fig. 2. Photoacoustic experimental arrrangement.

experiments were performed at a modulation frequency of 20 Hz. The aerosol particles, collected on 1.2 μm Millipore filter substrates, were mounted on a 1.5 mm-thick Pyrex backing with the particles facing the incident light beam. Experiments were also performed with the laser beam first incident on the filter substrate.

It is easy to show in the limit of low frequency light modulation [6] ($\leqslant$100 Hz) that the ratio of the photoacoustic signal to a reference sample for which the signal is saturated is given by:

$$S_{ph} = V/V_{sat} = 1 - \exp(-\alpha \ell).$$

This saturable behavior was observed for highly absorbing samples, and the sample which yielded the largest photoacoustic signal was used as the reference, V_{sat}. Note that such samples yield values of $\alpha \ell \geqslant 3$, as deduced from the optical attenuation measurements; hence the highest signal obtained from available samples is close to the actual saturation value.

The experimental setup for the optical attenuation measurements is described above. In this technique the signal, S_{op}, is defined as $1 - \exp(-x)$, where x is the optical attenuation of the sample and is given by $-\ell n\ I/I_0$, where I is the transmitted intensity of a loaded filter and I_0 is the transmitted intensity of a blank filter.

In Fig. 3 we present a plot of the normalized photoacoustic signal, S_{ph}, vs. S_{op} for a wide range of ambient samples and samples collected directly from combustion sources. The samples include urban particulates collected over a 24-hr period in Fremont and Anaheim, California; Denver, Colorado; New York City; and particles collected in a highway tunnel and from an acetylene torch. The least squares fit of the experimental points yields a correlation coefficient, r, of 0.98 and a slope of 1.03, which would be expected if both techniques measure the same optical property of the aerosol particles. Since the photoacoustic signal is proportional to the heat generated by absorption, we conclude that the optical attenuation method measures the light-absorbing component of the aerosol particles.

From a theoretical point of view, this result is somewhat surprising, since aerosol

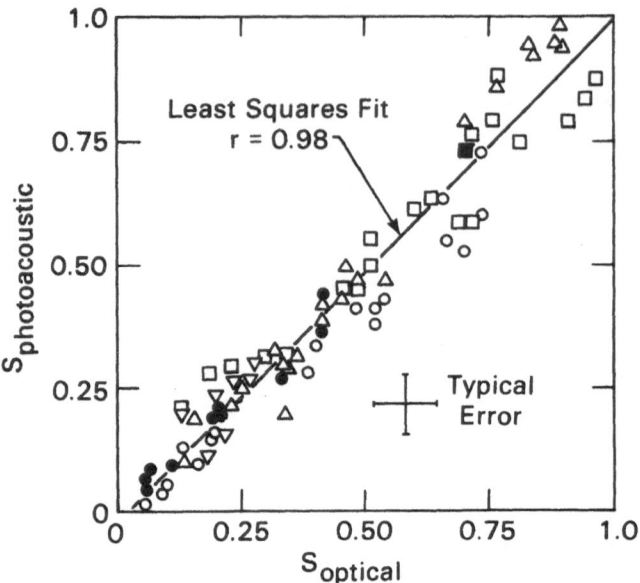

Fig. 3. Plot of S_{ph} vs. S_{op} for various samples: ▼ — Fremont; □ — Anaheim; ○ — Denver; ▲ — New York City; ■ — highway tunnel; ● — acetylene torch.

The solid line is a least squares fit of the data.

particles have a large scattering coefficient, which would be expected to contribute to the optical attenuation measurement and not to the photoacoustic signal. This is especially true where the absorbing component represents only a small fraction of the aerosol mass. In this paper, a simple model calculation will be presented which explains these observations and points out the critical role of the filter substrate as an almost perfect diffuse reflector in the technique. Similar considerations may also apply to the opal glass method used by Weiss *et al.* [7].

For this model calculation, we will assume that the particles and the filter media can be treated independently and consider the geometry shown in Fig. 4. A similar

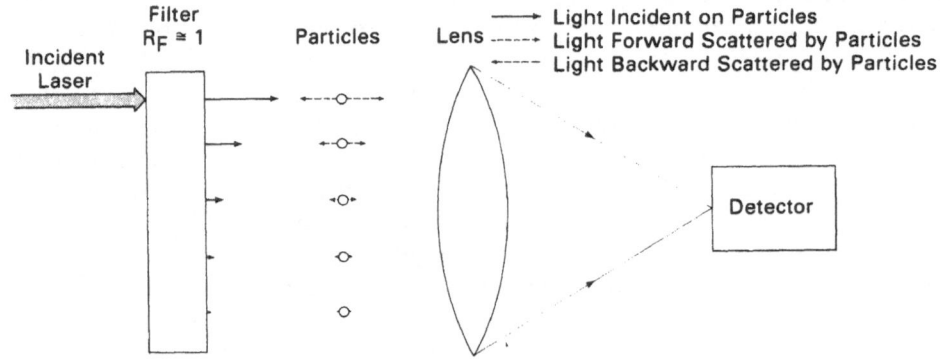

Fig. 4 Schematic of experimental arrangement used for model calculation.

treatment, where the light beam is first incident on the particles, gives identical results. After the light beam passes through the filter medium, it is incident on the particles with an intensity, I_0. The particles forward scatter a fraction of the incident light, backward scatter a fraction, and absorb a fraction. These components in the low loading limit are respectively given by $n_s \sigma_F I_0$, $n_s \sigma_B I_0$, and $n_A \sigma_A I_0$ where n_s is the number of scattering aerosol particles per unit area, n_A is the number of absorbing aerosols per unit area, σ_F is the forward scattering cross section, σ_B is the backward scattering cross section, and σ_A is the absorption cross section.

Since the optical attenuation technique only measures the forward scattering light, it would seem as if the backscattered light would be lost to the system and would contribute to the attenuation. However, the filter, in our method, is almost a perfect reflector. Under these circumstances, the backscattered light will be reflected in the forward direction and will again be incident on the particles. This process will continue until almost all the backscattered radiation is collected by the optics and so that it does not contribute to the optical attenuation. This result can be put in mathematical form [1], where I is the light detected by the collection optics and R_F is the reflectivity of the substrate:

$$I = I_0 (1 - n_s \sigma_B - n_A \sigma_A) + I_0 (1 - n_s \sigma_B - n_A \sigma_A) n_s \sigma_B R_F$$

$$+ I_0 (1 - n_s \sigma_B - n_A \sigma_A) n_s^2 \sigma_B^2 R_F^2 + \ldots I_0 (1 - n_s \sigma_B - n_A \sigma_A) n_s^n \sigma_B^n R_F^n$$

$$= I_0 \frac{(1 - n_s \sigma_B - n_A \sigma_A)}{(1 - n_s \sigma_B R_F)}$$

Consider several limits, for example, if $R_F \simeq 0$, which normally would be considered an ideal substrate, Eq. (1) reduces to

$$I = I_0 (1 - n_s \sigma_B - n_A \sigma_A).$$

Under these conditions, the backscattered radiation will contribute significantly to the optical attenuation and make the technique unsuitable for exclusively measuring the absorbing properties of the aerosol. In our method, however, $R_F \simeq 1$ and Eq. (1) becomes

$$I = I_0 \left(1 - \frac{n_A \sigma_A}{1 - n_s \sigma_B}\right) .$$

and the optical attenuation in the low loading limit is

$$ATN = I_0 - I = \frac{I_0 n_A \sigma_A}{1 - n_s \sigma_B} .$$

From this expression it is clear that a nonabsorbing aerosol will make almost no contribution to the optical attenuation which is consistent with our experimental

References p. 292

results. The magnitude of the optical attenuation is somewhat dependent on the scattering properties of the aerosol. In the low loading limit, however, this effect is small. For example, if the substrate has 50% coverage, and if the scattering cross section of the particles, σ_S, is twice the particle area, then $n_s\sigma_s \simeq 1$. If σ_B is about 20% of σ_S (the maximum value measured by Charlson and his coworkers [8]), then

$$1 - n_s\sigma_B \simeq .8;$$

so that even for this rather high loading, the error in the absorption measurement due to scattering of the aerosol is only about 20%. This treatment should only be viewed as giving physical insight into the LBL method for determining absorption coefficients and clearly is approximate since it assumes that the scattering properties of the particles are not affected by the filter substrate and neglects the penetration of the particles into the substrate. Future analysis will evaluate the significance of these effects.

The LBL laser transmission method has been compared to the integrating plate technique [5] developed at the University of Washington. Five sampling sites in the western part of Washington state were used in this study. The sites varied from a highly congested site in a highway tunnel to a very remote site on a western foothill in the Olympic Mountains. By sampling such diverse aerosol conditions, the range of absorption coefficients of the air samples was more than three orders of magnitude, from $\sim 10^{-7}$ m^{-1} to nearly 10^{-3} m^{-1}, with the total carbon concentration ranging from about 5 μg/m^3 (Mt. Octopus) to nearly 90 μg/m^3 (highway tunnel).

The results of these comparisons [3] are shown graphically in Fig. 5, which contains a plot of absorption coefficients, as determined by the IPM using a Nucle-

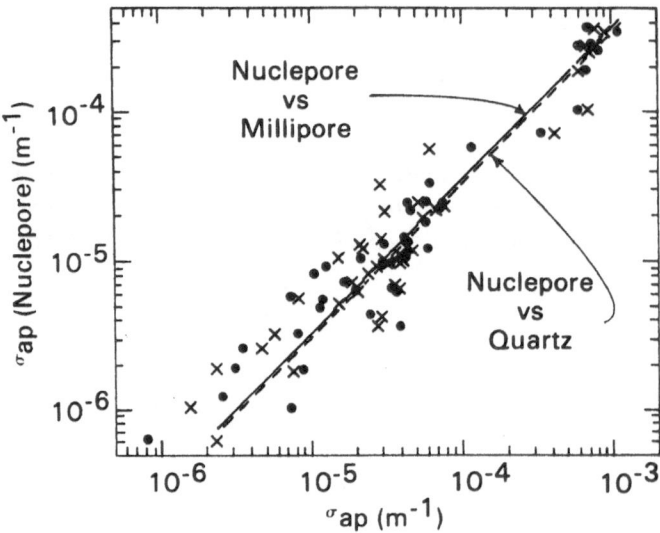

Fig. 5. Light absorption coefficient, σ_{ab} (m^{-1}), determined by the integrating plate method on Nuclepore filters versus σ_{ab} (m^{-1}) determined on Millipore (X) and quartz (O) filters

pore substrate, and the LTM using a Millipore substrate. Forty-four filters of each type were used in this comparison. The correlation coefficient between the two measurements is 0.95, with the absorption coefficient determined by the LTM being greater than that determined by the IPM by a factor of approximately 2.5. The light absorption coefficients determined by these two methods are comparable and highly correlated. The reason for the higher indicated absorption coefficient using the LTM could be due to several factors, including:
1. penetration effects in the Millipore substrate, which could lead to enhanced absorption due to multiple scattering effects within the filter medium itself;
2. possible pile-up at the holes or loss of particles in the holes of the Nuclepore substrate;
3. differences in the collection efficiency of the two substrates.

Studies are under way in both laboratories to assess the magnitude of these effects. It should be emphasized, however, that the differences found between these two techniques are small in comparison with the large uncertainties reported in the literature for the absorbing component of aerosol particles. These differences are also small compared to the range of more than three orders of magnitude observed in the value of σ_{ab}.

IDENTIFICATION OF OPTICALLY ABSORBING SPECIES IN URBAN AEROSOLS

Raman spectroscopy is a highly selective method of analysis, which until recently, has not been applied to the characterization of air pollution particulates [9-11]. The technique can often be used to make unambiguous identifications since different chemical species have characteristic vibrational modes and, therefore, characteristic Raman spectra. The Raman spectroscopy apparatus uses a Coherent Radiation argon ion laser producing 1 W of power at 514 nm. The laser beam is focused by a 75 mm focal length cylindrical lens to a spot 0.06 mm x 2 mm on the sample surface via a small mirror, and the backscattered radiation is collected and imaged by an f/l lens onto the slit of a 1 m Jarrell-Ashe double monochromator equipped with two 1180-grooves/mm gratings blazed at 5000 Å. The output of the spectrometer is detected by an FW130 photomultiplier which is cooled to $-20°C$ and used in a photon-counting mode. The pulses, after appropriate shaping, are counted and displayed on a multichannel analyzer. A computer-controlled grating drive made by RKB, Inc., allows a given spectral region to be scanned many times and added to the memory of the multichannel analyzer, greatly improving the signal to noise ratio. In order to minimize heating effects, the highly absorbing samples used in these experiments are rotated at 1800 rpm by a motor, which increases the area illuminated by the laser beam by a large factor with almost no loss in signal level. The focal spot of the laser is located approximately 5 mm below the axis of rotation so that the effective illuminated area is an annulus of 5-mm radius and 2 mm width, resulting in the low power density of approximately 1 W/cm².

The Raman spectra between 920 cm⁻¹ and 1950 cm⁻¹ of ambient air, automobile exhaust, and diesel exhaust particulates are compared with the spectra of activated carbon and polycrystalline graphite in Fig. 6. It is evident that the spectra of activated carbon, diesel exhaust, automobile exhaust, and the ambient samples are

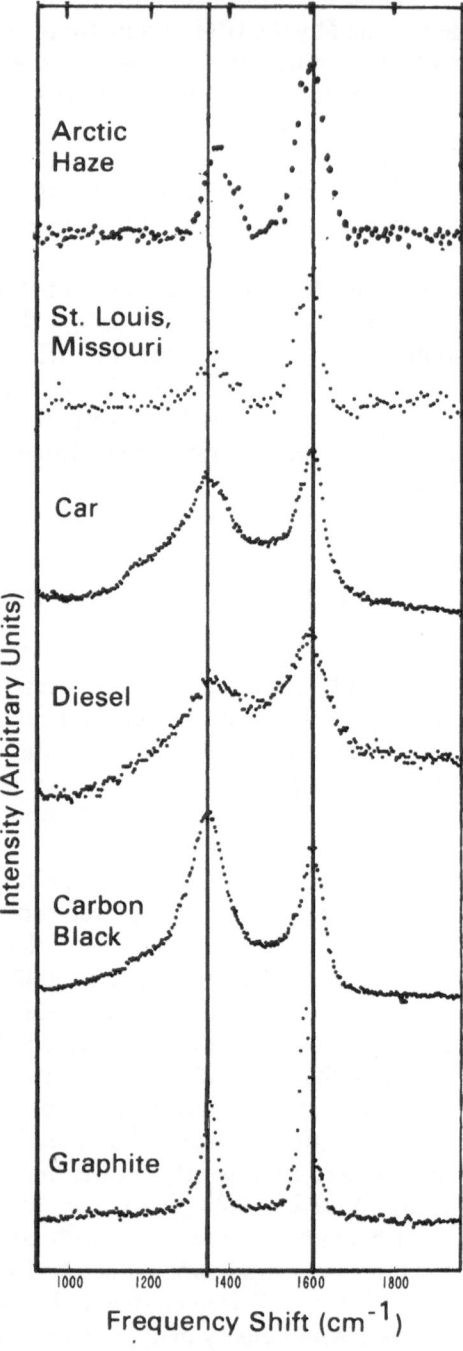

Fig. 6. Raman spectrum of an Arctic sample compared to those of urban particulates, various source emissions, carbon black, and polycrystalline graphite.

very similar, with the positions of the two Raman modes coincident to within ± 10 cm⁻¹ which is the estimated experimental error. The ambient samples shown were collected as part of the RAPS program in St. Louis, Missouri and in the Arctic region near Barrow, Alaska. However, the same Raman modes are also evident in every urban sample studies so far (New York, Buffalo, Berkeley, Anaheim, Fremont). Koenig *et al.* [12] have studied the Raman spectrum of activated carbon and have identified that the modes near 1600 cm⁻¹ and 1350 cm⁻¹ are due to phonons propagating within graphitic planes. The close correspondence of the spectra in Fig. 6 indicates the presence of physical structures similar to activated carbon in both source and ambient samples. These graphitic species are formed directly in combustion, and we shall use the term graphitic soot throughout the text to describe them.

Urban and combustion source particulates collected on various filter media have a grey or black appearance. The graphitic species identified by Raman spectroscopy is the most likely candidate to explain this coloration. To test this hypothesis, we have developed an optical absorption technique to quantitatively measure various properties of the absorbing species (see above).

Using this apparatus we have studied the temperature stability and solubility of the absorbing species in ambient and source particulate samples. Our results show that these species have high temperature stability [13] with only minimal oxidation up to 400°C and are essentially insoluble in a wide variety of solvents [7, 13]. We have also shown, using a spectrophotometer, that to within 20% over the visible spectral region, the optical attenuation has a 1/λ wavelength dependence characteristic of a constant imaginary index of refraction [7, 13].

All these results strongly suggest that the absorbing substances in urban and source particulate samples is graphitic soot. A direct substantiation of this hypothesis is provided by comparing the integrated intensity of the 1600 cm⁻¹ Raman mode with the optical attenuation of the same filter sample. These measurements have been done on three types of samples: acetylene soot samples, which were essentially pure carbon with only trace amounts of metallic impurities; highway tunnel samples; and ambient samples collected in Berkeley and Fremont in the San Francisco air basin and Anaheim in the Los Angeles air basin. The results, shown in Fig. 7, indicate that within experimental error there is a direct correspondence between the optical attenuation and the Raman intensity or graphitic soot content for all samples studied, despite widely different chemical compositions (e.g., for a given optical attenuation, the Pb and Fe concentrations vary by more than a factor of 100). The only reasonable explanation is that the optical attenuation is due to the graphitic soot content of the collected particulate.

In summary, we have shown that the species responsible for the high optical absorptivity of particulate samples has high temperature stability in air, is insoluble in a variety of solvents, and absorbs uniformly throughout the visible spectrum. We have also demonstrated that the amount of the absorbing species is directly proportional to the graphitic soot content as defined by Raman spectroscopy. All these results taken together indicate that the high optical absorptivity of both ambient samples collected in urban environments and various source particulate samples is due to the graphitic component of the aerosol.

References p. 292

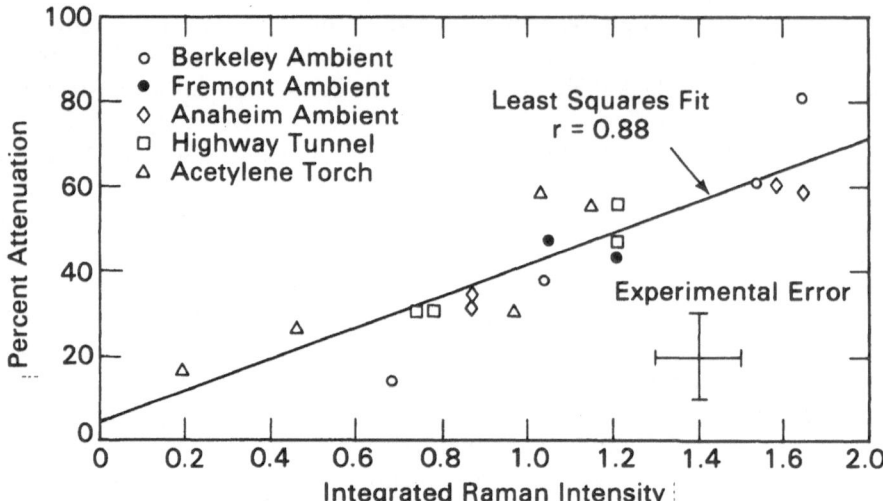

Fig. 7. Plot of integrated Raman intensity of the 1600 cm^{-1} mode versus percent optical attenuation at 633 nm for ambient, acetylene soot, and tunnel samples.

DETERMINATION OF GRAPHITIC OR BLACK CARBON CONTENT OF AEROSOLS BY COMBINING THERMAL ANALYSES WITH OPTICAL ABSORPTION MEASUREMENTS

Malissa [14] has reported a method of analysis for the carbonaceous component of atmospheric aerosol particulate material which involves measurement of the evolved gas during a temperature-programmed combustion in oxygen. We have extended this analysis by constructing an apparatus which simultaneously measures the optical transmission of the particulate matter collected on a filter. Since the optical absorptivity of this material has been shown to be due to a graphitic component [1], this combination of analytical techniques may provide a direct determination of the light-absorbing fraction of the sample.

A schematic representation of the apparatus used in our analysis of carbonaceous material is shown in Fig. 8. The particulate sample, collected on a prefired quartz filter, is placed in the quartz combustion tube so that its surface is perpendicular to the tube axis. The tube is supplied with purified oxygen and the excess oxygen escapes through an axial opening at the end of the tube. The remainder of the oxygen, together with gases produced during analysis, passes through a nondispersive infrared analyzer (MSA LIRA 202S) at a constant rate. Sample carbon may be evolved through volatilization, pyrolysis, oxidation, or decomposition. To ensure complete conversion of this carbon to CO_2, a section of the quartz tube, immediately outside the programmed furnace, is filled with a CuO catalyst, which is kept at a constant 900°C by a second furnace. This is necessary, especially when the first furnace is at low temperatures where volatilization and incomplete combustion are the dominant processes occurring.

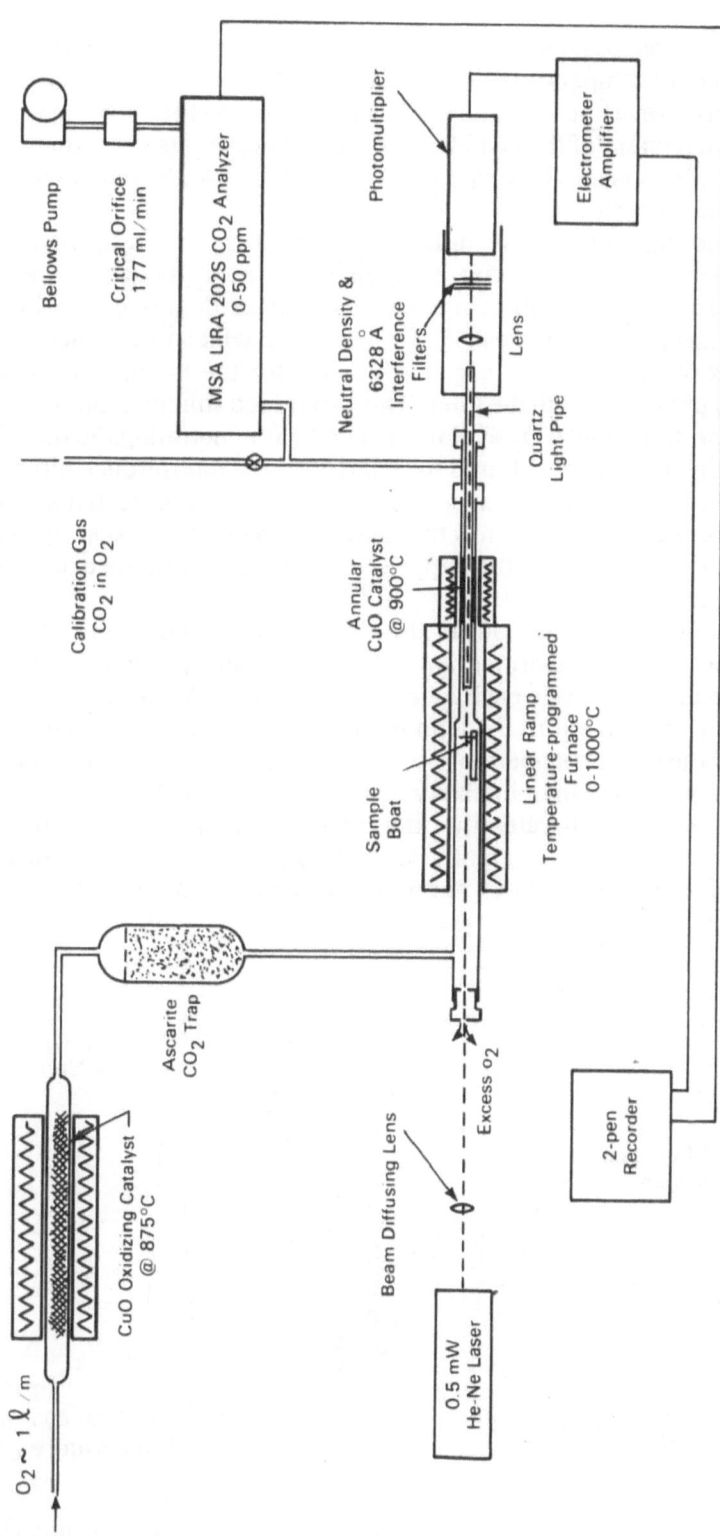

Fig. 8. Optical-thermal analysis apparatus.

The actual measurement consists of monitoring the CO_2 concentration as a function of the sample temperature. The result is a "thermogram," which is a plot of the CO_2 concentration vs. temperature. The area under the thermogram is proportional to the carbon content of the sample. The carbon content is determined by calibrating with CO_2 in oxygen. This calibration is crosschecked by analyzing samples of known carbon content.

The thermograms of ambient and source aerosol samples reveal distinct peaks or groups of peaks. One important component of the carbonaceous aerosol is the graphitic carbon, which is known to cause the black or grey coloration of ambient and source particulate samples [1]. To determine which of the thermogram peaks corresponds to this graphitic carbon, we monitor the intensity of a He-Ne laser beam which passes through the filter. This provides a simultaneous measurement of sample absorptivity and CO_2 evolution. The light penetrating the filter is collected by a quartz light guide and filtered by a narrow band interference filter to minimize the effect of the glow of the furnaces. An examination of the CO₂ and light intensity traces enables the identification of the thermogram peak or peaks which correspond to the black carbon because they appear concurrently with the decrease in sample absorptivity.

The potential of this method (in the CO_2 mode) is shown in Figs. 9-11, which illustrate the complete thermograms of several source samples and an ambient sample. The lower traces in each figure represent the CO_2 concentration, while the upper curves correspond to the light intensity of the laser light beam that reaches the detector during the temperature scan. Inspection of the thermogram shows that a sudden change in the light intensity occurs concomitantly with the evolution of a CO_2 peak. This demonstrates that the light-absorbing species in the sample are combustible and contain graphitic carbon. The carbonate peak in the ambient sample evolves at about 600°C, and since carbonate is not light absorbing, it does

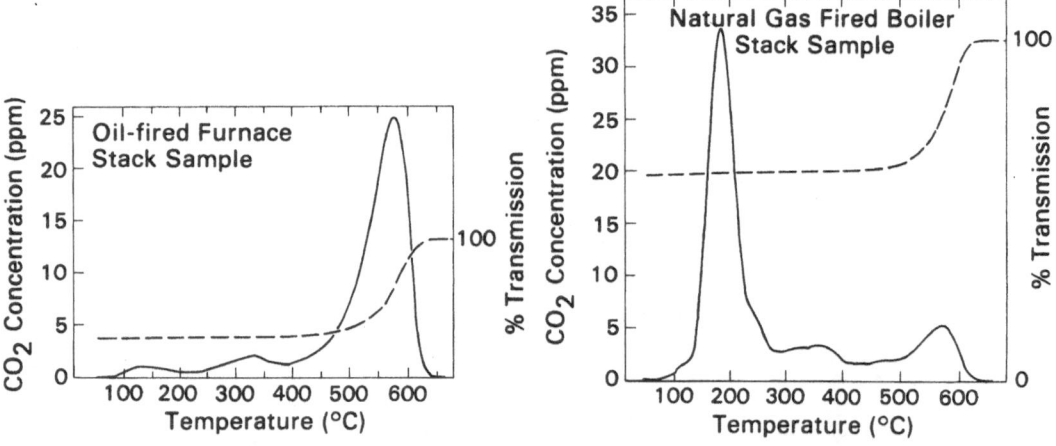

Fig. 9. Example of combustion thermograms of particulate material emitted from stationary sources. Dashed line is optical transmission.

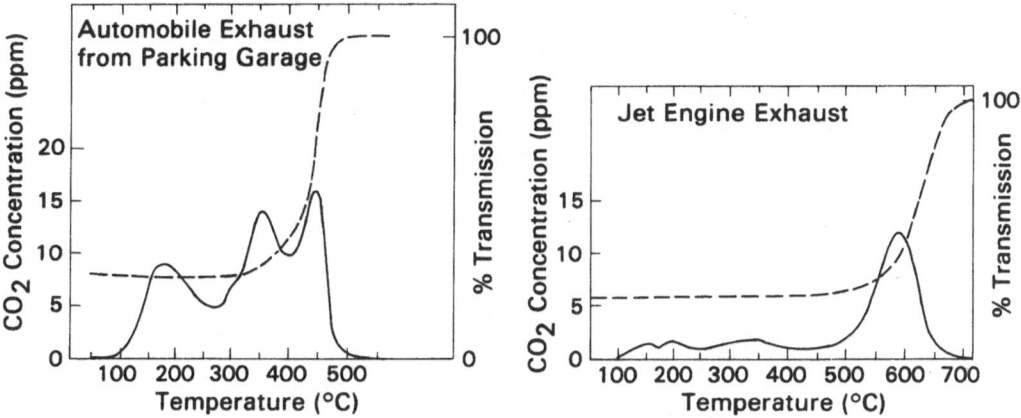

Fig. 10. Example of combustion thermograms of particulate material emitted from mobile sources. Dashed line is optical transmission.

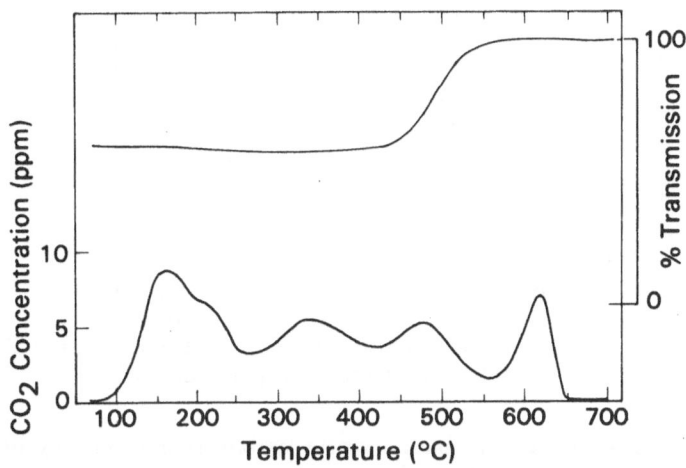

Fig. 11. CO_2 thermogram of Berkeley ambient aerosol particles (5/31/79).

not change the optical attenuation of the sample. In addition to black carbon and carbonate, the thermograms also show several distinct groups of peaks at temperatures below approximately 400°C. These peaks correspond to various organics which do not appreciably affect the optical absorption measurement.

Using this apparatus, we have done a preliminary study of the relationship between the graphitic content of the aerosol and its optical absorption coefficient. A comparison of the room-temperature optical attenuation measurement with the amount of carbon, represented by the high temperature peak which appears concurrently with the decrease in the sample absorptivity, is shown in Fig. 12. The points in the figure include the following source emissions: highway tunnel, parking garage, oil-fired furnace, natural gas boiler, motor scooter, jet engine, and an acetylene torch. The optical attenuation and the graphitic content show a good correlation with th least squares fit corresponding to a specific attenuation of 20. Further studies are under way to test the generality of these results.

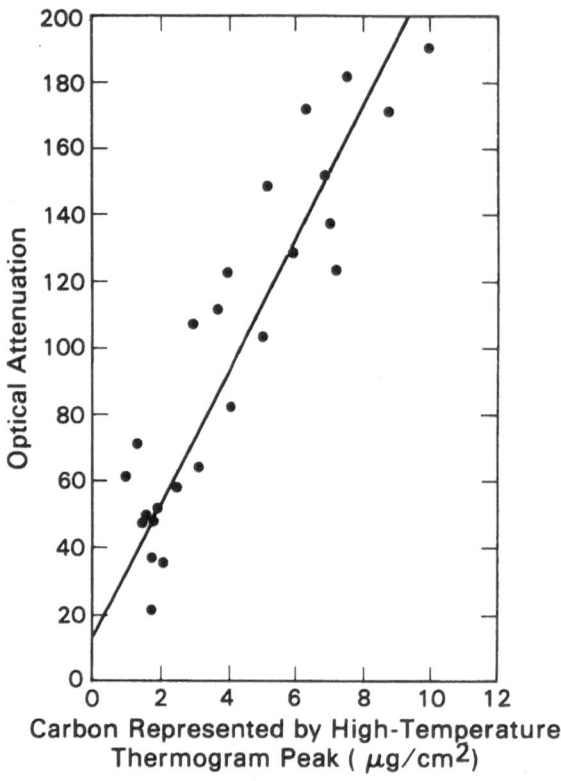

Fig. 12. Graph of optical attenuation versus carbon represented by high-temperature thermogram peak. Line is best fit to points.

SOOT IN THE ARCTIC

Recent studies in the Arctic [15, 16] show the presence of large aerosol concentrations which significantly affect the transfer of light through the atmosphere and lead to the phenomenon of Arctic haze, which was first reported by Mitchell [17]. In particular, the observation of substantial concentrations of particulate sulfur and vanadium at the NOAA-GMCC sampling station near Barrow, Alaska has attracted considerable attention [15, 16]. Questions have been raised as to the sources, and climatic impacts of these aerosols. In order to gain a better understanding of these issues, a study of the physical and chemical properties of the carbonaceous aerosol at Barrow was initiated in October 1979.

Recent studies of the urban aerosol indicate the presence of substantial graphitic carbon concentrations. These graphitic species — identified by a variety of modern analytical techniques, which include Raman spectroscopy [1], photoacoustic spectroscopy [2], thermal analysis [4], are very effective absorbers of visible radiation and are responsible for the high optical absorption coefficients which have recently been observed in urban air [3, 7]. The impacts of these highly absorbing particles on a regional or global scale have not been assessed so far, but they could be important, especially over regions like the polar icecaps with a high surface albedo. Furthermore, graphitic carbon may offer an attractive and convenient tracer for anthropogenic activity.

An aerosol sampler was constructed to collect parallel 47-mm quartz fiber and Millipore filter samples at a flow rate of ~ 1.5 cfm. The sampler had two chambers — a lower chamber which contained the pumps and an upper chamber for collecting the aerosol samples. The upper chamber was warmed by several thermostatically controlled heatlamps, while temperature control in the lower chamber was achieved by using a thermostatically controlled fan and heatlamps. The exhaust of the pumps was vented below the sampling platform. To minimize local contamination, the aerosol sampler at Barrow was controlled by a wind sensor which was built by the University of Rhode Island [15]. This allowed samples to be collected only when the wind was from the clean air sector, between 0° North and 130° Southeast. Results obtained with and without the wind controller suggest that there is no significant influence from local sources. Approximately 50 filter pairs have been collected at sampling time intervals ranging from 2 days to 1 week. The quartz filters were used to determine the total carbon content of the aerosol by a combustion method [18], and the Millipore substrate was used to determine the optical absorption coefficient of the aerosol by the LBL laser transmission method. The absorption coefficients reported here are consistent with the optical constants of graphitic carbon and are expected to have an accuracy of better than a factor of 2. The Millipore substrate was also analyzed by the X-ray fluorescence technique to determine the concentration of elements with $Z > 11$. The quartz blanks had a carbon loading of $< .5 \ \mu g/cm^2$, while typical filter deposits corresponded to 15 $\mu g/cm^2$. Selected filters have been analyzed by Raman spectroscopy, thermal analysis, and solvent extraction techniques. The results of these analyses are described below.

Raman spectroscopy is a highly selective method of analysis which has been used to identify large concentrations of graphitic carbon in urban particulates [1]. This technique has been applied to the analysis of several samples collected at Barrow. The results are shown in Fig. 6, where the spectrum of an Arctic sample collected in December 1979 is compared to that of urban particulates, various source emissions, and carbon black. All these spectra show the presence of two intense Raman modes located at 1350 and 1600 cm^{-1}, which have been identified as being due to phonons propagating within graphitic planes [12]. This result shows that graphitic structures, similar to carbon black from combustion processes, are present in the Arctic aerosol. Furthermore, the intensity of these modes indicates that, just as in urban samples, the graphitic structures represent a major component of the aerosol.

The filters collected at Barrow from fall through late spring have a grey or black

Reference p. 292

appearance similar to that found for urban particulates. For the urban samples, this optically absorbing component has been identified as graphitic carbon [1]. Preliminary measurements on a limited number of samples show that the absorption coefficient of the optically absorbing species in the Arctic has a $1/\lambda$ wavelength dependence, is insoluble in a wide range of solvents, and has a high temperature stability with an oxidation threshold of approximately 500°C. These results are consistent with the properties of graphitic carbon and strongly suggest that indeed the optically absorbing species at Barrow is graphitic in nature.

The seasonal variation of the optical absorptivity (or graphitic carbon concentration) of the Barrow aerosol is shown in Fig. 13. The absorption coefficient changes

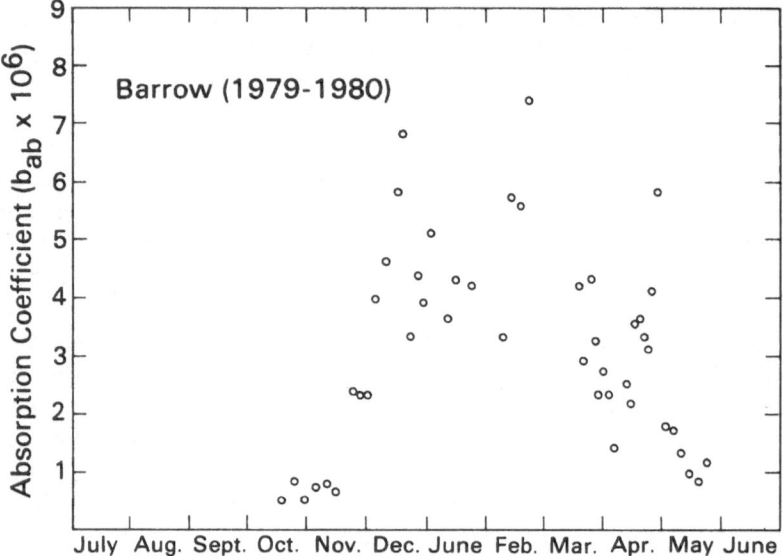

Fig. 13. Seasonal variations in the optical absorption coefficient in m^{-1} at Barrow from October 1979 to May 1980.

by more than an order of magnitude from mid-October to early January and remains at a relatively high level throughout most of February, March, and April. It decreases substantially in May. The magnitude of the absorption coefficient during the winter and spring is comparable to that found in urban environments (Table 1). The peak values in February are only about a factor of 10 less than the average absorption coefficients in New York City and a factor of 3 less than those found in Berkeley, California, and Denver, Colorado. This absorption coefficient is large enough to produce significant optical effects. For example, an absorption coefficient of 5×10^{-6} m^{-1} extended over a pathlength of 10 km would produce an optical thickness of 0.05 which is consistent with the large optical depths and low single scatter albedos suggested by the work of Shaw and Stamnes [16]. This would be a large perturbation on the transfer of optical radiation through the atmosphere when the sun's irradiance is at a high level, as it was during part of these pollution episodes. In order to assess the importance of these effects, more detailed measurements of the seasonal variation in the optical properties (visible and infrared) as well as the vertical and horizontal extent of the aerosols need to be determined.

TABLE 1

Comparison of Absorption Coefficients and Carbon Content
of Aerosols at Urban Locations and in the Arctic

Site	Date	Number of Samples	Average Absorption Coefficient x 10⁵ (m^{-1})	Average Total Carbon ($\mu g/m^3$)	Average Graphitic Carbon as Percent of Total Carbon[a]
Barrow, Alaska	12/79-4/80	33	.4	1.2	24
Argonne, Illinois	1/79-3/80	438	2.8	8.1	22
Gaithersburg, Maryland	1/79-3/80	381	2.1	6.1	22
Denver, Colorado	11/78-5/79	141	2.4	9.8	16
Anaheim, California	8/77-1/80	852	4.9	16.6	19
Fremont, California	7/77-3/80	924	3.4	12.0	18
Berkeley, California	6/77-4/80	998	2.1	6.7	20
New York, New York	11/78-4/80	439	6.4	15.2	27

[a]Calculated from optical constants in Ref. 13.

References p. 292

The pollution episodes observed at Barrow are not a local phenomenon but appear to be areawide with similar seasonal variations occurring at widely spaced sites across the Arctic [19, 20]. This is illustrated in Fig. 14, where a comparison of the monthly variations in the blackness of the filter deposits collected at Barrow and

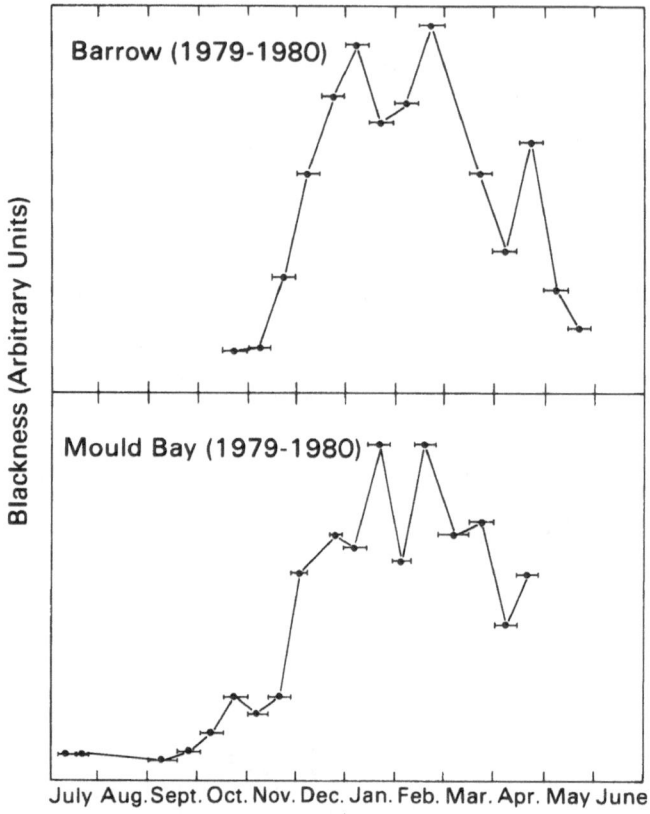

Fig. 14. A comparison between the seasonal variation in the blackness of filter deposits collected at Barrow, Alaska, and Mould Bay, Canada, in 1979 and 1980. Mould Bay samples were provided by L. A. Barrie and P. Hoff of the Canadian Atmospheric Environment Service.

Mould Bay, Canada, are shown. (The samples from Mould Bay were provided by Dr. L. A. Barrie and Dr. P. Hoff of the Canadian Atmospheric Environment Service.) These sampling stations are widely separated, yet the initial onset and duration of the episodes observed at these two sites are almost identical.

Using the optical constants of graphitic carbon determined from our urban studies [21], we can make an estimate of the graphitic carbon concentration at Barrow.

These results are shown in Fig. 15, which is a plot of the graphitic carbon concentration as a fraction of the total carbon content of the aerosol for various periods of time during the year. The results show a strong enrichment of the graphitic fraction

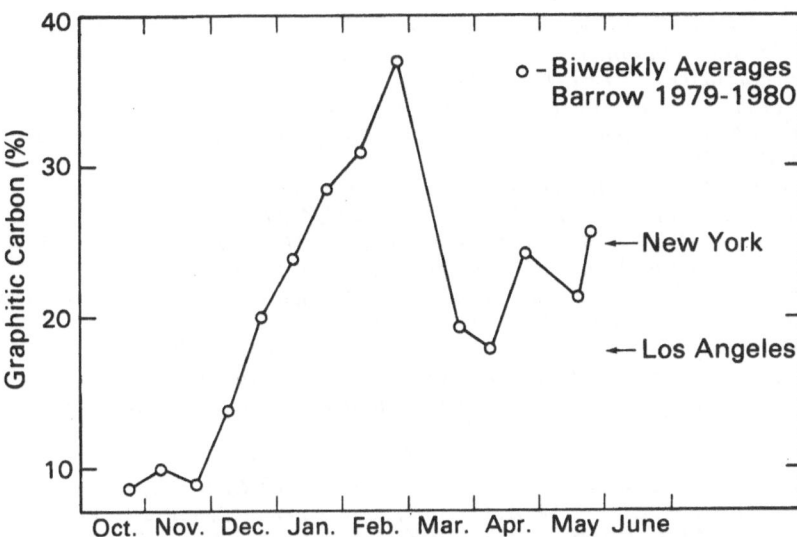

Fig. 15. Seasonal variation of graphitic carbon as a percentage of the total carbon content of the aerosol at Barrow from October 1979 to May 1980. These values are compared to the average values obtained in New York City and Los Angeles.

of the carbonaceous aerosol from early to late winter. The concentration of graphitic carbon in late February is almost 40 % of the carbonaceous mass. Remarkably, this percentage is higher than that found in urban centers like New York City and Los Angeles [22]. After its peak in February, the fraction of graphitic carbon seems to level off to a value which is more typical of those found in urban environments. The interpretation of these interesting features in terms of transport, atmospheric chemistry, and deposition processes is complicated and will have to await more detailed analyses and further measurements.

In summary, these observations indicate that significant graphitic concentrations can develop at remote locations. If one ignores the possible contribution of natural burning processes (e.g., forest fires), which are expected to be small during these times of the year in the northern hemisphere, this component can be attributed directly to anthropogenic combustion processes. The source and the climatic effects of these highly absorbing species are uncertain and will have to await more systematic measurements and careful modeling.

ACKNOWLEDGEMENT

This work was supported by the Biomedical and Environmental Research Division of the U.S. Department of Energy under contract No. W-7405-ENG-48 and by the National Science Foundation.

Reference p. 292

REFERENCES

1. H. Rosen, A. D. A. Hansen, L. Gundel and T. Novakov, Appl. Opt., Vol. 17 (1978), p. 3859.
2. Z. Yasa, N. M. Amer, H. Rosen, A. D. A. Hansen and T. Novakov, Appl. Opt., Vol. 18 (1979), p. 2528.
3. M. Sadler, R. J. Charlson, H. Rosen and T. Novakov, "An intercomparison of the integrating plate and the later transmission methods for determination of aerosol absorption coefficients," LBL-11176 (1980).
4. R. Dod et al., "Application of thermal analysis to the characterization of nitrogenous aerosol species," LBL-10735, Ch. 8 from LBL Energy and Environment Division Annual Report, 1979 (1980).
5. C. I. Lin, M. Baker and R. J. Charlson, Appl. Opt., Vol. 12 (1973), p. 1356.
6. A. Rosencwaig and A. Gersho, J. Appl. Phys, Vol. 47 (1976), p. 64.
7. R. E. Weiss, A. P. Waggoner, R. J. Charlson, D. L. Thorsell, J. S. Hall and L. A. Riley, "Studies of the optical, physical and chemical properties of light absorbing aerosols," in Proceedings, Conference on Carbonaceous Particles in the Atmosphere, Lawrence Berkeley Laboratory Report LBL-9037, (1979), p. 257.
8. R. E. Weiss, R. J. Charlson, A. P. Waggoner, M. B. Baker, D. Covert, D. Thorsell, and S. Yuen, "Application of directly measured aerosol radiative properties to climate models," in Atmospheric Aerosols: Their Optical Properties and Effects, NASA Report CP-2004.
9. H. Rosen and T. Novakov, Nature, Vol. 266 (1977), p. 708.
10. H. Rosen and T. Novakov, Atmos. Environ., Vol. 12 (1978), p. 923.
11. G. J. Rosasco, E.S. Etz and W. A. Cassate, Appl. Spectros., Vol. 29 (1975), p. 396.
12. F. Tuinstra and J. L. Koenig, J. Chem. Phys., Vol. 53 (1970), p. 1126.
13. H. Rosen, A. D. A. Hansen, L. Gundel and T. Novakov, "Identification of the graphitic carbon component of source and ambient particulates by Raman spectroscopy and an optical attenuation technique," in Proceedings, Conference on Carbonaceous Particles in the Atmosphere, Lawrence Berkeley Laboratory Report LBL-9037, (1979), p. 49.
14. H. Malissa, H. Puxbaum and E. Pell, Z. anal. Chem., Vol. 282 (1976), p. 109.
15. K. A. Rahn and R. J. McCaffrey, "On the origin and transport of the winter Arctic aerosol," Annals N.Y. Acad. Sci., Vol. 338 (1980), p. 486.
16. G. E. Shaw and K. Stamnes, Ibid. (1980), p. 533.
17. M. Mitchell, J. Atmos. Terr. Phys., Spec. Suppl, (1956), p. 195.
18. P. K. Mueller, R. W. Mosley and L. B. Pierce, "Carbonate and non-carbonate carbon in atmospheric particulates," Second International Clean Air Congress, Proceedings (New York, Academic Press, 1971).
19. L. A. Barrie, R. M. Hoff, S. M. Duggupaty, Atmos. Environ. (In press), paper presented at Second Symposium on Arctic Air Chemistry, Kingston, Rhode Island, 6-8 May 1980.
20. B. Ottar, Ibid (In press).
21. A. D. A. Hansen et al., "The use of an optical attenuation technique to estimate the carbonaceous component of urban aerosols," LBL-10735, Chap. 8 from LBL Energy and Environment Division Annual Report, 1979 (1980).
22. H. Rosen, A. D. A. Hansen, R. L. Dod and T. Novakov, Science, Vol. 208 (1980), p. 741.

DISCUSSION

J. Heintzenburg (Stockholm University)

Is it allowable to deduct the size of the graphitic component in the Arctic aerosols from the Raman Spectrum?

H. Rosen

It is for the crystalite size, but that is not the actual particle size. The crystallite size would be something like 35 to 40 angstroms.

M. Shelef (*Ford Motor Company*)

What are the two active Raman modes?

H. Rosen

One of them is reasonably well understood. According to the work of Tuinstra and Koenig;* that one is around 1600 wave numbers and it corresponds to an E_{2g} phonon propagating within the plane of the graphitic structure. The other Raman mode is really not well understood. There has been a considerable amount of work on it but we just do not know. Some have suggested that it may have to do with a Raman mode that is normally inactive in a graphite lattice but that is made active by the finite size of the graphitic structures.

M. Shelef

Would you observe this kind of mode in a large polynuclear aromatic molecule?

H. Rosen

No. The Raman spectra of polynuclear aromatics which are available in the literature are quite distinct from these. We did one experiment on Vilanthaone which is the largest aromatic structure we could find and it also happened to be black in appearance. This molecule also had a distinct Raman spectra.

G. Hidy (*Environmental Research & Technology, Inc.*)

Are you saying that the carbon particles in Raman scattering have roughly ten to twenty crystals?

H. Rosen

No. It would be something like this. Visualize these little crystallites in the following fashion. You have a planar structure which aromatic rings all coupled together. Then there is another plane underneath and so on. The only thing that the Raman spectrum is sensitive to is the structure within the plane. So, I am talking about the crystallite size within the plane. The combustion particles coming directly out of the exhaust may come out as 50 angstrom crystallites. When they get out into the air, they would coagulate and grow into larger particles which would typically be the size of that found in the urban aerosol.

---

*F. Tuinstra and J. L. Koenig, J. Chem. Phys. Vol. 53, p. 1126 (1970).

G. Hidy

These are crystallites which are basically the same size as the particle size.What I am wondering is, if these are spherical particles, will this additional stress on the crystal structure lead to an additional Raman mode?

H. Rosen

That is possible. The extra Raman mode that we have observed may be due to a distortion of the crystal strucutre. Also, if you look carefully at those Raman modes, there is a shift between polycrystalline graphite and carbon black, and indeed if you look very carefully in the ambient samples you can detect small shifts, much less than 1 %, in the vibrational frequency which may be due to some sort of disordering in graphitic structure. That would be very interesting to look at because you would expect more order to come out of the higher temperature combustion processes than lower temperature ones.

G. Hidy

What is the amount of graphitic carbon in the Barrow sample relative to Los Angeles and New York? Is it similar to the ratio from the Russian samples? I am just thinking it might be different because we burn things more efficiently in the United States.

H. Rosen

I would love to have samples to do that kind of analysis.

SESSION IV
AMBIENT MEASUREMENTS

Session Chairman

G. R. HILST

Electric Power Research Institute
Palo Alto, California

PARTICULATE CARBON AT VARIOUS LOCATIONS IN THE UNITED STATES

G. T. WOLFF, P. J. GROBLICKI, S. H. CADLE and R. J. COUNTESS

Environmental Science Department
General Motors Research Laboratories
Warren, Michigan

ABSTRACT

Particulate elemental and organic carbon concentrations were determined on filters collected between 1972 and 1980 at ten United States' sites representing urban, suburban, rural, and remote areas. The results showed that particulate elemental carbon is ubiquitous with mean concentrations ranging from 1.1 micrograms per cubic meter at the remote site in South Dakota to 13.3 micrograms per cubic meter in a congested area in New York City. About 80% of the elemental carbon mass consists of particles with a diameter of less than 2.5 micrometers. Particulates in this size range are responsible for most pollutant-related visibility reductions. Since it appears that elemental carbon is the only light-absorbing particulate species, the specific light-absorption coefficient for elemental carbon was calculated to be 12.7 m²/g while the specific light-scattering coefficient was 3.2 m²/g. Using these coefficients, the contributions of elemental carbon to the observed visibility reduction at the various sites are estimated. These range from 6 to 38%. Also discussed are the seasonal and diurnal variations of particulate elemental and organic carbon as well as the contribution from secondary organic particulates. In addition, an updated carbon-source apportionment, based on recent analytical developments, is presented for the Denver area.

INTRODUCTION

During the past few years, the potential importance of particulate elemental carbon has been recognized because of its role in atmospheric optics [1-3] and chemistry [4, 5]. In addition, it can also constitute a significant fraction of the respirable particulate fraction [6].

Although all particles between 0.1 and 1 micrometers scatter light, elemental carbon is the major particulate light-absorbing specie in the atmosphere [2] and this has two consequences. First, on a unit mass basis, it is the most efficient visibility-reducing particulate. For example, during the winter in Denver, elemental carbon comprises 15% of the fine particulate mass (diameter less than 2.5 micrometers), but it is responsible for 38% of the visibility reduction [3]. Second, the heating or cooling of the atmosphere will be sensitive to the amount of elemental carbon because of the light-absorption properties of this particulate specie [7].

Elemental carbon may also be important in atmospheric chemistry because of its catalytic activity. Laboratory studies have shown that elemental carbon promotes the catalytic oxidation of sulfur dioxide to sulfate [4] and acts as a site for the synthesis of amine, amide, and nitrile compounds from adsorbed ammonia and nitric oxide [5].

Despite these important characteristics of elemental carbon, measurements of ambient concentrations (and also source emission rates) are sparse. Accordingly, we have initiated research programs to determine the ambient concentrations and the emission rates from various sources. In this paper we focus on several fundamental questions concerning particulate elemental carbon . . . What are the ambient concentrations in urban and rural areas? . . . How is the material distributed between the fine and coarse particle size modes? . . . How does the concentration vary temporally? . . . What is its effect on visibility? . . . What are its sources?

The principal reason for the sparsity of elemental carbon data has been the lack of a suitable analyzer which could routinely determine the amounts of elemental carbon on filters. Consequently, before the above questions could be addressed, such an analyzer was developed and is discussed in the methodology section of this paper. Further details are contained in Cadle *et al.* [8]. Once this analyzer was developed, both particulate elemental and organic carbon measurements were incorporated into ongoing field programs [9-11]. In addition, carbon was also measured on stored filters collected in previous field studies.

METHODOLOGY

Site Locations — The locations where the carbon samples were collected are shown in Fig. 1 and described in Table 1. Urban sites were located in New York City, Washington, DC, Downey, CA, and Denver, CO. Suburban sites were located in Pleasanton and Pomona, CA and Warren, MI. Abbeville, LA and Luray, VA were rural sites, while the site near Pierre, SD was a remote site. For the first five sites listed in Table 1, carbon determinations were made during 1980 on stored high-volume sampler filters. At the remaining sites, sequential samplers or a dichotomous sampler were operated for the specific purpose of obtaining carbon data. For these sites, the carbon analyses were generally conducted within two months of sample collection. For the earlier sites, however, filters were stored up to seven years and it is possible that some loss of organic carbon particulate occurred. It is not expected that elemental carbon would be lost during storage.

TABLE 1

Information on Ambient Monitoring Sites

Site and Date	Type of sampler[a]	Sampling interval (h)	No of Samples	Site Description
1. New York City 2/10-3/6/72	Hi-Vol	24	21	East 45th St. and Lexington Avenue — heavy traffic area.
2. Washington, DC 6/9-6/28/72	Hi-Vol	24	18	Corner of 1st and D Streets — next to parking lot several blocks from busy thoroughfares.
3. Pleasanton, CA 8/13-9/5/72	Hi-Vol	24	24	Alameda County Fairgrounds — removed from local traffic sources.
4. Downey, CA 9/8-10/2/72	Hi-Vol	24	22	Grounds of Rancho Los Amigos Hospital — several blocks from industrial or traffic sources.
5. Pomona, CA 10/3-11/19/72	Hi-Vol	24	39	Los Angeles County Fairgrounds — minimal local traffic; about 3 km to freeway.
6. Pierre, SD 7/13-9/7/78	ERT SFS	4	59	40 km W-NW of Pierre on a grain farm — remote; no local sources; no detectable influence from Pierre.
7. Denver, CO 11/8-12/20/78	ERT SFS	4	231	In an unused parking lot of the Mile High Kennel Club in Commerce City — frequently downwind of Denver and major highways.
8. Abbeville, LA 8/5-9/11/79	GMR SFS	4	142	15 km N of Gulf Coast in rural field — infrequent traffic on road 50 m to S; no known local sources of particulate C.
9. Warren, MI 1979-1980	GMR SFS	24	79	Roof of GM Research Laboratories — many traffic and industrial sources within 2 km.
10. Luray, VA 7/14-8/15/80	Dicot	8	20	16 km N of Luray in rural field in Shenandoah Valley — no known local sources.

a Hi-Vol is a high-volume sampler.

ERT SFS is a pair of Environmental Research & Technology, Inc., sequential filter samplers operating at 60 liters per minute. One sampler collected particulates <30 µm diameter, while a second, fitted with a cyclone, collected particles <2.5 µm.

GMR SFS is a General Motors sequential filter sampler which collected total particulates and operates at about 20 liters per minute.

Dicot is a Beckman dichotomous sampler with two size cuts: ≤2.5 µm and 2.5 µm-15 µm.

References p. 314.

GMR Carbon Measurement Sites

1. New York City, NY	2/10-3/6/72
2. Washington, DC	6/9 -6/28/72
3. Pleasanton, CA	8/13-9/5/72
4. Downey, CA	9/8 -10/2/72
5. Pomona, CA	10/3-11/19/72
6. Pierre, SD	7/13-9/7/78
7. Denver, CO	11/8-12/20/78
8. Abbeville, LA	8/5 -9/11/79
9. Warren, MI	1979-1980
10. Luray, VA	7/14-8/15/80

Fig. 1. Locations and dates of carbon measurements.

Carbon Analyses — Filter samples were analyzed for apparent organic (Cao) and apparent elemental carbon (Cae) using thermal methods. The automated analyzer designed for this purpose has been described [8]. Initially, all samples were analyzed using a two-step pyrolysis-oxidation procedure. A punch from the filter was dropped into a furnce held at 650°C. A helium stream swept the pyrolyzed carbon over an oxidation catalyst where oxygen was added and the carbon was converted to CO_2. The CO_2, measured with a nondispersive infrared CO_2 analyzer (NDIR), corresponds to apparent organic carbon, Cao. After the pyrolysis step, oxygen was introduced into the sample compartment to remove the apparent elemental carbon as CO and CO_2 which were also passed over the oxidation catalyst and detected as CO_2 with the NDIR analyzer. Since carbonization or charring of organics during the pyrolysis step of the analyses can cause large errors with some samples, we have referred to the results as apparent organic, Cao, and apparent elemental carbon, Cae. Johnson *et al.* [12] have reported that air oxidation of the sample at a temperature which does not remove elemental carbon minimizes carbonization. We have obtained similar results [13]. Therefore, a representative number of filters from each site were reanalyzed using a 4-step thermal procedure. One punch from the filter was analyzed for total carbon. A second punch was heated in air at 350°C for 10 minutes in a furnace in order to remove as much of the organic carbon as possible. This second piece was then analyzed using the pyrolysis-oxidation procedure described above to determine elemental carbon. Organic carbon was determined by difference. The difference in the elemental carbon measured by the two procedures was calculated. Large variations in this difference occurred between samples collected at the same site, but the 4-step procedure always gave the lower value. This indicates that carbonization was occurring in the 2-step procedure. However, linear-regression analysis between the elemental carbon determined by these two methods for Denver Hi-Vol samples and Warren Lo-Vol samples showed correlations of 0.99 and 0.98, respectively. Therefore, it was concluded that valid corrections could be made to the average elemental carbon values generated by the 2-step thermal method. The correction factors are given in Table 2. All correction factors are simple averages except for Warren, which was determined by linear regression. Luray filters are not included since all these samples were analyzed by the 4-step method. It is possible that carbonization of organics is still occurring in this new analytical method, thus, we retain the 'apparent' nomenclature, but designate the corrected values as Cao' and Cae'.

There are several possible explanations for the large site-to-site variation in the carbonization error indicated in Table 2. Most natural products such as pollen or wood smoke carbonize and could be present in varying amounts in these total particulate samples. Also, it appears that most carbonization occurs with the higher molecular weight compounds which are extractable in polar solvents [13]. Finally, some of our recent work with pure compounds on different quartz and glass fiber filters shows carbonization can be affected by filter type.

TABLE 2
Change in Apparent Elemental Carbon after Oxidation

Site	Number of samples reanalyzed	Avg. % decrease in apparent elemental carbon after oxidation @ 350°C	Standard Deviation
Pierre, SD	10	3	12
Abbeville, LA	11	10	22
Warren, MI	20	11	9
Denver, CO	27	18	10
Washington, DC	7	18	14
New York, NY	6	32	7
Pomona, CA	6	40	21
Downey, CA	7	37	9
Pleasanton, CA	7	30	12

Duplicate and sometimes triplicate analyses were run on all Hi-Vol samples to help account for uneven loadings of the filters which had been folded in half for storage. Differences between duplicates averaged 10%. Repeat analyses on Lo-Vol

References p. 314.

filters generally showed better than 5% agreement. Blanks were determined from the edges of the Hi-Vol filters. The average blank for a given site was subtracted from each sample. Some Hi-Vol samples had been stored for several years in Mylar folders before analysis. It is assumed that any pickup of organics from the folder are accounted for by the blank measurements. We do not know if any organic material was lost through volatilization during storage.

Impactor Samples — At the Denver site, the particulate size distribution of carbon was determined by using impactors which operated on an 8-hour sampling schedule. The impactors were a modified version of the 8-stage impactor designed by Mercer *et al.* [14]. Instead of a single jet per stage, the lower stages of our units have multiple circular jets arranged symmetrically in an annulus about the axis of flow. At each stage, particulate material is impacted onto a removable 22 mm diameter glass cover slip. A 25 mm diameter final filter is housed in the bottom of the unit. Studies performed in this laboratory indicate that the equivalent aerodynamic cutoff sizes for the upper 7 stages of the impactor operating in Denver at a flow rate of 4 L/min^{-1} are: 7.8-, 3.6-, 2.0-, 1.0-, 0.52-, 0.28-, and 0.16-micrometers diameter, with the final filter catching all material less than 0.16-micrometer diameter. For the impactor data, fine particulate is defined as that material which is less than or equal to 2.0-micrometers diameter.

The impactors used for collecting carbon samples had uncoated coverslips since we could not find a suitable coating that was free of carbon. We minimized the particle bounce problem by removing all particles larger than 2.0-micrometers diameter with a 10-mm Dorr Oliver cyclone preseparator [15]. The final filter for these impactors was a 25-mm diameter Gelman Micro-quartz filter. All filter and glass coverslips used for sampling carbon were heated at 500°C for several hours before use in order to reduce the carbon blank. Carbon data was also obtained for all particles larger than 2.5 micrometers diameter with the sequential filter sampler [6]. Further details on the impactors are contained in reference [16].

Particulate Light Absorption Measurements — Measurements of light absorption by particulates were made during recent field studies. To accomplish this, particulate matter was collected on 0.4-micrometer pore-size 47 mm Nuclepore filters. A sample flow rate of 4 to 10 L/min maintained for 4 hours provided a deposit of particulate matter on a filter which had a total area of 12.8 cm^2.

Two simultaneous samples were taken in Denver, i.e., total particulate was collected on an open-faced filter holder, and fine particulate was collected with an in-line filter holder preceded by a 10 mm Dorr Oliver cyclone which had a 50% cutoff size of 2.2 micrometers at the 4 L/min flow rate [15].

A modification of the Integrating Plate Method (IPM) of Lin *et al.* [17] was used for the measurement of the absorption coefficient (bap) of the Denver aerosol. In this method, particulate collected on a filter is supported on an opal glass plate and placed in a light beam. The opal glass serves to direct light scattered by the particles back to the detector. The amount of light transmitted through the filter and particles is compared with that transmitted through the blank filter. The difference in transmission is ascribed entirely to light absorption by the particles. We modified the method of Lin *et al.* by placing the clean side of the filter on the opal glass [18] and by reducing the optical beam size and scanning the filter across the beam, thus allowing the use of the unexposed filter edge as the reference. Measurements were made using 550 nm light. Additional details of the measurement method and a comparison of results with the original method are given in reference [3].

RESULTS AND DISCUSSION

Summary of Measurements — Average data from all of the sites are summarized in Table 3. As expected, higher concentrations of particulate elemental carbon, Cae', were observed in the urban areas, with the highest levels being recorded in downtown New York City. As one moves from the urban areas to the remote areas, a distinct gradient is observed. All the urban areas have Cae' levels greater than the suburban areas, while the suburban values exceed the rural values which are greater than those observed at the remote site.

This pattern, however, is not as consistent for the particulate organic, Cao', concentrations. Although the highest levels were seen in New York City (19.8 micrograms per cubic meter), the next highest levels were found at rural Abbeville

TABLE 3
Summary of Mean Ambient Particulate Carbon Measurements

Site	Particulate carbon concentration ($\mu g/m^3$)			
	Cae' [a]	Cao' [b]	Cae'/TSP [c]	Cae'/Ct [d]
Urban				
New York City	13.3	19.8	0.09	0.40[e]
Washington	6.5	5.1	0.11	0.56[e]
Denver (total	5.4	10.4	0.06	0.34
Denver (fines)	4.4	7.6	—	0.35
Downey	4.1	5.9	0.06	0.41[e]
Suburban				
Warren	3.7	8.6	0.06	0.29
Pleasanton	3.2	6.4	0.03	0.33[e]
Pomona	3.6	8.0	0.04	0.31[e]
Rural				
Abbeville	1.7	10.8	0.04	0.14
Luray	1.7	7.7	—	0.18
Remote				
Pierre				
(total)	1.1	5.1	0.08	0.18
(fines)	0.8	3.4	—	0.19

[a] Corrected elemental carbon.

[b] Corrected organic carbon.

[c] Total suspended particulate mass.

[d] Total particulate carbon.

[e] Possible loss of Cao' in these samples due to long storage time.

(10.8 micrograms per cubic meter), Denver (10.4 micrograms per cubic meter), Warren (8.6 micrograms per cubic meter), and Pomona (8.0 micrograms per cubic meter). The lowest values (5.1 micrograms per cubic meter) were observed in Washington, DC and at our remote site in South Dakota. The South Dakota site, however, had appreciable coarse mode (diameter greater than 2.5 micrometers) Cao', which may have been due to nearby grain farming operations.

The last column in Table 3, elemental carbon/total carbon (Cae'/Ct), is analogous to the optical attenuation/Ct ratio that Rosen *et al.* [19] used as an indicator of secondary the highest ratios occur in the urban areas, and systematically decrease as one moves out to the suburban and then to the rural areas. The deviation from this pattern at the remote site may be due to larger uncertainties associated with the low concentration of both Cae' and TSP.

The last column in Table 3, elemental carbon/total carbon (Cae'/Ct), is analogous to the optical attenuation/Ct ratio Rosen *et al.* [19] used as an indicator of secondary organic production, i.e., the lower this ratio the higher the amount of secondary organic particulate present. For primary emissions, this ratio should be nearly constant. For example, our Denver source-estimates [20] which will be summarized in a subsequent section, indicate that the overall ratio of Cae'/Ct for primary emissions in Denver is 0.37. Consequently, the mean ratio of 0.35 observed in the ambient samples suggests that secondary organic production was minor during the Denver study. Furthermore, since the ratio for motor vehicle and natural gas emissions was 0.36 and since these sources are important contributors of carbon particulate emissions in many other urban areas, a ratio of about 0.36 will probably be a reasonable ratio for bulk primary emissions in other areas. Hence, primary organic particulate appears to be dominant (over secondary organic particulate) at all of the urban sites listed in Table 3. Although some loss of organic particulate may have occurred during storage of the New York, Washington, DC, and Downey samples, the magnitude of the loss probably would not account completely for the high ratios observed at those sites.

The ratios for all the surburban sites are lower than 0.36 and for the rural and remote sites they are a factor of 2 or more lower than 0.36. This could be due to two factors other than the presence of secondary organic particulate material: organic material from biogenic sources at the suburban and rural sites, or emissions from sources with a low Cae'/Ct ratio. The first factor is more likely to be important at the sites where the samples were not size-segregated. However, since biogenic material is primarily in the coarse-size fraction (more than 2.5 micrometers), it could not account for the low ratios for the fine particulate fraction (less than or equal to 2.5 micrometers) in Luray (0.15) and Pierre (0.19). Since most of the fine particulate matter at the rural and remote sites was due likely to the transport of aged emissions from various upwind directions, their chemical composition should be a composite of the significant upwind sources. Unless these upwind sources differ significantly from those in Denver, one would expect a ratio of about 0.36. Consequently, secondary production of organic particulate appears to be a logical explanation for the lower ratio at the rural and remote sites. Other factors such as time of year and ozone levels may influence the Cae'/Ct ratio and will be discussed in the section on temporal patterns.

Particle Size — Particle size data for ambient elemental carbon was obtained at both the Denver and South Dakota sites. On the average in Denver, 82% of the Cae' mass was in the fine particulate mode (less than or equal to 2.5 micrometers) compared to 74% in South Dakota. For organic carbon, the percentages in the fine modes were 71% and 67%, respectively.

Using 8-stage impactors, described by Countess *et al.* [16], 120 detailed size distributions of fine carbon were obtained during the Denver study. A cyclone was used at the inlet of the impactor to remove particles less than 2.2 micrometers diameter. The mean distributions are reproduced in Fig. 2. For elemental carbon, the temporal variations in the size distribution were extremely small as the mass median aerodynamic diameter for the fine fraction was 0.28 micrometers, with a standard deviation of the geometric mean of only 0.03 micrometers. The respective values for organic carbon were 0.32 and 0.06 micrometers. In addition, the distribution of both species in the fine fraction appears to be log-normal.

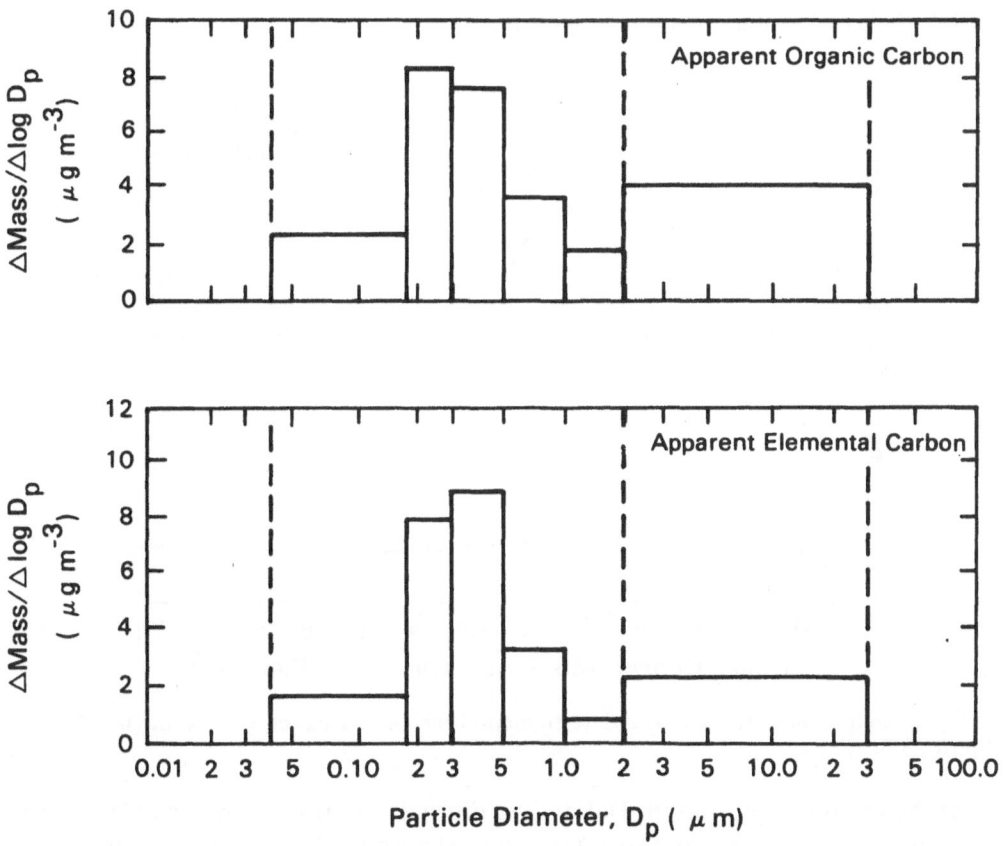

Fig. 2. Mean size distributions of organic and elemental carbon in Denver.

Temporal Patterns — Information on temporal patterns is of interest because it can provide information on sources and chemistry. Our data sets contain two types of temporal variations. In Denver, South Dakota, and Louisiana, 4-hour sampling intervals were used so diurnal variations could be examined. In Warren, samples were collected in the fall of 1979 as well as the winter and late spring of 1980 so that seasonal variations could be analyzed.

Diurnal Variations – The mean diurnal patterns at the three sites are shown in Fig. 3. For the urban site, Denver, the diurnal pattern is determined by the local patterns in diurnal emissions and atmospheric ventilation. The maximum between 8-1200 h is due to maxima in emissions from motor vehicles and natural gas emissions [20]. In addition, at 800 h the ventilation is generally at a daily minimum and does not begin to improve until after 1000 h. The best mixing occurs in the afternoon and this reflected by a Cae′ minimum. From this time on, the Cae′ steadily increases to a second maxima between 1800-2400 h. This second peak has been attributed to wood burning emissions during the evening hours when the nocturnal inversion is becoming reestablished [20].

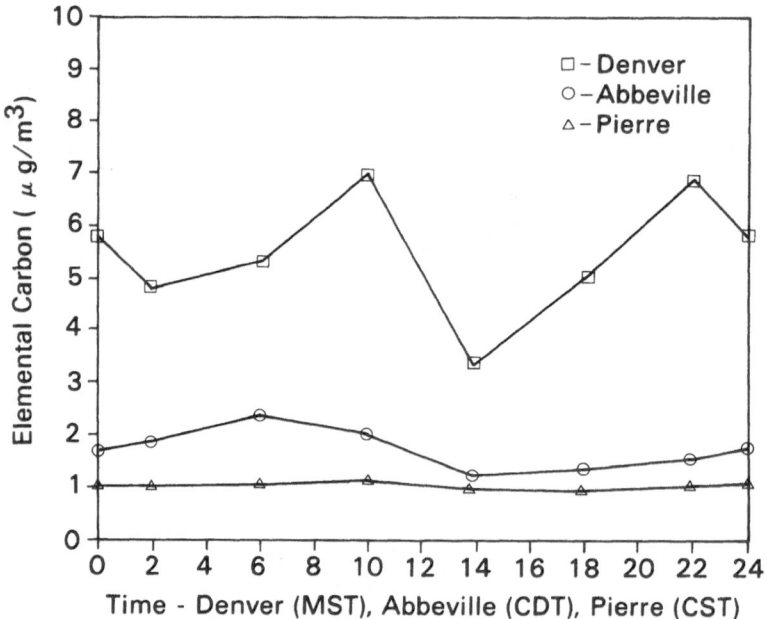

Fig. 3. Diurnal patterns of particulate elemental carbon at Denver, CO, Abbeville, LA and Pierre, SD.

At the remote site in South Dakota, no diurnal variation is preceivable. This is expected in areas far removed from sources. The Abbeville profile is a combination of the other two. The Cae′ maximum and minimum occur during the worst and best periods of ventilation, respectively, but the absolute magnitude of the variations are small compared to Denver. Consequently, there appears to have been a minor influence from local sources on Cae′ concentrations in Abbeville.

The diurnal patterns for Cao' are similar to those in Figure 3, but there are some subtle differences which are more discernible when the Cae'/Ct ratio is examined (Fig. 4). In both Denver and Abbeville, the ratio attains a minimum value during the time of day with maximum photochemical activity. Although these differences are small, they are statistically signficant at the 95% confidence level in Denver and the 90% level in Abbeville. In Denver, particulate nitrate, NO_3^-, which is formed photochemically, typically peaked between 12-1600 h [20]. In Abbeville, a broad ozone, O_3, maximum occurred between 12-1900 h. Partitioning the days into cloudy and sunny days (defined as the days in the lower and upper quartiles of ultraviolet light intensity) also shows a consistent pattern. In Denver, the ratio during the 12-1600 h period for the fine fraction on sunny days averaged 0.30, while on cloudy days it was 0.33. For Abbeville, the values were 0.12 and 0.14, respectively. Consequently, the variations in the 12-1600 Cae'/Ct ratio are consistent with photochemical production of secondary Cao'. In both cases, however, the magnitude of local secondary organic particulate seems to be quite small and would account for only a small fraction of the total observed Cao'. For example, the increase in Cao' concentrations on sunny afternoons as compared to cloudy afternoons was 0.7 micrograms per cubic meter in Denver and 0.2 micrograms per cubic meter in Louisiana.

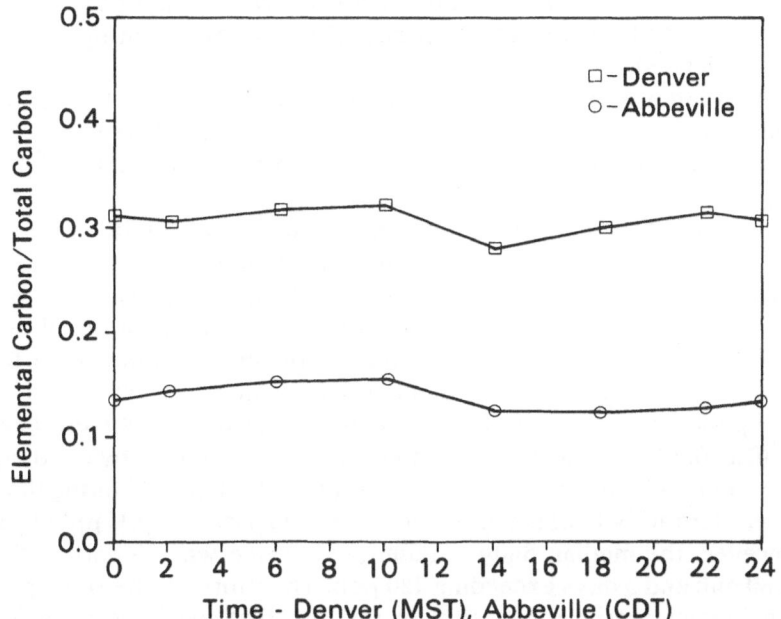

Fig. 4. Diurnal patterns of the ratio of particulate elemental carbon to total particulate carbon at Denver, CO and Abbeville, LA.

Seasonal Variations – Carbon data from Warren, MI, segregated by season, are presented in Table 4. The mean Cae' concentrations for each of the three seasons are essentially identical, and only slight differences were observed in the maxima. An examination of local and synoptic scale meteorological data, however, shows that

References p. 314.

WOLFF, GROBLICKI, CADLE, COUNTESS

TABLE 4
Carbon Data from Warren, MI, Averaged by Season

Season	Dates	Mean Cae′	Max Cae′	Mean Cao′	Mean TSP	Mean Cae′/Ct
		μg/m³				
Fall	10/5-10/31/79	3.6	8.7	11.2	54.6	0.21
Winter	1/22-2/18/80	3.8	12.6	6.6	57.5	0.36
Spring	5/20-6/13/80	3.7	7.9	8.0	90.7	0.30
Mean	all seasons	3.7	—	8.6	66.7	0.29

there was a significant seasonal difference in the type of meteorology which accompanied the higher (greater than or equal to 6 micrograms per cubic meter) concentration days. In the fall, the highest levels occurred with a southwest air flow associated with the backside of high-pressure systems and the intrusion of warm humid maritime tropical air into the Great Lakes region. These days also experienced the highest TSP values and most restricted visibility due to haze. Since previous studies have shown that these meteorological systems are effective in transporting certain pollutants, including fine particulate matter, long distances [11,21,22], it is likely that a significant fraction of the Cae′ observed on those days in Warren was not emitted locally. During the spring, similar conditions occurred on the higher Cae′ days when high ozone levels were also experienced.

In the winter, intense nocturnal inversions accompanied by weak pressure gradients appear to be the conditions most conducive for high Cae′ concentrations. Consequently, the winter concentrations appear to be dominated by local emissions.

In contrast to the Cae′ concentrations, the mean Cao′ and TSP values did vary considerably from season to season (see Table 4). Obviously, the Cae′/Ct ratio also showed considerable variation: fall, 0.21 winter, 0.36, and spring, 0.30. Based on our previous discussions, this could be interpreted as an indication of negligible secondary Cao′ production in the winter and significant production in the fall. As a result, the data was examined for possible relationships between Cae′/Ct and an indicator of photochemical activity, ozone. First, mean values were examined. There was little photochemical acitivity in the fall, as evidenced by the daily maximum ozone values which ranged up to 53 parts per billion (ppb). During the winter, there was even less activity as the maximum ozone (O_3) was only 35 ppb. During the spring, however, the median daily maximum O_3 value was 55 ppb, with 6 days exceeding 80 ppb and 3 days exceeding 120 ppb. Therefore, it does not appear that local photochemical activity can explain the fact that the lowest Cae′/Ct ratio occurred in the fall.

Since the greatest variations in O_3 were in the spring, this data set was used to examine possible statistical relationships between Cae′/Ct and O_3. A graph of these two parameters (Fig. 5) also indicates there is no relationship between O_3 and secondary Cao′ production.

In summary, we found no evidence for a relationship between O_3 and secondary Cao′ production at Warren, MI. This agrees with the results of Rosen et al. [19].

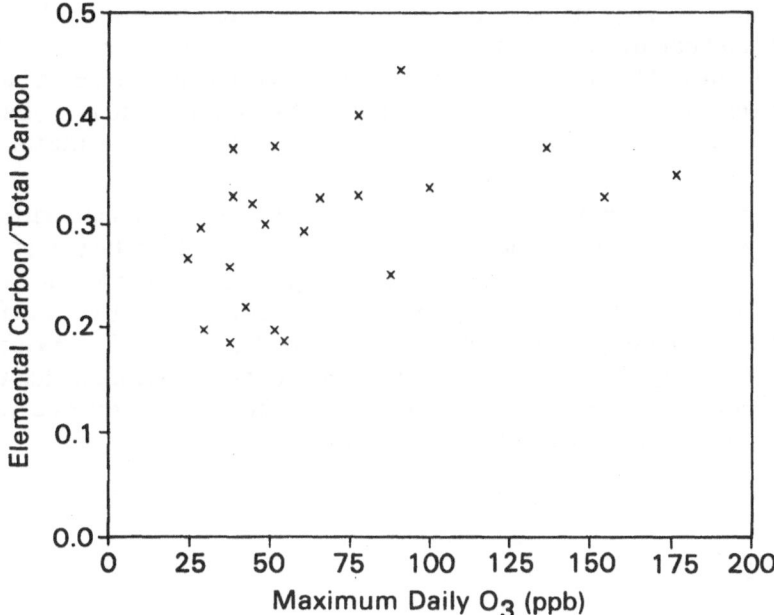

Fig. 5. Plot of maximum daily ozone versus Cae′/Ct in Warren, MI during the 1980 spring sampling period.

The only consistent relationship which was observed at all the sites was the decrease in the ratio as one moved from urban to suburban to rural areas, which seems to indicate that an aged air mass may have lower Cae′/Ct ratios. It is possible then, that the seasonal difference observed in Warren is due to a seasonal variation in the mean age of particulates in the prevailing air masses.

Contribution of Cae′ to Visibility Reduction — As previously mentioned. Cae′ is the most effective visibility-reducing specie because it both absorbs and scatters light. The purpose of this section is to develop a relationship between Cae′ and extinction so that the fractional contribution of Cae′ to visibility reduction at the sites examined in this paper can be estimated.

Visibility is inversely proportional to the extinction coefficient [23], bext, which is the sum of several terms:

$$bext = bsp + bap + bsg + bag.$$

The first term, bsp is the extinction due to scattering of light by particulate matter and it is measured with an integrating nephelometer operating at ambient temperature and humidity conditions. In many previous studies, the practice of operating at ambient conditions was not followed for various reasons, so those values are only approximations of the actual bsp. A detailed discussion of this appears in Groblicki *et al.* [3]. All nonambient nephelometer data used in this report will be so indicated. The second term, bap, is the extinction coefficient due to light absorption by particulate matter, and was determined by the integrating plate method [17] described in the experimental section, or it was predicted from the relationship between Cae′ and bap derived later. Rayleigh scattering of light by gases, bsg, is a constant term and will be neglected in our analysis because it represents baseline extinction in a clean, particulate-free atmosphere. Absorption due to NO_2, bag, will also be neglected, because it typically represents only a few percent of the total light extinction [3].

Based on our Denver analyses, Cae′ appears to be the only significant particulate light-absorbing specie [3]. Consequently, it is assumed that all of the bap is due to Cae′. A plot of bap versus Cae′ (total) for 305 data points obtained in Denver, Warren, and Louisiana is shown in Fig. 6. The slope of the regression line forced through zero corresponds to a specific bap/Cae′ ratio of 12.7 m^2/g. From Groblicki *et al.* [3], the mean bsp/Cae′$_f$ ratio is 3.2 m^2/g. Since the scattering is due exclusively to the fine Cae′ [3], we will assume that 80% of the Cae′ is in the fine fraction for the sites where this percentage was not determined.

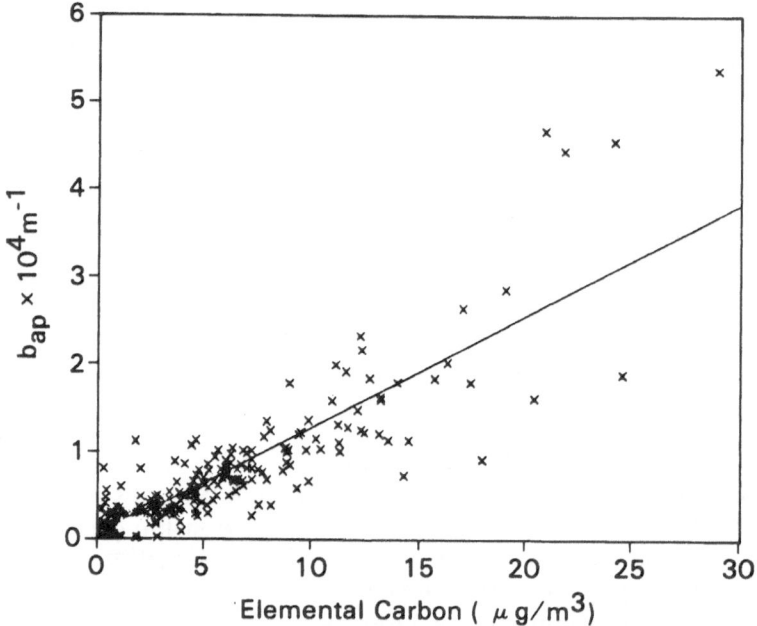

Fig. 6. Relationship between bap and Cae′.

The results of these calculations for sites where bsp data was available are shown in Table 5. For those sites where the ambient air was heated, New York City and Abbeville, the percentages in the last column are overestimates because the bsp value does not include the influence of water associated with the particulates. A comparison of heated and ambient nephelometers in the Denver study [3] suggests that the bsp for the heated air may be underestimated by about a factor of 1.6. If this is the case, the percent of extinction due to Cae′ would decrease to 35% in New York City and to 6% in Abbeville, LA. In Downey, Pleasanton, Pomona, and Pierre the nephelometer was in an air-conditioned laboratory. Since the ambient temperature fluctuated above and below the laboratory temperatures, the mean bsp value

TABLE 5

Contribution to the Visibility Reduction Due to Elemental Carbon at Various Locations

Site	Mean Cae′	Mean Cae′ f	Calculated contribution from Cae′ bap	Calculated contribution from Cae′ bsp	Measured bsp	Total bext	% bext due to Cae′
	— μg/m³ —				— x 10⁻⁴m⁻¹ —		
New York	13.3	10.6[a]	1.7	0.3	2.5 [b]	4.2	48
Denver	5.4	4.4	0.7 [c]	0.1	1.5	2.2	36
Downey	4.1	3.3[a]	0.05	0.1	2.2 [d]	2.7	22
Pleasanton	3.2	2.6[a]	0.4	0.1	0.9 [d]	1.3	38
Pomona	3.6	2.9[a]	0.5	0.1	2.8 [d]	3.3	18
Abbeville	1.7	1.4[a]	0.22	0.04	1.25[e]	1.47	18
Pierre	1.1	0.8	0.14	0.03	0.50d	0.64	27

[a] Assume 80% of Cae′ is in fine fraction.

[b] In heated room; therefore, an underestimate.

[c] Actual measurement of bap.

[d] Air-conditioned laboratory.

[e] Heated; therefore, underestimated.

may be fairly representative of the actual bsp. Even with all of these shortcomings, however, the data indicate that elemental carbon contributes significantly to the reduced visibility at all of the sites examined.

Sources — In light of the significant impact of particulate carbon on visibility, it is important to determine the sources of this carbon. Unfortunately, few carbon emission measurements have been made, especially at high altitude, so the uncertainty becomes large when these measurements are extrapolated to an entire urban area. Nevertheless, we attempted a carbon emission inventory for the Denver area for the November-December 1978 period. We estimated that for most sources, the uncertainty is about a factor of 2 [20]. The information (or techniques) which was used to construct the carbon inventory included: measurement of tracer species, the Colorado emissions inventory, fuel usage data, vehicle miles traveled (VMT) data, published emission factors, emission tests conducted by GMR[24], and radiocarbon dating. Except for three differences, the procedures employed are described in detail in reference [20]. Consequently, only these differences will be discussed here.

The first difference is that the ambient Denver data used in this report was corrected for carbonization. Second, we also corrected the light-duty vehicle emissions for carbonization. For noncatalyst vehicles, 12% of the Cae, and for catalyst-equipped vehicles, 15% of the Cae, was due to carbonization. The third change involves the Cae′ emission rate from natural gas combustion. The fraction of Cae′ in the total particulate emissions was calculated by assuming it to be equal to the ratio of bap for total natural gas emissions to bap for Cae′. Originally, we used a value of 4.22 m²/g [25] for the total and 8.3 m²/g [26] for the Cae'. Since the 8.2 m²/g value was determined in a laboratory from acetylene flame emissions, we felt it would be more appropriate to use the actual bap/Cae' ratio of 14.5 m²/g observed in Denver's ambient air. This reduces the fraction of Cae' in natural-gas particulate emissions from 0.51 to 0.29.

The results of the Cae' and Cao' budget calculations are shown in Table 6 and 7. In general, no single source dominates, but, combined wood-burning and motor vehicles account for 74% of the Cae' and 55% of the Cao'. Natural gas and fuel-oil combustion account for most of the remainder.

TABLE 6
Estimated Source Contributions of Fine Elemental
Carbon (Cae'$_f$) in the Denver Area.

Source	Daily emission rate, kg/day	% of total Cae'
Light-duty catalyst vehicles	39	2
Light-duty noncatalyst vehicles	144	5
Diesel vehicles	632	24
Natural gas	292	11
Fuel oil	217	8
Coal	19	1
Aircraft	199	7
Wood	1113	43
Sum of known sources		101

TABLE 7
Estimated Source Contributions of Fine Organic
Carbon (Cao'$_f$) in the Denver Area

Source	Daily emission rate, kg/day	% of total concentration
Light-duty vehicles	655	15
Diesel trucks	538	13
Natural gas	714	17
Fuel oil	400	9
Aircraft	199	3
Wood	1158	27
Sum of known sources		84
Unknown sources		16

SUMMARY AND CONCLUSIONS

The data presented in this report demonstrate the ubiquitous nature of particulate elemental carbon. Mean concentrations range from 1.1 micrograms per cubic meter at the background site in South Dakota to 13.3 micrograms per cubic meter in a congested area in New York City. In general, the highest concentrations were observed in the urban areas with a decreasing concentration trend as one moves from urban to suburban to rural to remote areas.

In Denver, 82% of the Cae' was in the fine particulate fraction (i.e., diameter less than or equal to 2.5 micrometers) while 74% was in the fine fraction in South Dakota. Impactor measurements indicate that the mass median aerodynamic diameter for the fine fraction is about 0.28 micrometers.

The diurnal variations of Cae' are a reflection of the diurnal variation in local emission rates and local mixing conditions. At the South Dakota site, where there were no appreciable local emissions, the profile was flat. No seasonal variations in Cae' were observed at the one site where seasonal data were available, i.e., Warren, MI.

Significant site-to-site and seasonal variations as well as small diurnal variations in the ratio of elemental carbon to total carbon, Cae'/Ct, were observed. Since the collective ratio for major carbon sources is about 0.36, ratios less than 0.36 probably indicate secondary organic particulate production. An examination of this hypothesis led to two conclusions. First, there does not appear to be a direct relationship between ozone, an indicator of photochemical activity, and Cae'/Ct. Second, there appears to be a qualitative relationship between Cae'/Ct and the age of a given air mass, which indicates that the amount of secondary organic particulate increases with age.

Using data from Denver, Abbeville, and Warren, the specific light absorption coefficient per unit mass of elemental carbon was determined to be 12.7 m^2/g. With several assumptions, it was then possible to estimate the percentage of the extinction (or visibility reduction) due to Cae'. At all sites, the contribution was significant. It ranged from a low of 6 to 18% at Abbeville to 34 to 48% in New York City. The ranges are due to measurement uncertainties rather than variations.

In Denver, we employed source-estimation procedures which were able to account for essentially all of the ambient Cae'. The major sources include: wood burning, 43%, motor vehicles, 31%, and natural gas, coal, and fuel-oil combustion, 20%.

While this report is one of the first papers to present such a large Cae' data base, it illustrates how little is known about ambient levels, trends, variations, and sources of carbon particulate, especially the elemental carbon fraction. Questions concerning the importance of secondary organic particulate have also been raised. In addition, with the present trend towards the use of fuels rich in carbon emissions (i.e., wood and diesel fuels), it is expected that these issues will receive continued attention.

ACKNOWLEDGMENTS

The authors are grateful to the members of the Environmental Science Department who assisted in the data collection and chemical analysis. They included: Martin Ferman, Nelson Kelly, Carolina Ang, Carrie Masters, Gerald Morris, Patricia Mulawa, Martin Ruthkosky, William Scruggs, David Stroup, and Jerome Zemla. In addition, Denise Pierson provided computer programming and data management skills.

References p. 314.

REFERENCES

1. *A. P. Waggoner and R. J. Charlson, in "Denver Air Pollution Study, 1973, Vol. II," p. 35, EPA-600/9-77-001, (1977).*
2. *H. Rosen, A. D. A. Hansen, L. Gundel, and T. Novakov, Appl. Optics, Vol. 17 (1978), p. 3859.*
3. *P. J. Groblicki, G. T. Wolff and R. J. Countess, Atmos. Environ, Vol. 15 (1981), p. 2473.*
4. *T. Novakov, S. G. Chang, and A. B. Harker, Science, Vol. 186 (1974), p. 259.*
5. *S. G. Chang and T. Novakov, Atmos. Environ., Vol. 9 (1975), p. 495.*
6. *R. J. Countess, G. T. Wolff, and S. H. Cadle, J. Air Pollut. Control Assoc. Vol. 30 (1980), p. 1194.*
7. *R. A. Reck, Science, Vol. 186 (1974), p. 1034.*
8. *S. H. Cadle, P. J. Groblicki, and D. P. Stroup, Anal. Chem., Vol. 52 (1980), p. 2201.*
9. *N. A. Kelly, G. T. Wolff, M. A. Ferman, Atmos. Environ. (In Press); Also available from General Motors Res. Labs as GMR-3598).*
10. *G. T. Wolff, P. J. Groblicki, R. J. Countess, and M. A. Ferman, "The Design of the Denver 'Brown Cloud' Study", General Motors Research Laboratories, Warren, MI, Publication GMR-3050, (1979).*
11. *G. T. Wolff, N. A. Kelly, and M. A. Ferman, Science, Vol. 211 (1981), p. 703.*
12. *R. L. Johnson, J. J. Shah, and J. J. Huntzicker, presented at Conference on Sampling and Analysis of Toxic Organics in the Atmosphere, Boulder, CO, August (1979).*
13. *S. H. Cadle and P. J. Groblicki, (These proceedings) p. 89.*
14. *T. T. Mercer, M. I. Tillery, and G. J. Newton, J. Aerosol Sci., Vol. 1 (1970), p. 9.*
15. *T. Chan and M. Lippmann, Environ. Sci. Technol., Vol. 11 (1977) p. 377.*
16. *R. J. Countess, S. H. Cadle, P. J. Groblicki, and G. T. Wolff, J. Air Pollut. Control Assoc., Vol. 31 (1981), p. 247.*
17. *C. I. Lin, M. Baker, and R. J. Charlson, Appl. Optics, Vol. 12 (1973), p. 1356.*
18. *R. E. Weiss, A. P. Wagonner, D. L. Thorsell, J. S. Hall, L. A. Riley, and R. J. Charlson, in Proceedings, "Carbonaceous Particulate in the Atmosphere," LBL-9037, Lawrence-Berkely Laboratories, Berkeley, CA, (1979), p. 257.*
19. *H. Rosen, A. D. A. Hansen, R. L. Dod, and T. Novakov, Science, Vol. 208 (1980), p. 741.*
20. *G. T. Wolff, R. J. Countess, P. J. Groblicki, M. A. Ferman, S. H. Cadle, and J. L. Muhlbaier, Atmos. Environ, Vol. 15 (1981), p. 2485.*
21. *G. T. Wolff and P. J. Lioy, Environ. Sci. Technol. Vol. 14 (1980), p. 1257.*
22. *G. T. Wolff, Ann. N. Y. Acad. Sci., Vol. 338 (1980), p. 379*
23. *W. E. K. Middleton, Vision Through the Atmosphere, University of Toronto Press, Toronto, Ontario, Canada, (1963).*
24. *J. L. Muhlbaier and R. L. Williams, (These proceedings), p. 185.*
25. *J. L. Nolan, M. S. Thesis, Department of Civil Engineering, University of Washington, Seattle, WA, (1977).*
26. *D. M. Roessler and F. R. Faxvog, J. Opt. Soc. Amer., Vol. 69 (1979), p. 1699.*

DISCUSSION

W. White, *(Washington University)*

Do you have any thoughts on why your specific absorption is so much higher than the value that Roessler and Faxvog* calculated from Mie theory and measured in the laboratory?

---

*D. M. Roessler and F. R. Faxvog, J. Opt. Soc. Amer., Vol. 59, p. 1699 (1979).

G. Wolff, *(General Motors Research Laboratories)*

There are certain assumptions that went into their calculation. If they used different assumptions they could get a higher specific adsorption. There have been some other photoacoustic measurements at Ford that have come out as high as 17 m²/g. Also, there is a possibility that there is something else besides elemental carbon that is contributing to the absorption and this would cause an apparent increase in the specific absorption coefficient. In addition, it is a function of particle size.

W. White

As I recall, their largest value for any size was half of your value.

G. Wolff

Their largest was about 8m²/g. We are considerably higher than that but there have been other estimates higher than ours.

D. Roessler, *(General Motors Reseach Laboratories)*

The calculation from Mie Theory was based on spheres. There are all sorts of problems with that calculation. It gives a surprisingly good agreement with the ambient aerosol measurements that George was talking about. We have done some measurements on exhaust and we can give you any number you like. Our experimental measurements give numbers from less than 5 m²/g to 15 m²/g.

J. Daisey, *(New York University)*

Just a comment on your ratios, George, I think you have to be a little careful with those for this reason. I noticed in the beginning you had some four-hour samples and 24-hour samples. We have done short term studies and found that sampling-time effects the organic carbon which is consistent with Bruce Appel's work. It depends on the fraction, but for non-polar materials, it looks as though the longer you sample, the higher the apparent concentration of organic carbon that is observed.

G. Wolff

We are aware of this problem and we will be examining it more closely.

OPTICAL MEASUREMENTS OF AIRBORNE SOOT IN URBAN, RURAL AND REMOTE LOCATIONS

R. E. WEISS and A. P. WAGGONER

University of Washington
Seattle, Washington

ABSTRACT

The Integrating Plate Method and integrating nephelometry were used to measure the light absorption and scattering extinction coefficients at 15 urban and rural locations. Graphitic carbon is highly absorbing and for most ambient aerosols is probably the dominant absorbing material. Average absorption coefficients and albedos for single scattering were (2.7-11.8) x 10^{-5} m^{-1} and 0.5 to 0.65 respectively for the urban areas and (0.6-3.7) x 10^{-5} m^{-1} and 0.73-0.87 for the rural areas. For sites where the submicron sized aerosol only was analyzed, average specific absorption was (1.4-1.8) m^2/g and (0.7-1.0) m^2/g for the urban and rural sites respectively. These correspond to an average graphitic carbon content of about 20 percent for the urban areas and 10 percent for the rural areas. Inversions of the albedo for single scattering give similar graphitic carbon fractions for the submicron sized aerosol.

INTRODUCTION

Possible increases in ambient levels of airborne soot due in part to increases in the number of diesel vehicles has increased interest in the effects of aerosol soot on human health and visual air quality. Germane to understanding the effects and atmospheric cycles of soot is the development of techniques to identify and quantify soot separately from other forms of carbon. Techniques currently used are based on combustion properties or Raman spectra [1]. However, the light absorption properties of soot may also provide a useful measurement method. Soot is probably unique as an effective absorber among the persistent and ubiquitous components of ambient aerosol, and the fraction of soot should be quantifiable by comparing the absorption per unit mass of the aerosol with that of pure soot.

Reported here are seven years of light absorption and scattering extinction measurements taken at a variety of sites that include urban, rural and remote locations.

References p. 324

Using either the absorption per unit mass for the sub 2.5 μm particles (σ_a/ρ_f) or the albedo for single scattering ($\widetilde{\omega} \equiv \sigma_{sp}/(\sigma_{sp}+\sigma_a)$), a range of average soot concentrations for the fine particles is inferred for each site. The single scattering albedo, $\widetilde{\omega}$, is also the fraction of visibility reduction due to scattering and $1 - \widetilde{\omega}_o$ is the fraction due to absorption. The Integrating Plate Method (IPM) [2, 3] was used for the absorption measurements and scattering was measured using an integrating nephelometer. Both measurements were made between 520 - 550 nm.

INTEGRATING PLATE METHOD

The method is based on comparing the light transmission through a Nuclepore filter that is sparsely covered with aerosol particles to that for the same filter without particles. Scattering by the particles is minimized by the approximate refractive index matching between the particles and the filter substrate so that the change in filter transmittance is caused by absorption only. Milky glass is placed behind the filter (away from the light source) with the intent of transmitting an isotropic light flux from scattered and transmitted light through the filter. The filter should be oriented such that the particles are toward the light source, thus reducing multiple reflections through the particle layer. A neutral density filter between the filter and the milky glass or an absorbing milky glass will also help reduce errors due to reflections. The geometry is shown in Fig. 1. The method has been tested by comparing long path extinction measurements with combined measurements of absorption and scattering extinction at two sites: one with low fractional absorption (σ_a/σ_e) and a site with high σ_a/σ_e [4]. The combined scattering and absorption extinction coefficient agreed to within $\pm$ 10 percent with the long path extinction measurements.

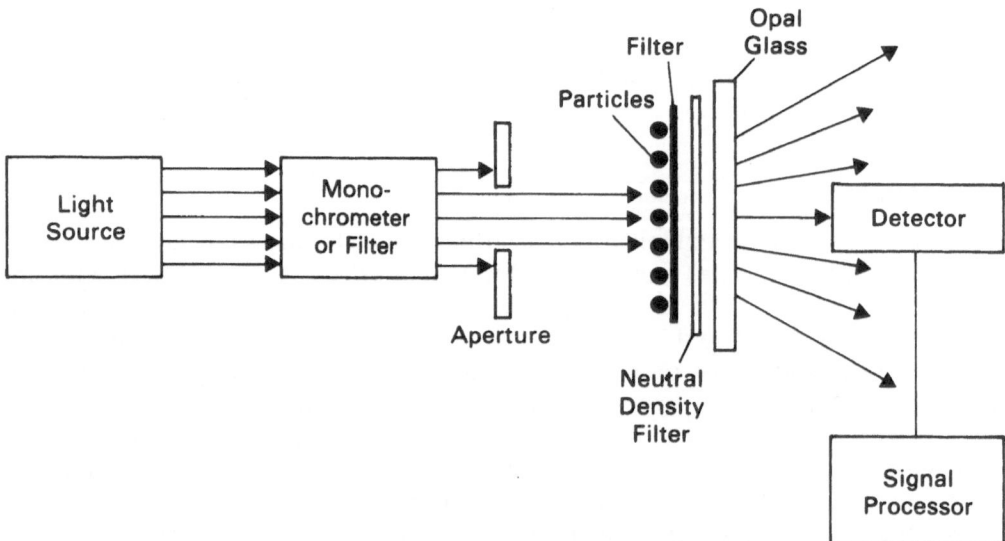

Fig. 1. Schematic of integrating plate method (IPM).

The filter material is flat, hydrophobic and transparent. Since the particles sit "on" the surface rather than "in" the filter material as with other filter types such as glass fiber, the optical depth is known. The optical depth is equivalent to that of the suspended particles in a column whose length, x, is determined by the volume of air sampled and the surface area of the filter. The absorption coefficient is calculated from the volume of air sampled and the change in transmission, I, caused by the absorbing particles on the filter, or

$$\frac{dI_\lambda}{I_\lambda} = - \sigma_{a\lambda} \, dx$$

where $x \equiv V/A_f$,
          $V \equiv$ volume of air sampled,
          $A_f \equiv$ area of filter used for collecting particles, and
          $\lambda \equiv$ wavelength.

IMAGINARY REFRACTIVE INDEX CALCULATION AND SPECIFIC ABSORPTION

If only sub 2.5 μm particles are collected (using a prefilter) and they are assumed to absorb light in proportion to their volume [3, 5], then the average imaginary part of the complex refractive index can be estimated by assuming that the particles are homogeneous, with uniform electrical polarizability.

For small, uniform spheres, the absorption cross section is proportional to the size parameter and the polarizability of the particle.

For small spheres the absorption cross section, $S_{a\lambda}(r)$ is

$$S_{a\lambda} (r) = -4\pi K \, Im[\alpha]$$

where $\alpha \equiv$ polarizability, m = refractive index
and $K \equiv 2\pi/\lambda$

For small isotropic spheres

$$\alpha = r^3 [\frac{m^2 - 1}{m^2 + 2}], \quad m = n - in' \quad .$$

The absorption coefficient for a distribution of particles

$$\sigma_{a\lambda} = \int_0^\infty S_{a\lambda} (r) \, n(r) \, dr, \quad n(r) \, dr \equiv \text{number concentration of particles}$$

References p. 324

then,

$$\sigma_{a\lambda} = -3K \, \mathrm{Im}[\frac{m^2-1}{m^2+2}] \quad \int_0^\infty V_p(r) \, n(r) \, dr, \text{ where } V_p(r) = \frac{4}{3}\pi r^3$$

or

$$\sigma_{a\lambda} = -3K(\frac{\rho_f}{\rho_o}) \, \mathrm{Im}[\frac{m^2-1}{m^2+2}] \quad .$$

Since

$$\int_0^\infty V_p(r) \, n(r) \, dr = V_F, \text{ the particle volume concentration,}$$

$$V_F = \rho_F/\bar{\rho}_o, \text{ where } \rho_F \equiv \text{mass concentration of the fine mode}$$

$$\text{and } \bar{\rho}_o \equiv \text{average particle mass density}$$

and

$$\mathrm{Im}[\frac{m^2-1}{m^2+2}] \overset{\sim}{=} 6\overline{nn'} \, / \, (n^2+2)^2 \text{ for } \bar{n} \gg n', \, m = n\text{-}in'$$

therefore,

$$\bar{n}' \overset{\sim}{=} - \frac{\sigma_{a\lambda}}{\rho_f}(\bar{n}^2 + 2)^2 \bar{\rho}_o/18K\bar{n} \quad .$$

Assumptions must be made for $\bar{n}$ and $\bar{\rho}_o$. The imaginary part, $\bar{n}'$, is not strongly dependent on $\bar{n}$ in the region $\bar{n} = 1.5$ to 1.6; a value of 1.55 was used here. A traditional value of $\bar{\rho}_o = 1.5$ g/cm³ was taken as the average particle density. The quantity, $\sigma_{a\lambda}/\rho_f$, is determined from the IPM and fine particle mass concentration.

SPECIFIC ABSORPTION

The ratio of absorption extinction to fine particle mass (σ_a/ρ_f) is directly related to the soot concentration in the fine particle mode provided that soot is the dominant absorbing material. An estimate of the soot concentration can be determined from the ratio of specific absorption for the fine particle aerosol to that of pure soot. Measured values of σ_a/ρ_f for pure aerosol phase soot seems to be somewhat source dependent with values ranging from about 7 to 11 m²/g [3, 6]. These values are about 50 percent higher than would be expected from calculations based on the optical and physical properties of the bulk material. This discrepancy has not been resolved. The range of soot fractions for each site calculated from the ambient measurements of σ_a/ρ_f (Table 1) are determined by using 7 and 11 m²/g as the absorption efficiency for pure soot.

For comparison purposes or for periods when fine particle mass was not measured, the soot content was estimated from the albedo for single scattering, $\omega \equiv \sigma_{sp}/(\sigma_{sp} + \sigma_a)$. The calculations were done for log normal distributions of spherical particles having layered refractive indices. Each particle is assumed to contain a soot core surrounded by a non-absorbing shell; the ratio of the core volume to total particle volume is kept constant throughout the distribution. The refractive index used for the core is 1.57-.48i and 1.525 - 0i for the shell. The range of soot concentration values listed in Table 1 is determined by using a mass mean diameter of either 0.2 or 0.4 μm and a standard deviation, $\sigma = 2$. Our measurements of size distribution indicate that this is approximately the range of mass mean diameter for the types of sites listed in Table 1.

MEASUREMENTS

Absorption and scattering extinction measurements were made at 18 sites including urban and rural areas and areas remote from major man-made air pollutant sources. An integrating nephelometer was used to measure the aerosol light scattering extinction; the IMP was used for the absorption measurements. All measurements were done on the dry aerosol. The nephelometer inlet was heated several degrees above ambient and the filters were maintained at below 50 percent RH for at least 24 hours prior to the weighing and optical measurements. The filters were weighed on a Cahn 4700 Electrobalance, having a measurement precision of about 2 μg. Flow rates through the filters were adjusted to limit the mass concentration on the filter to less than about 20 μg/cm², or a total of about 80 μg on 25 mm diameter filters.

Fine particle mass was measured at 11 of the sites by removing the larger particles from the sample airstream using either a dichotomous virtual impactor with a 50 percent size cut at 2.5 μm [7] or a modified Bendix model 18 cyclone with a size cut near 1 μm.

Listed in Table 1 are the results for each of the 18 sites. Included are average values for the absorption coefficient (σ_a), the albedo for single scattering ($\widetilde{\omega}$), absorption per fine particle mass (σ_a/ρ_f) and the mass fraction of soot in the fine particle mode inferred from either the σ_a/ρ_f or $\widetilde{\omega}$ measurements.

DISCUSSION

Particle absorption was of the order of 10 percent of scattering extinction or less in clean background areas and nearly equal to σ_{sp} in urban areas (Table 1). The

TABLE 1

Site	$\bar{\sigma}_a$ (10^{-5} m^{-1})	$\tilde{\omega} = \sigma_{sp}/(\sigma_{sp} + \sigma_a)$	$\overline{\sigma_a/\rho_f}$ (m^2/g)	f(soot) from σ_a/ρ_f	f(soot) from $\tilde{\omega}$
Urban-Industrial					
1. Seattle, WA (Duwamish)	2.7	0.65	1.8	.16-.26	.13-.20
2. Portland, OR	8.6	0.56	1.6	.15-.23	.21-.29
3. St. Louis, MO (Univ.)	5.0	0.60	—	—	.18-.24
4. Denver, CO (Henderson)	4.9	0.50	—	—	.26-.35
5. Denver, CO (Trout Farm)	6.0	0.54	—	—	.22-.31
6. Phoenix, AZ	11.8	0.64	1.4	.13-.20	.15-.20
7. Denver, CO (1978)	6.4	0.61	—	—	.17-.23
Rural-Residential					
8. Seattle, WA (Maple Leaf)	1.0	0.79	0.7	.06-.10	.06-.10
9. St. Louis, MO (Wash. Univ.)	3.7	0.76	—	—	.09-.11
10. Tyson, MO (1973)	1.9	0.78	—	—	.08-.10
11. Tyson, MO (1975)	1.2	0.81	1.0	.09-.14	.06-.09
12. Milford, MI	1.7	0.73	1.0	.09-.14	.09-.12
13. Hall Mt., AR	0.6	0.87	0.7	.06-.10	.04-.06
14. Puget Island, WA	0.6	0.80	0.7	.06-.10	.06-.09
Remote					
15. Anderson Mesa (near Flgstf.)	0.12	0.94	0.3	.03-.04	~.03
16. Mauna Loa Obs., HI	0.005**	0.95**	—	—	~.02
17. Mesa Verde, CO	0.12	0.91	0.4	.03-.05	.03-.04
18. Abastumani Obs., USSR	0.57	0.89	0.5***	.04-.07	.04-.05

Average Measured Optical Properties and Inferred Soot Concentrations. The f(soot) is the inferred mass fraction of soot in the fine particle mode aerosol. The range of f(soot) values using σ_a/ρ_f are determined by using either 7 or 11 m^2/g as the absorption efficiency for pure soot; the range of values inferred from $\tilde{\omega}$ are for either 0.2 or 0.4 μm as the mass mean diameter of the size distribution (Figure 2).

* One measurement

** Size cut with cyclone, $D_p \approx 1$ μm, all others with dichotomous separator, $D_p \approx 2.5$ μm

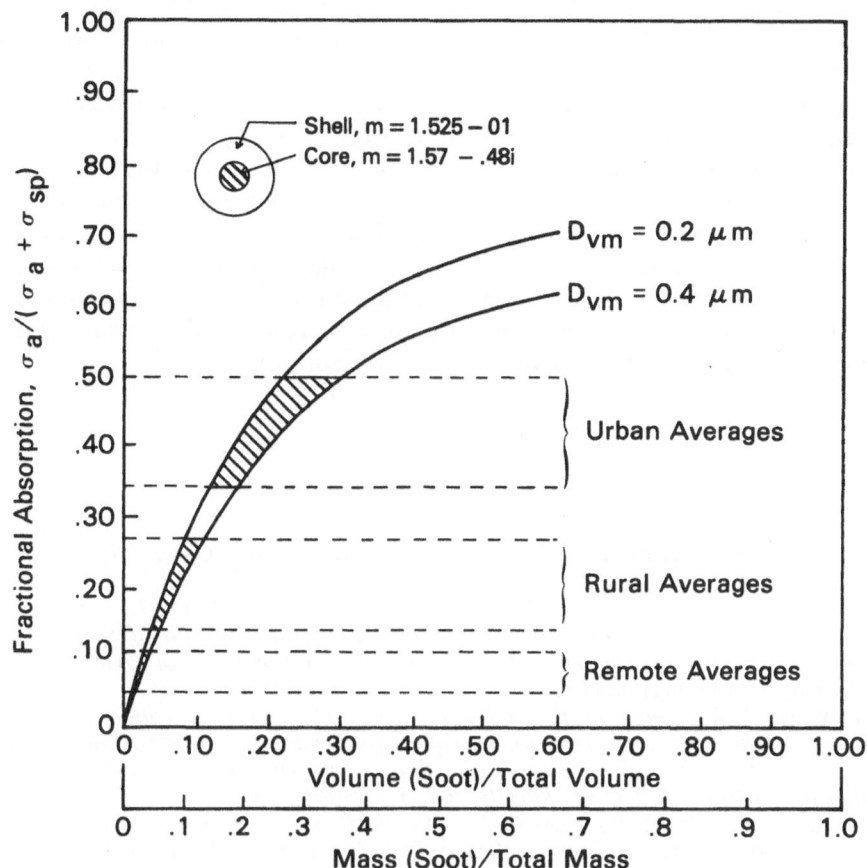

Fig. 2. Fractional absorption, $1\text{-}\widetilde{\omega} \equiv \sigma_a (\sigma_a + \sigma_{sp})$, for spheres with absorbing soot cores and non-absorbing shells for two different log normal size distributions, $D_{vm} = 0.2$ and 0.4 μm, $\sigma = 2$ and $\lambda = .55$ μm. The refractive index of the core is taken as m_c F27 $1.57\text{-}.48i$; the shell as $m_s = 1.525\text{-}0i$; the density of the core as $\rho_c = 2$ g/cm³ and the shell, $\rho_s = 1.5$ g/cm³. The shaded areas represent the range of measured $\sigma_a/(\sigma_a + \sigma_{sp})$ values.

amount of absorption per mass depends on the molecular composition and the size distribution of the particles [3, 5, 8]. The most important contributor to this absorption appears to be graphitic carbon (in the form of soot) [1, 4, 9]. Sources of this highly absorbing soot include the combustion of liquid fuels, particularly in diesel engines, and wood burning, whereas efficient coal combustion may not be a major source [10].

If soot is the dominant absorbing material in the fine particles at all the sites listed in Table 1 then aerosol soot fractions are site class dependent, varying from about 20 percent in industrial/urban areas to 10 percent or less in other areas. In urban areas soot probably dominates light extinction over any other single aerosol component (other than water at high RH).

References p. 324

Visual air quality will be more sensitive to changes in suspended soot concentrations than any other aerosol component. For example, soot in the fine aerosol has an extinction efficiency per unit mass of about 10 - 15 m²/g (including scattering) compared with 3 - 5 m²/g for non-absorbing materials such as sulfates. Therefore an incremental increase in soot concentration would require a three-fold increase in sulfate (for example) to produce a similar adverse effect on visibility.

Increases in the diesel fleet fraction and in the amount of wood burning for the purposes of conserving crude oil may cause substantial increases in aerosol optical absorption due to graphitic carbon. We believe that the optical method may be useful in providing a quick and easy determination of soot concentration. Our finding that about 20% of urban fine particle mass is graphitic carbon by optical measurements is in agreement with chemical measurements.

REFERENCES

1. *H. Rosen, A. D. A. Hansen, L. Gundel and T. Novakov, App. Opt., Vol. 17 (1978), pp. 3859-3861.*
2. *C. Lin, M. Baker, and R. J. Charlson, App. Opt., Vol. 12 (1973), pp. 1356-63.*
3. *R. E. Weiss, "The Optical Absorption Properties of Suspended Particles in the Lower Troposphere at Visibile Wavelengths," Ph. D. Dissertation Dept. of Civil Engineering, University of Washington, (1980).*
4. *R. E. Weiss, A. P. Waggoner, R. J. Charlson, D. L. Thorsell, J. S. Hall, and L. A. Riley, "Studies of the optical, physical and chemical properties of light absorbing aerosols," in Proc. Conference on Carbonaceous Particles in the Atmosphere, Lawrence Berkeley Laboratory LBL-9037, CONF-7803101, UC-11, (1978).*
5. *R. W. Bergstrom, Beitz. Phys. Atm., Vol. 46 (1973), pp. 223-234.*
6. *D. M. Roessler, and F. R. Faxvog, J. Opt. Soc. Am. Vol. 69 (1979), pp. 1699-1704.*
7. *B. W. Loo, J. M. Jaklevic, and F. S. Goulding, "Dichotomous Virtual Impactors for Large Scale Monitoring of Airborne Particulate Matters," in Fine Particles, edited by B. Y. H. Liu, Academic Press, New York, (1972), pp. 332-350.*
8. *A. P. Waggoner, M. B. Baker and R. J. Charlson, App. Opt., Vol. 12 (1973), p. 896.*
9. *W. R. Pierson and P. A. Russell, Atmos. Environ., Vol. 13. (1929), pp. 1623-1628.*
10. *G. T. Wolff, R. J. Countess, P. J. Groblicki, M. A. Ferman, S. H. Cadle, and J. L. Muhlbaier, Atmos. Environ. Vol. 15 (1981), p. 2485.*

DISCUSSION

P. Mueller, *(Electric Power Research Institute)*

What is the standard deviation on the correlations between the nephelometer and the fine particle mass?

A. Waggoner, *(University of Washington)*

The ratio of fine mass to bsp was $0.32 \pm 0.02 \, gm^{-2}$.

H. Gerber, *(Naval Research Laboratory)*

Would you elaborate on your technique for measuring the mass.

A. Waggoner

We take a Cahn model 4700 electro-balance into the field. There is disagreement between our group and Bob Stevens' group at EPA on the precision. We normally measure the filter within one day before collection and weigh it within one day after in the field. Keeping the time short helps the accuracy of the measurement. We used 25-mm diameter 0.4-micron pore size millipore filters. On those filters we get reproducibility of the mass or of the tare weight of the filter within plus or minus two micrograms on each measurement. We collect the fine particle mass using either a Liu and Jaklevic design dichotomous sampler, one of the original two-stage samplers with a 2.5 micron cut size or a 1-micron cut with a modified Bendix cyclone. In our opinion, the cyclone often has advantages for fine particle mass because you can adjust the cut point very easily, whereas adjustments on a dichotomous sampler are very difficult or impossible.

T. Hansen, *(Lawrence Berkeley Laboratory)*

Typically, what kind of loadings can you get on these filters before they start to clog appreciably?

A. Waggoner, *(University of Washington)*

We use flow regulators, active flow control devices, on the filters and we are typically pulling about a quarter of an atmospheric pressure drop across the filter initially. That works out to be on the order of 10-20 liters per minute on a 25-millimeter filter. We have a constant flow and it varies with the ambient levels. For example, in Seattle the average fine particle aerosol concentration is about 15 micrograms per cubic meter. In Luray, Virginia, it was probably closer 50. We typically end up with 50 to 100 micrograms on the filter by varying the flow rate and sampling time.

R. Charlson, *(University of Washington)*

There was another light absorption measurement which Alan did not mention because there was not a scattering coefficient to go with it. On the west coast of the state of Washington, in a very remote area, the absorption was a few times ten to the minus six per meter. This fits exactly with the other data that Alan mentioned.

BLACK AND WHITE EPISODES, CHEMICAL EVOLUTION OF EURASIAN AIR MASSES, AND LONG-RANGE TRANSPORT OF CARBON TO THE ARCTIC

K. A. RAHN

University of Rhode Island
Kingston, Rhode Island

C. BROSSET

Swedish Water and Air Pollution Research Laboratory
Gothenburg, Sweden

B. OTTAR

Norwegian Institute for Air Research
Lillestrom, Norway

E. M. PATTERSON

Georgia Institute of Technology
Atlanta, Georgia

ABSTRACT

Long-range transport of pollution aerosol from central Europe to southern Scandinavia has been recognized for a decade. Depending on the trajectory, high concentrations of sulfate in southern Sweden are accompanied by greater or lesser carbon, and have become known as black or white episodes, respectively. It is now known that, when conditions are right, pollution aerosol (including elemental carbon) can be transported from Eurasia to the Arctic. A simple model of long-range transport of SO_2, $SO_4^=$, and trace metals has been constructed to account for the large changes between Eurasia and the Arctic. This paper presents a broadened interpretation of black and white episodes which, in combination with atmospheric measurements at other locations in Eurasia and environs, allows this transport model to be checked empirically. It also allows the atmospheric lifetime of carbon to be compared with lifetimes of other primary aerosol pollutants. The results confirm most features of the model, and suggest that the European black and white episodes are highly related, early stages of systematic atmospheric aging, later stages of which can be observed as far away as the Arctic. The lifetime of elemental carbon appears to be similar to that of other primary, submicron elements, but this is still very uncertain.

References pp. 339-340.

TRANSPORT OF AIR POLLUTANTS TO SCANDINAVIA
AS BLACK AND WHITE EPISODES

The transport of polluted air masses from the highly industrialized regions of central Europe to Scandinavia has been known for over a decade [1-3]. This transport is highly episodic, with each episode having high concentrations of submicron aerosol particles (including SO_4^{-2}, H^+, etc.) and sometimes high concentrations of gaseous pollutants such as SO_2 and NO_2. Episodes occur when air moves directly from European source regions to Scandinavia. The major environmental effects, which have been the driving force for the European Organization for Economic Cooperation and Development's Long Range Transport of Air Pollutants (OECD/LRTAP) studies, are the acidification of precipitation and its effects on the environment [1, 4]. Changes in SO_2 and SO_4^{-2} during transport have been extensively measured and modeled in the OECD/LRTAP program [3, 5].

Although the greatest attention has been paid to the various forms of sulfur being transported, an entire suite of aerosols and gases is carried to Scandinavia as well. Important constituents of the aerosol include Fe, Mn, V, Pb, Cd, etc., and, of course, elemental C. Although the latter may have natural sources such as forest fires, natural sources should not be important during winter, when most of the pollution episodes occur in Scandinavia. Elemental C, or more properly the darkness which it imparts to samples of aerosol, has long been paid special attention in urban studies because it can be measured rapidly, inexpensively and reproducibly. Typically, the mass of the aerosol in a given region is estimated from the grayness of a filter, by using region-specific empirical calibration curves which relate grayness to the total mass of aerosol. An example of this procedure is the OECD technique [6], described by Brosset and Åkerström [7], which uses reflectance measurements on filters to estimate grayness. Unfortunately, total masses derived in this way are often referred to as "soot" a historical misnomer which can be confused with the modern meanings of "soot", i.e., graphitic carbon with or without the other mass closely associated with it [8].

Soot was rarely measured in remote regions however. Some of the first investigations were those of Brosset and co-workers, who have considered both soot and sulfate in Scandinavia since the earliest studies of long-range transport to that region. Brosset and Nyberg [2] found that concentrations of each were linearly related in winter episodes to those which Rodhe *et al.* [9] and Nord *et al.* [10] later showed to be associated with air flow from central Europe. Those episodes were also accompanied by high concentrations of nitrate and metals such as Mn, Fe and V. The particles were of moderate acidity, with much more or most of the H^+ neutralized by NH_3. Occasionally, high concentrations of SO_2 were also observed [11, 12].

Subsequent measurements by Brosset revealed, however, that the high concentrations of sulfate could be accompanied by drastically reduced soot when air masses were transported eastward to Sweden over the North Sea. Such episodes, which occurred primarily during summer, were termed "white episodes" by Brosset, to distinguish them from the "black episodes" of winter [11]. The white episodes were also characterized by much higher concentrations of hydrogen ions and much lower concentrations of metals and nitrate than were found in black episodes. Although the source of the sulfate, soot, etc. of white episodes could not be documented with certainty because of long over-water trajectories, it was presumed to be Europe as well.

Later, another degree of complexity was added to black and white episodes when Brosset discovered that white episodes could occur with transport from the east (eastern white episodes) as well as with transport from the west (western white episodes) [12]. Properties of particles in the eastern white episodes were nearly the same as those in western white episodes. Interestingly, eastern white episodes were not found during the summer; the first cases were observed during February and March (1976). Trajectory analysis by the Norwegian Institute for Air Research showed that air masses associated with eastern white episodes had passed over either eastern Europe or the western USSR some days earlier.

It should be noted that Brosset's classification of episodes as black or white was never intended to be rigid. Rather, the two types of episodes were deemed only to represent particles of "essentially different genetic origin", and the existence of transition stages was recognized [12]. This point is important for the discussion below.

TRANSPORT OF CARBON AND OTHER POLLUTANTS FROM EURASIA TO THE ARCTIC

Recently, a series of studies has demonstrated that aerosols and gases are transported from Eurasia to the Arctic regularly during the winter half-year [13-19]. Constituents to which the most attention has been paid include SO_2, sulfate, V and Mn. But the Arctic aerosol during winter also contains relatively large amounts of sooty carbon to the point that all filter samples become noticeably gray. This was a surprise, because the aerosols in other comparably remote regions, such as the equatorial Atlantic and Pacific Oceans, Mauna Loa, and the Antarctic, are essentially colorless. (By contrast, summer samples in the North American Arctic are nearly colorless, and those from the Norwegian Arctic are only light gray.) Winter transport to the Arctic is controlled near the surface by the Icelandic low-pressure and Asiatic high-pressure systems and is episodic, as is that from central Europe to Scandinavia. At 5,000 to 10,000 km, it is the longest routine transport of pollution aerosol presently known.

One of the outstanding chemical features of the Arctic aerosol is its high secondary/primary ratio relative to near-source aerosol. For example, the winter sulfate/V ratio in the Arctic is about 3×10^3, a full order of magnitude higher than the values of 0.3×10^3 found in eastern North America and Eurasia during winter [14]. Because the winter Arctic aerosol appears to be derived in large measure from midlatitude polluted areas such as eastern North America or Eurasia, the study of it and its precursors offers an ideal opportunity to observe systematic chemical and physical evolution of polluted air masses on a scale not previously possible. Physical properties of interest include the particle-size distribution of the aerosol as well as its scattering and absorption of light. Chemical properties of interest include particle/

gas ratios such as SO_4^{-2}/SO_2, secondary/primary ratios within the aerosol such as SO_4^{-2}/V, and primary/primary ratios within the aerosol for elements of different particle sizes, such as Mn/V. Of particular interest for this paper is the behavior of elemental carbon relative to other primary submicron elements such as Mn and V.

A simple description of certain systematic chemical changes during transport from Eurasia to Barrow, Alaska has been offered by Rahn and McCaffrey [14]. It starts with the mean chemical properties of polluted air masses of central Europe during winter, which are assumed to reasonably represent the true precursors of Arctic pollution. These air masses are then allowed to age for times up to 20 days, which should be long enough to include all reasonably direct transport to the Arctic. During this time, gases are allowed to react, particles and gases are removed, and the air masses are steadily diluted by external air with or without aerosol or trace gases. The rates of removal, oxidation and dilution decrease smoothly with time*. Calculations are carried out in one-day time steps.

An example of the results of Rahn and McCaffrey [14] is shown in Fig. 1. The k's are rate constants for removal of particles, oxidation of SO_2, and wet and dry removal of SO_2, in units of d^{-1}. These results show that the aerosol at Barrow was compatible with a European precursor, provided that rate constants decrease by roughly an order of magnitude during transport and that large-scale dilution was only a factor of 6. The shapes of the curves have a number of interesting features, such as rapid changes during the first days followed by slower changes, and a maximum in the secondary sulfate after 2-3 days. These features, which were consequences of the particular forms of evolution chosen for the rate constants, seemed generally reasonable, because polluted air masses ought to be most reactive when they are fresh and have high concentrations of reactants such as SO_2, and because SO_2 is known to be dry-deposited rapidly near its source, particularly in summer [20]. Eurasian air moving northward during winter should be no exception. The Arctic during winter is cold, dark and dry. These conditions may inhibit later transformation and removal of the pollutants [14, 15]. In addition, there are both theoretical reasons and observational evidence that the rate of dilution of polluted air masses by surrounding air decreases with time [3, 21].

It would be highly desirable to check the aging model of Rahn and McCaffrey empirically. The most direct way to do this would be through a Lagrangian experiment in which Eurasian air masses were actually followed on their way to the Arctic. No such experiment has been performed, however, nor is one likely to be undertaken in the near future because of the high costs as well as the logistical and political complexities. Failing this, a pseudo-Lagrangian approach may be taken in which atmospheric measurements at different sites downwind of different source regions within Eurasia are compared. Such an approach is possible provided the age of the aerosols is known and the various parent air masses are similar. The success of such a venture can never be predicted theoretically, because it is not known whether the parent air masses within Europe are sufficiently similar or whether the ages of the aerosols can be deduced with sufficient accuracy. The only way to test these assumptions is to perform the experiment and look for coherence

_____

*The rate "constants" referred to here include terms which vary with meteorological factors such as temperature, humidity, sunlight intensity, rate of precipitation, etc.

of the results. This paper presents such an experiment and evaluates its results.

At the time that Rahn and McCaffrey [14] was written (late 1978), it was not possible to attempt to verify Fig. 1 with a pseudo-Lagrangian approach, because atmospheric data were available only for the initial and final points of the trajectory. Somewhat later, data from northern Norway and the Norwegian Arctic became available. These data were, however, judged to represent conditions after the critical, first few days during which the most rapid changes occurred. New data, from earlier stages of aging, i.e., nearer the source, were needed to complete the picture.

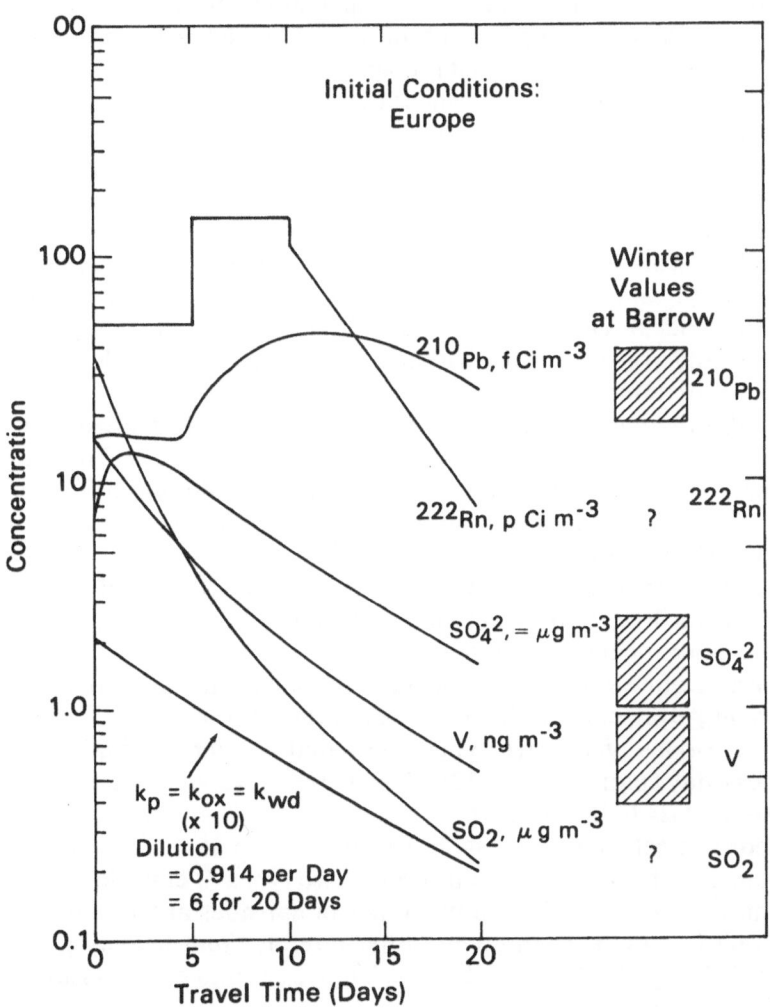

Fig. 1. Simulated aging of a polluted air mass as it travels from Europe to the Arctic during winter (after Rahn and McCaffrey [14]).

References pp. 339-340.

BLACK AND WHITE EPISODES AS RELATED TO
A SIMPLIFED THEORY OF ATMOSPHERIC AGING

We have recently become aware that black and white episodes, as observed by Brosset and co-workers at Onsala in southern Sweden, can provide the important missing data on the first days of aging of Eurasian air masses. In combination with data from northern Norway, the Nowegian Arctic, Barrow, and source regions in Eurasia, they offer a chance to observe details of the progressive aging of Eurasian air masses on a time scale of 0-20 days, and hence, to check the simulation of Rahn and McCaffrey [14]. The result is the first verified model of large-scale chemical changes during true long-range transport of polluted air masses. Exposition of this idea, together with its consequences and implications, forms the rest of this paper.

The key point in the development of this new approach was a broader interpretation of black and white episodes that recognizes and stresses their basic similarities rather than the more obvious differences that have been discussed in the literature to date. Black and white episodes are surprisingly similar in several ways:

1. The sulfate/soot and sulfate/Mn ratios, which are secondary/primary ratios that have been used to distinguish black from white episodes, are not always as different as one might imagine. (Here we restrict ourselves to eastern white episodes because they occur during winter, the season of maximum transport to the Arctic, and they have more easily defined source areas than do western white episodes). Data from Brosset [12] show that there is typically a factor-of-two range of each ratio within a given type of episode, whereas mean sulfate/soot ratios are a factor of three different between the episode types and sulfate/Mn ratios are only a factor of two different. In fact, sulfate/Mn ratios overlap considerably between the two types of episodes. Thus, the secondary/primary ratios of black and eastern white episodes differ from one another by a factor of three at most and the variation within each class is of the same order as the separation between classes.

2. A factor-of-three change in the secondary/primary ratio of sulfate/V is predicted from Fig. 1 to occur during the first days of aging, when the rate constants are decreasing smoothly.

3. Black and white episodes can occur at the same site during the spring. Sometimes they even come right after one another. For example, during February 1976, episodes of the two types alternated regularly, sometimes occurring only one day apart. Brosset [12], in his Table 11, reports four such alternations between 14-24 February 1976.

4. Meteorological maps for 14-24 February 1976 reveal surprisingly similar placements of synoptic systems for the two types of episodes.

5. Air-mass trajectories for the two types of episodes at Onsala [12] are not so different in length. Both come rather directly from heavily polluted areas: central Europe to the south for black episodes and eastern Europe/western USSR for eastern white episodes.

6. Black and eastern white episodes must be traceable back to a similar polluted precursor. Because each type of episode has about the same final concentration of sulfate (15-20 μg m^{-3}), each must have started with approximately the

same concentration of SO_2, and thus roughly the same concentrations of other primary pollutants such as Mn, Fe, etc.

7. White episodes may be derived from black episodes, because the initial conditions for a white episode, 10-15 μg m^{-3} SO_2 and low concentrations of primary particles [12], are characteristic of a partially aged polluted air mass, which initially could have been a black episode or something related to it.

In view of these considerable similarities between black and eastern white episodes, we have adopted the following working hypothesis. *Black and eastern white episodes represent progressive stages of atmospheric aging of similar parent air masses which may come from different regions within Eurasia. The main variable causing the differences between black and eastern white episodes is time, or extent of aging; this alone is enough to explain many of their apparently unrelated features.* According to this view, there is a gradual decrease in the blackness of a pollution aerosol relative to its total mass. In the beginning, near the source, primary constituents like soot are of major importance, and the aerosol is very black. As the aerosol ages, these primary constituents decrease in concentration monotonically, whereas secondary substances like sulfate increase monotonically relative to the primary constituents, and may even pass through a maximum in absolute concentration. Thus, the relative blackness of an aerosol decreases steadily during aging.

Here we wish to acknowledge that relating black and eastern white episodes to nothing more than time, or extent of aging, is an obvious oversimplification. It is, however, deliberate on our part, to determine just how little is required to explain the essence of black and white episodes, namely the sulfate/soot or secondary/primary ratio. The goal of this paper is to create a first level of explanation only by keeping the number of variables as small as possible. In this way, we hope to reveal the fundamental process of atmospheric aging and its most basic results. The results, discussed below, suggest that this approach has merit. In the future, after the successes and shortcomings of this first approach have been treated, a second, somewhat more sophisticated, approach will be taken to explain the remaining properties of black and white episodes by adding additional variables to this basic theory.

AN EMPIRICAL, PSEUDO-LAGRANGIAN AGING DIAGRAM FOR EURASIAN AIR MASSES

It is not difficult to assign approximate times of aging to black and eastern white episodes. The presumed source for black episodes is central Europe. From southern Sweden to central Europe is 700-800 km, or roughly 1 day's travel at 6-7 m s^{-1}, the speed commonly cited for long-range transport in Europe [3]. The source for eastern white episodes is presumably eastern Europe and the western USSR. From southern Sweden to major pollution sources in the western USSR is roughly 1500 km, or about double the distance to central Europe. Because of the slower travel speeds expected for trajectories from the east, which are generally associated with smaller pressure gradients near high-pressure areas, we take a travel time of three

References pp. 339-340.

days from these eastern sources. This figure is confirmed by the 850-mb trajectories shown by Brosset [12], which take three days to arrive in southern Sweden from eastern source areas. To complete the picture, we assign approximate travel times of 5 days to northern Norway (Skoganvarre and Jergul) under normal "direct" northward flow, and 10 days to the Norwegian Arctic (Spitsbergen and Bear Island) via the "return-flow" pathway [15]. We thus have data for Eurasian air masses at aging times of 0 days (central Europe), 1 day (Onsala, black episodes), 3 days (Onsala, eastern white episodes), 5 days (northern Norway), 10 days (Norwegian Arctic), and 20 days (Barrow). Fig. 2 shows the sources, the measurement sites, and the presumed transport paths. We stress that the transport times to northern Scandinavia and the Arctic are still imperfectly known, and will probably be adjusted in the future. The values used here are first approximations, for demonstration purposes only.

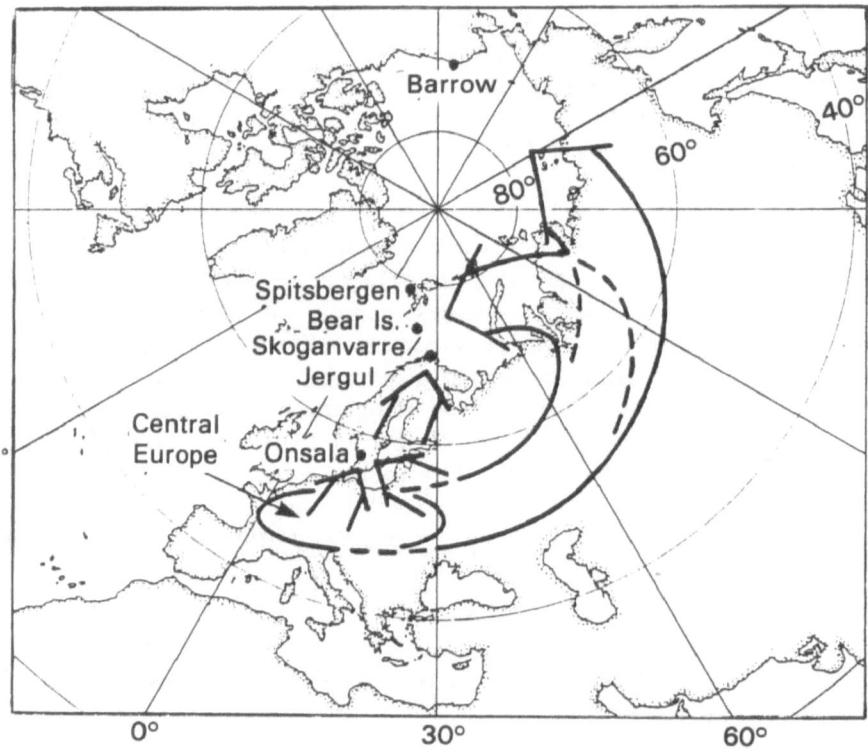

Fig. 2. Locations of measurement sites for aged air masses discussed in this paper, and principal atmospheric pathways from Eurasian source regions to each site.

The available data on SO_2, SO_4^{-2}, C, V and Mn for these six sites have been plotted as an empirical, pseudo-Lagrangian aging diagram in Fig. 3. Several remarks about the construction of this figure are in order. Non-marine sulfate was determined from total sulfate, whenever possible, by using Na to subtract marine sulfate, according to the following formula:

$$SO_{4_{nonmarine}}^{-2} = SO_{4_{total}}^{-2} - Na\left(\frac{SO_4^{-2}}{Na}\right)_{seawater}$$

Similarly, crustal Mn and V have been removed by calculating them from Al:

$$Mn_{noncrustal} = Mn_{total} - Al\left(\frac{V}{Al}\right)_{crust}$$

$$V_{noncrustal} = V_{total} - Al\left(\frac{V}{Al}\right)_{crust}$$

The calculated noncrustal Mn and V will both be primarily submicron, and so can be compared reliably to the submicron nonmarine SO_4^{-2}.

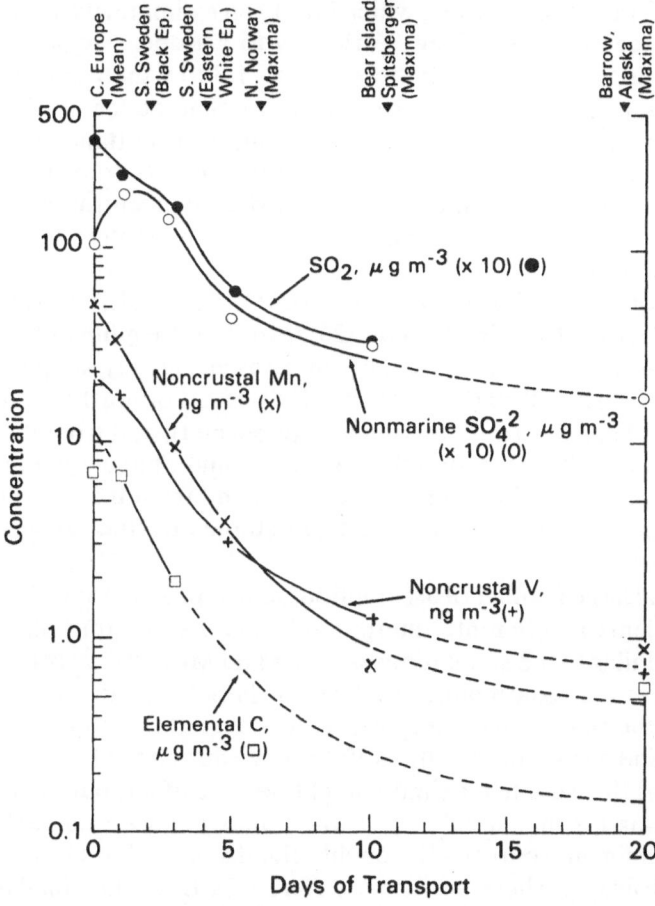

Fig. 3. Empirical, pseudo-Lagrangian aging diagram för polluted air masses travelling between Eurasia and the Arctic during winter.

References pp. 339-340.

We attempted to normalize the entire plot to the maximum mean 24-h concentrations found at each site during episodes. This was made difficult, and probably only partially successful, by the different sampling times at the different sites, which ranged from one day to one week. When 24-h data were available, typical winter maxima were chosen. When only longer-period data were available, a factor was determined which related these maxima to 24-h maxima. For central Europe, winter mean values were used because we could not assume that transport, especially to the Arctic, was automatically preceded by higher-than-normal concentrations at the source.

Elemental carbon for black and eastern white episodes was derived by multiplying Brosset's "soot" data for 1975-76 [12] by 0.2. This factor represents the approximate mass fraction of carbon in the fine-particle mode of typical urban aerosols, and was derived from data on elemental C in the New York aerosol of the late 1970's [22] and the mass of the accumulation mode there during the summer of 1976 [23]. The corresponding ratio in Charleston, West Virginia, which is in an area generally richer in sulfate than is New York, was 0.13 in the summer of 1976 [24]. The value for elemental C in central Europe, which is only an estimate, was derived by scaling the annual concentrations in New York City to European regional winter conditions. The scaling factors were 1/3 to regional, 1.3 to winter, and 37/16 to Europe (based on SO_2 in Europe and North America). The concentration of elemental carbon at Barrow was derived from broad-band transmittance measurements of filters taken during March 1978. The transmittance was converted to absorption which was used to estimate the elemental carbon concentration. The calculated March carbon values were then multiplied by scaling factors to convert them to winter episodic values.

Other data for central Europe came from the OECD EMEP study for 1977-78 [25] (sulfate and SO_2) and from Hoste et al. [26] (Mn, V in Belgium in 1972-73). Data for black and eastern white episodes came from Brosset [11, 12], supplemented by data from the Swedish EMEP station at Rörvik [25]. Data for northern Norway in 1971-72 came from Skoganvarre (K.A. Rahn, unpublished), and from the EMEP site at Jergul during 1977-78 [25]. Data for Bear Island and Spitsbergen in 1977-78 came from the Norwegian Institute for Air Research and from the University of Rhode Island. Barrow data of 1977-78 came from studies by the University of Rhode Island.

In spite of the large number of assumptions and approximations used to construct Fig. 3, it presents a coherent, interpretable picture of atmospheric aging. It is surprisingly similar to the simulations of Rahn and McCaffrey [14] shown in Fig. 1, and reproduces their basic features such as: the initial rapid decreases followed by a slackening in the rates of decline, the greatest changes in composition within the first 4-5 days and a maximum in the sulfate curve during the first 2-3 days. We first considered that this sulfate maximum might be an artifact, but the OECD Lagrangian calculations for individual days show a maximum for sulfate about 800 km downwind of its main sources [3], roughly the distance that Onsala is down wind from central Europe. There are some differences from the simulations, though: in Europe during the black episodes the peak in SO_4^{-2} occurs earlier; SO_2 parallels SO_4^{-2} and the rest of the aerosol after 3-4 days (it becomes more nearly inert than in the simulations); and the curves are flatter in the last 10 days then expected (less dilution, due to shorter travel distances than expected). New information from Fig. 3 is that Mn decreases faster than V probably because the particle size of Mn is larger. In addition, the similar rate of C and Mn decay suggest that they are somehow associated in the aerosol and that elemental C has a larger particle size than V. Few data on the particle size of C are available to compare with V, other than experience with cascade impactors that C is generally submicron. The association of C and Mn in the aerosol is compatible with the study of Linton *et al.* [27], which showed that Mn is preferentially volatilized during combustion, then condenses onto the surface of pre-existing particles such as fly ash or soot.

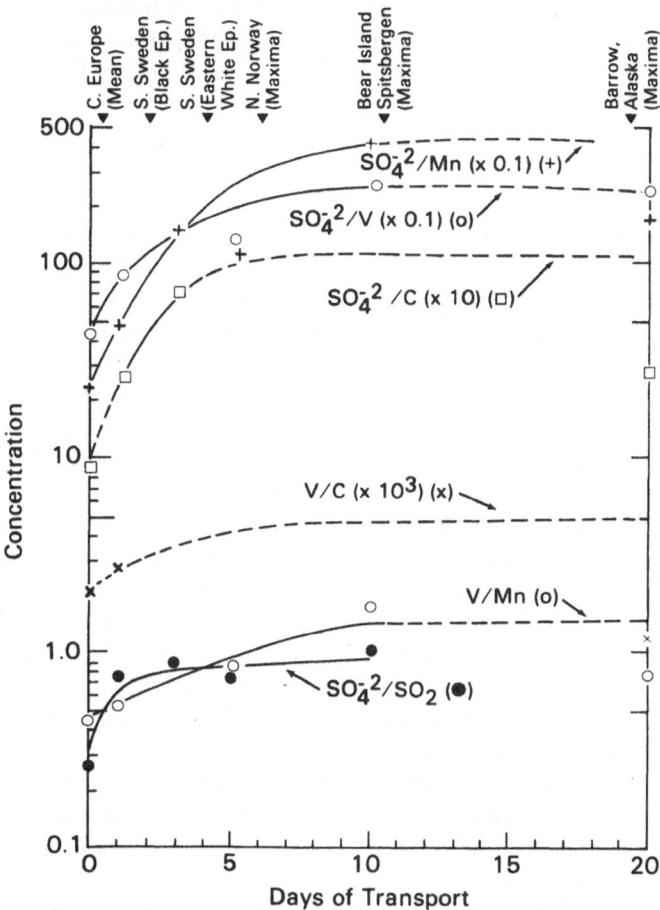

Fig. 4. Evolution of various secondary/primary and primary/primary ratios during winter transport of polluted air masses from Eurasia to the Arctic. All data from Fig. 3.

The progressive primary-to-secondary transformation of the aerosol can be seen from the various ratios plotted in Fig. 4. Note that the SO_4^{-2}/SO_2 ratio rises more rapidly initially than SO_4^{-2}/Mn, SO_4^{-2}/V, or SO_4^{-2}/C. This is due to the rapid initial conversion of SO_2 to SO_4^{-2}. Depending on the indicators chosen, the secondary/primary ratio of the aerosol increases by factors of 4 to 20 during aging. The

References pp. 339-340.

338   RAHN, BROSSET, OTTAR, PATTERSON

approximate factor-of-three increase from black to eastern white episodes is compatible with that predicted from Fig. 1. The V/Mn ratio increases by about a factor of 4 between Eurasia and the Norwegian Arctic. This increase is in the same direction but larger than the factor of 2 predicted by Rahn [18] if particles down to radius 0.3 μm were removed. The reason for this discrepancy is not yet clear.

Two other interesting features of Figs. 3 and 4 are the excesses of Mn (2x) and elemental C (4x) at Barrow compared to values extrapolated from the other points. The Mn at Barrow is actually higher in concentration than it is in the Norwegian Arctic. This excess Mn at Barrow has been noted previously [18] and appears as well at Mould Bay, NWT, also in the North American Arctic [17]. The North American Arctic would thus seem to be, influenced at least partially by an additional source of aerosol which seems to emit enriched Mn, C, and probably other elements as well. This source is probably in the central U.S.S.R.

CONCLUSIONS

A number of basic conclusions are suggested by the above analysis:

1. Large-scale aging of polluted air masses can be observed in and around Eurasia during winter, once the principal pathways are understood. The regularity of the empirical pseudo-Lagrangian aging plot (Fig. 3) and the similarity to the simulated aging (Fig. 1) suggest that aerosols at different aging times have common precursors, and are produced by a highly unified and regular process of atmospheric aging. The main variable in this process, at least as far as these species are concerned, is time alone. This confirms our working hypothesis stated above.

2. Black and eastern white episodes represent early stages of this aging process, and have ages of roughly 1 and 3 days, respectively. The major difference between black and eastern white episodes is time, or degree of aging, which is reflected in the secondary/primary (sulfate/soot) ratio of the aerosol. Although there may well be differences in sources or aging between black and eastern white episodes, or between any of the other aerosols discussed here, these other differences do not have to be invoked to explain the basic patterns of the empirical aging diagram.

3. The rate of aging decreases rapidly during transport, and reaches very small values in the cold, dry and dark Arctic night. The major transformation from primary to secondary character of the Eurasian aerosol is largely completed within the first 4-5 days.

4. The relative atmospheric lifetime of elemental C compared to other primary pollutants such as V or Mn is not yet clear. In the first days of transport, it would appear to be similar to them, and to Mn in particular. Subsequent behavior is obscured by the apparent additional source influencing Barrow. Further data at intermediate points are needed to resolve this question.

An intriguing question for the future is the extent to which the large-scale features of atmospheric aging revealed here are general, i.e., can be found downwind of other major polluted areas. Eastern North America and eastern Asia are the two other source regions near which such studies might be carried out. In each of these cases, however, the downwind transport is over oceans and at lower latitudes, where meteorological and chemical conditions, hence atmospheric aging, may be quite different from the Eurasian case.

ACKNOWLEDGEMENTS

This work was supported in part by the Office of Naval Research Contract N00014-76-C-0435. Aerosol samples at Barrow were collected by observers at the NOAA/GMCC Clean-Air Observatory, and analyzed by T. J. Conway at the Rhode Island Nuclear Science Center.

REFERENCES

1. S. Odén, Nederbördens och luftens försurning, dess orsaker, förlopp och verkan i olika miljöer. Statens Naturvetenskapliga Forskningsrad, Ekologikommitéen. Bull. No. 1, Stockholm, 1968.
2. C. Brosset and A. Nyberg, in Proceedings of the Second International Clean Air Congress, Washington, 6-11 December 1970, H. M. England, W. T. Beery, eds., Academic Press, New York and London, (1971), p. 481.
3. OECD, The OECD Programme on long range transport of air pollutants. Measurements and findings. OECD, Paris, 1977.
4. B. Ottar, Atmos. Environ., Vol. 12 (1978), p. 445.
5. A. Eliassen, Atmos. Environ., Vol. 12 (1978), p. 479.
6. OECD, Methods of Measuring Air Pollution: Report of the Working Party on Methods of Measuring Air Pollution and Survey Techniques. OECD Publication No. 17913, January (1965), p. 15.
7. C. Brosset and Å. Åkerström, Atmos. Environ., Vol. 6 (1972), p. 661.
8. H. Rosen, A. D. A. Hansen, L. Gundel and T. Novakov, in "Carbonaceous Particles in the Atmosphere," ed. T. Novakov, Lawrence Berkeley Laboratory Report LBL-9037, Berkeley, CA 94720, (1979), pp. 49-55.
9. H. Rodhe, C. Persson and O. Åkesson, Atmos. Environ., Vol. 6 (1972), p. 675.
10. J. Nord, A. Eliassen and J. Saltbones, Advances in Geophysics, Vol. 18B (1974), p. 137.
11. C. Brosset, Ambio, Vol. 5 (1976), p. 157.
12. C. Brosset, Atmos. Environ., Vol. 12 (1978), p. 25.
13. K. A. Rahn and R. J. McCaffrey, in Papers Presented at the WMO Symposium on the Long-Range Transport of Pollutants and its Relation to General Circulation including Stratospheric/Tropospheric Exchange Processes, Sofia, 1-5 October 1979, WMO No. 538, Geneva, p. 25
14. K. A. Rahn and R. J. McCaffrey, Ann. NY Acad. Sci., Vol. 338 (1980), p. 486.
15. K. A. Rahn, E. Joranger, A. Semb and T. J. Conway, Nature, Vol. 287 (1980), p. 824.
16. S. Larssen and J. E. Hanssen, in Papers Presented at the WMO Technical Conference on Regional and Global Observation of Atmospheric Pollution Relative to Climate, WMO No. 549, Geneva, 1980.
17. L. A. Barrie, R. Hoff and S. Daggupaty, The influence of mid-latitudinal sources on haze in the Canadian Arctic, Atmos. Environ. (In press).
18. K. A. Rahn, The Mn/V ratio as a tracer of large-scale sources of pollution aerosol for the Arctic. Atmos. Environ. (In press).

19. K. A. Rahn, *Relative importances of North America and Eurasia as sources of Arctic aerosol. Atmos. Environ. (In press).*
20. J. A. Garland, *Atmos. Environ.,* Vol. 12 (1978), p. 349.
21. R. A. Scriven and B. E. A. Fisher, *Atmos. Environ.,* Vol. 9 (1975), p. 49.
22. R. J. Countess, G. T. Wolff and S. H. Cadle, *J. Air Pollut. Control Assoc.,* Vol. 30 (1980), p. 1195.
23. M. Lippman, M. T. Kleinman, D. M. Bernstein, G. T. Wolff and B. P. Leaderer, *Ann. NY Acad. Sci.,* Vol. 322 (1979), p. 29.
24. E. S. Macias, R. Delumyea, L.-C. Chu, H. R. Appleman, C. D. Radcliffe and L. Staley, in *"Carbonaceous Particles in the Atmosphere,"* ed. T. Novakov, Lawrence Berkeley Laboratory Report LBL-9037, Berkeley, CA 94720, (1979), p. 70
25. J. Schaug, H. Dovland and J. E. Skjelmoen, *EMEP/CCC Report 3/80, Data Report October 1977-September 1978,* Chemical Co-ordinating Centre, Norwegian Institute for Air Research, P.O. Box 130, N-2001 Lillestrom, Norway, 1980.
26. J. Hoste, R. Dams, C. Block, M. Demuynck and R. Heindryckx, *Study of National Air Pollution by Combustion-Final Report 1974. Part I: Inorganic Composition of Airborne Particulate Matter. Instituut voor Nucleaire Wetenschappen, Rijksuniversiteit Gent, Belgium, 1974.*
27. R. W. Linton, A. Loh, D. F. S. Natusch, C. A. Evans Jr. and P. Williams, *Science,* Vol. 191 (1976), p. 852.

DISCUSSION

D. Stedman, *(University of Michigan)*

Ken, where did the data points on Figure 4 come from? In particular, are they averages?

K. Rahn

They are averages whenever possible. Central Europe is an average. All other points are attempts to simulate peaks of episodes of one day duration at various places. It is always difficult to assemble data from many different studies and try to make them coherent. Some used here were one-day samples, some were two-day samples, and some were weekly samples. I have tried to reduce them all to one day to simulate episodic pulses. These data come from many different studies at various institutions including our laboratory, Dr. Ottar's laboratory and Dr. Brosset's laboratory.

R. Klimisch, *(General Motors Research Laboratories)*

Do you have any nitrate data? Is there evidence that nitrate is being lost along the transport path?

K. Rahn

There is relatively little nitrate at the end of these paths. This has been measured by the Canadians in Mould Bay and by Dr. Ottar's laboratory in Spitsbergen, and the nitrate is miniscule compared to sulfate. The inference is that it is probably lost because of the highly acidic sulfate particles. There is relatively more nitrate in summer than there is in winter but we are speaking of the winter case here.

G. Wolff, *(General Motors Research Laboratories)*

Referring to the data that I presented this morning*, when we look at the ratio of elemental carbon to total carbon, we see a decrease as we move away from urban areas or other source areas. One hypothesis is that as the emissions age, there is an increase in the secondary organic particulate. Do you have any organic carbon measurements to compare to your elemental carbon so we can explore this hypothesis further?

K. Rahn

Joan Daisey finds typically 1 μg/m^3 extractable organics in the Barrow aerosol during spring. During this same period, I believe that Hal Rosen finds 0.2-0.4 μ/gm^3 elemental carbon in the aerosol. This is after transport from midlatitudes.

H. Rosen, *(Lawrence Berkeley Laboratory)*

The data that I have, if anything, indicates the presence of more primary carbon in the Arctic than in urban locations. That is, the ratio of graphitic carbon to total carbon is higher in the Arctic than in typical U.S. urban locations.

G. Wolff

George Hidy brought up the point earlier that it may not be proper to compare U.S. and Arctic ratios directly because we tend to have more efficient combustion in this country. The ratio of organic to elemental carbon emissions may be different in Europe.

K. Rahn

I agree. That is the main reason we have not been able to draw any conclusions about source areas from this ratio.

H. Rosen

The thing that is surprising is that the graphitic fraction, which is definitely primary, is about 50% of the carbon during February. So, at most, you have 50% secondary carbon.

---

*Wolff, G. T., Groblicki, P. J., Cadle, S. H. & Countess R. J., These Proceedings, p. 297.

J. Daisey, *(New York University)*

We looked at the ratio of the nonpolar organics, which tend to be primary, to the vanadium during the spring season at Barrow. It looked like they were enriched by an order of magnitude over what they were in European and United States cities. This suggests that you may be seeing a shift in the vapor and particulate phase equilibria as you go to those colder regions. People have suggested that the Arctic acts as a cold finger and condenses the organics.

D. Atkinson, *(Environmental Protection Agency)*

What is the synoptic situation that brings the air flow from Europe to Barrow.

K. Rahn

It changes somewhat from early winter to late winter. In early winter, the flow at the surface in Barrow comes primarly from the northeast. The flow is dominated by the very large Icelandic low pressure system in combination with the Asiatic high. Transport is essentially straight across the pole from the Siberian side to the Alaskan side. Later in the spring, it changes as a high-pressure cell develops over the Arctic.

ATMOSPHERIC PARTICULATE CARBON OBSERVATIONS IN URBAN AND RURAL AREAS OF THE UNITED STATES

P. K. MUELLER*, K. K. FUNG, S. L. HEISLER
D. GROSJEAN and G. M. HIDY

Environmental Research & Technology, Inc.
Westlake Village, California

ABSTRACT

Observations of airborne particles containing carbon have been attempted by a variety of different methods. These have been aimed at detailed characterization of compounds for research purposes, as well as simple measures for air monitoring. The latter focused on solvent extractables prior to 1970, but more recently have evolved into techniques such as the thermal or catalytic oxidation to carbon dioxide. In this paper, a survey of recent measurements for noncarbonate carbon and elemental carbon are reported using the oxidation technique. Urban data for communities in New England, Houston, Denver and Los Angeles are given. Rural data from nine stations in the greater northeastern United States and the California San Joaquin Valley also are presented for comparison. The observations indicate the presence of noncarbonate carbon at concentrations above $1\mu g/m^3$ at most locations. The urban and rural sites in California are particularly rich in carbonaceous material. Winter data in Denver taken in 1978 indicate a substantial fraction of carbon in the aerosol is elemental in nature. In urban sites, as much as half of the noncarbonate carbon present can be elemental in character. Carbon is widespread in airborne particles, and is potentially important to public health considerations and visibility impairment. Therefore, its molecular composition should be better characterized and it should be considered for routine monitoring with other major particulate constituents in the United States.

INTRODUCTION

Particulate carbon in urban atmospheres has been the object of continuing attention for more than three decades. Early concern originated from observations of the

---

*Present address: Electric Power Research Institute, Palo Alto, CA

References pp. 367-368.

carcinogenic activity of organic extracts of atmospheric particulate matter in laboratory animals [1,2]. These observations led to a large number of studies of the ambient levels of polycyclic aromatic hydrocarbons (PAH) and of the carcinogenic activity of individual PAH including benzo(a)pyrene (BaP). Reflecting the general concern for public health, the National Air Surveillance Network (NASN) was initiated in 1953, and expanded to approximately 300 stations in 1957. The NASN included measurements of benzene-soluble organics (BSO) and, later by 1966, of benzo(a)pyrene. These studies of particulate polycyclic organic matter have been the object of a comprehensive review prepared by the National Academy of Sciences [3], whose recommendations for future research included the need for more systematic investigations of PAH and other hazardous organics in urban particulate matter, of the relationships between ambient levels of PAH and carcinogenic activity, and of possible interactions between PAH and other pollutants.

In the early seventies, renewed interest in the fundamental aspects of photochemical smog chemistry led to a number of investigations of secondary carbonaceous aerosols; i.e., those formed in the atmosphere by chemical reactions involving gas phase precursors including hydrocarbons, oxides of nitrogen, ozone and free radicals. Thus, smog chamber studies were directed at elucidating the chemical [4] and physical [5, 6] pathways leading to the formation of secondary organic aerosols, while systematic studies began to unravel the complex molecular composition of secondary organics in urban aerosols [7-9]. These studies led to the identification of a host of polyfunctional oxygenates including dicarboxylic acids as end products of hydrocarbon photooxidation, the determination of organic aerosol formation rates in photochemical smog, a better understanding of the major roles of ozone and OH chemistry, and increased knowledge of condensational growth as an important pathway for organic aerosol formation.

At the same time, a third group of studies focused on the issue of the relative importance of primary and secondary particulate carbon, which has major implications for control strategies. Following the carbon distribution presented by Mueller et al. [10] for Pasadena, CA, aerosols, the nature and levels of particulate carbon in ambient air have been the object of extensive field studies including the major Aerosol Characterization Experiment (ACHEX) [11]. In turn, the need for better methods of measuring particulate carbon in urban air has led to the development of new techniques including nuclear methods and several combustion methods [12-14]. While studies of photochemical aerosols continued to generate new kinetic and mechanistic information [15], Novakov and coworkers stressed the importance of elemental carbon (soot) and its possible role in the heterogeneous formation of particulate sulfate[16-18].

Current interest in particulate carbon encompasses all areas related to the above issues as well as to new ones. Recent-health related work has focused on mutagenic activity of ambient organic particles [19] and on possible degradation of PAH by reaction with gas phase pollutants, perhaps promoted by filter substrates [20]. Studies of secondary organic aerosols have considered those formed from aromatic hydrocarbons [21] and are now including heterogeneous as well as homogeneous processes. Among the new issues involving carbon aerosols, visibility and emissions from diesel powered motor vehicles are receiving much attention from environmental scientists and regulatory agencies [22]. Optical properties of carbon aerosols are being intensively investigated while assessments are being made of the air quality impact of light-duty diesel vehicles whose use is projected to increase significantly in the near future [22]. This latest surge of interest in carbon aerosols is well illustrated by the Denver Haze Study [23-24], which included extensive measurements of total, elemental and organic carbon in a large urban area, corresponding optical measurements, and the application of a modified chemical element balance method to the specific case of carbon and its sources, including diesel engine exhaust.

Of major importance in terms of air pollution control strategies is the relative abundance of primary and secondary carbon particles. While emissions of primary particles are obviously reduced by direct source control, control of secondary particles requires a different strategy aimed at reducing emissions of their specific hydrocarbon vapor precursors. However, application of cost-effective strategies is greatly hampered by our limited knowledge of the distribution of carbon containing particles in the atmosphere. Although the molecular composition of particulate organics has been investigated in studies involving mass spectrometry and other modern analytical methods [8, 25], these studies have been limited to date to a few urban and rural areas.

With the advent of more recently developed simple methods to measure total particulate carbon and/or some fraction of the total carbon [10, 12, 14], the analytical tools required to generate an improved data base concerning the distribution of particulate carbon in the atmosphere are now available. In this paper, we describe data obtained using a method based on sample combustion, reduction of the evolved carbon dioxide to methane, and quantitation using a flame ionization detector. Measurements performed on this basis include total carbon, total organic carbon, carbonate, volatile organic carbon, water-soluble carbon and/or organic solvent-soluble carbon. Selected examples derived from recent field studies are presented in order to illustrate the applications of the method to the determination of the concentration, size distribution and temporal variations of atmospheric particulate carbon and to establish the occurrence and variability of the ambient mass concentrations of the particulate carbonaceous material in various environmental settings. Such information provides guidance for the design of experiments, perspective on the relative importance of various subsets of the material in relation to their properties and, in principle, it provides the ability to relate ambient concentrations to emissions.

The information presented in this article is derived from studies conducted in agricultural and urban areas of California, in Denver, Houston, and in urban and nonurban northeastern USA. Results of some of these studies have been previously discussed [26]. Although these studies were limited in sampling duration, the diversity in geographical experience provides insight into the variability of particulate carbon in air over the United States.

References pp. 367-368.

METHOD OF DETERMINING CARBON

The direct filter combustion method was initially employed to measure organic carbon in particulate matter sampled in Los Angeles [10] with thermal conductivity detection of the evolved CO_2. This method has subsequently been made more sensitive by converting CO_2 to CH_4 for detection with a flame ionization detector (FID), and has been reinvestigated to achieve measurements of both total and organic carbon in ambient air samples. Operating parameters investigated include the combustion temperature, the source of oxygen (from a MnO_2 oxidizer and/or by direct addition of oxygen to the carrier gas) and the resulting analytical problem of separating oxygen from the evolved CO_2 prior to hydrogenation and flame ionization detection. Both combustion temperature and the source of oxygen are critical for unambiguous separation of organic and elemental carbon. The poisoning of the oxidizing agent, MnO_2, by repeated acidification to remove carbonates prior to analysis has to be monitored continually. To investigate these factors, we have modified a Dohrmann DC-50 carbon analyzer to include an oxygen supply and the following features (Fig. 1):

- addition of valve V1 to allow introduction of oxygen (10 ml/min for 3 minutes) to the pyrolysis zone;
- addition of valve V2 to direct the flow stream either through a Spherocarb column for the separation of O_2 from CO_2 [14] or directly to the reduction zone; and
- addition of valve V3 to allow venting of the effluent from the Spherocarb column or passage of the effluent to the nickel catalyst.

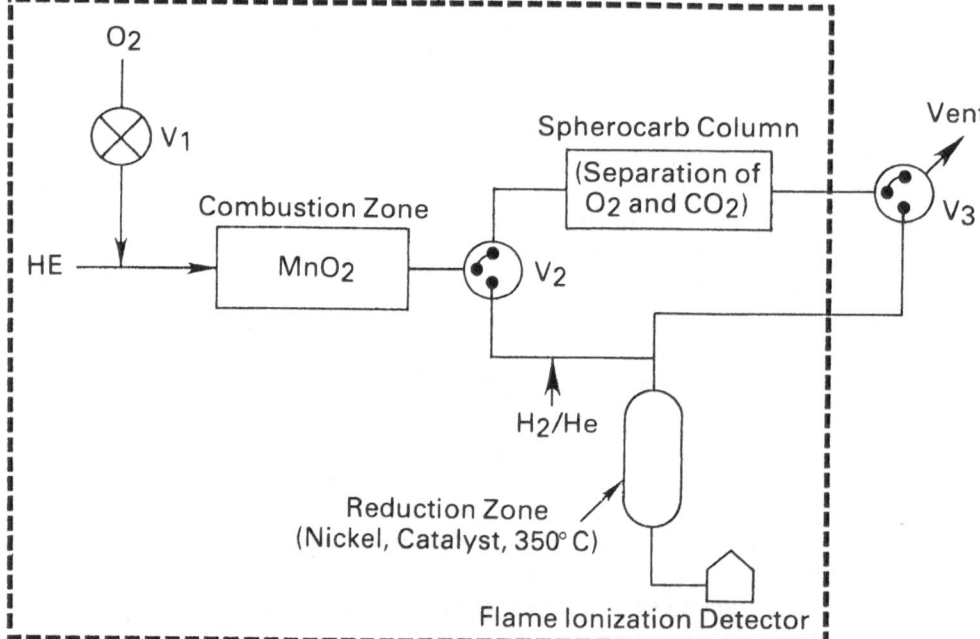

Fig. 1. Schematic flow diagram for ERT-modified Dohrmann DC-50 carbon analyzer.

The investigations using the modified carbon analyzer have led to the optimization of conditions for the speciation of carbon. Thus, for routine analysis, carbonaceous materials are determined by oxidation to CO_2 at a selected high temperature using MnO_2 as the oxidizing agent. The CO_2 is swept through a hydrogen-enriched nickel catalyst at 350°C and is reduced to CH_4 for detection with a flame ionization detector. The temperature at which sample oxidation is carried out depends on the form of carbon that is being determined. At 850°C, elemental and organic carbon are oxidized completely, while at 550°C the oxidation rate of elemental carbon by MnO_2 is so slow that only organic carbon is effectively converted to CO_2. Carbonate carbon, which liberates CO_2 upon heating, is removed from the sample by treatment with acid prior to the carbon determination. Thus total noncarbonate carbon is determined at 850°C and organic carbon at 550°C. Elemental carbon is then determined by difference.

The selection of 550°C for the oxidation of organic carbon is worth noting. It has been observed that elemental carbon is resistant to MnO_2 oxidation at 500°C. The oxidation may be observed at <550°C and just becomes measurable at 550°C. Meanwhile, the oxidation efficiency of organic compounds decreases quite noticeably going from 550° to 500°C, Hence, the selection of 550°C is optimal.

In the analysis of filter samples by direct combustion, a disc of 0.07 cm^2 area is removed from the filter and loaded into a platinum boat containing MnO_2. Thirty microliters of approximately 0.02 N HC1 are added to the disc. The boat is then advanced to a volatilization zone at 115°C to remove the water, CO_2 and any volatile organic material (a negligible amount in filter samples) present. The volatile organic material is trapped with a sorbent (Chromosorb 101 in a column) while water and CO_2 are vented to the atmosphere. The volatile organic material (VOC) is recovered upon heating the sorbent to 130°C while purging with helium. The helium carries the volatile organics to the reduction zone containing a hydrogen enriched nickel catalyst. The methane formed is directed to a flame ionization detector for detection and quantitation. Next, the boat is advanced to the high temperature oxidation zone. The CO_2 resulting from the sample oxidation is likewise converted to methane and detected by FID. In the instrument's normal mode of operation, this signal is integrated and accumulated with the VOC to yield a single carbon value for the sample.

When solvent-extractable organics are determined, a disc of 1.4 cm^2 removed from the filter is extracted ultrasonically with 0.5 mℓ of solvent (water, diluted aqueous acidic solution, benzene, or a mixture of dichloromethane and methanol depending on the nature of the analysis) in a glass stoppered centrifuge tube for one-half hour. After centrifugation at 2500 rpm for 30 minutes, 30 mℓ of the supernatant is loaded into the platinum boat containing MnO_2 for carbon determination. For aqueous extracts, the sample is analyzed at 850°C in a fashion analogous to a filter sample. For organic solvent extracts, the solvent is removed by placing the platinum boat over a hot plate at 50 to 60°C for about three minutes. The boat is then introduced into the analyzer and the carbon in the residue is determined as in direct combustion of filter samples, with the oxidation temperature set at 850°C. Table 1 summarizes the analytical conditions for the various carbon analyses.

Calibration of the carbon analyzer involves routinely nulling with acidified de-

TABLE 1
Analytical Conditions for Various Carbon Determinations

Analytical Conditions	Total Noncarbonate (TONC)[a]	Organic (OC)[a]	Elemental (EC)[a]	Solvent-extractable
Oxidizer	MnO_2	MnO_2	MnO_2	MnO_2
Oxidation Temp.	850°C	550°C	850°C	850°C
Reduction Temp.	350°C	350°C	350°C	350°C
Reduction Catalyst	Ni/H_2	Ni/H_2	Ni/H_2	Ni/H_2
Mode	Direct combustion of filter after acid pretreatment	Direct combustion of filter after acid pretreatment	Difference between TONC and OC	Residue from an aliquot of filter extract
Fiber Media	Teflon coated glass fiber (TFG)[27] Quartz fiber (QAST)[23, 31] Gelman AE-Glass fiber (GF)[29]	Quartz fiber (QAST)[23] Glass fiber (GF)[30]	Quartz fiber (QAST)[23] Glass fiber (GF)[30]	Teflon coated Glass fiber (TGF)[28]

(a) TONC = total noncarbonate carbon, OC = organic carbon, and EC = elemental carbon.

ionized distilled water and spanning with 180 and 360 ppm standard solutions of potassium hydrogen phthalate (KHP). Periodically, the methanizer efficiency is verified with injections of CO_2. Quartz filter discs loaded quantitatively with submicron graphite particles are also employed for checking the condition of the MnO_2. Exhaustion of MnO_2 may be indicated by:

- tailing of the methane peak in the recorder trace;
- grey appearance of the disc after analysis:
- brownish appearance of the MnO_2 which is characteristic of MnO formation;
- lowered instrument response to periodic tests with standardized filter discs impregnated with graphite.

With a properly functioning analyzer, calibration with KHP is adequate regardless of the form of carbon that is being determined.

The precision of the instrument is assessed by performing replicate analyses with standards and samples. The precision of the analysis is related to the amount of sample used in the determination. When the amount of carbon in the analyzed sample is smaller than 1.5 μg, the standard deviation is limited by the precision of the analyzer, and is about 0.03 μg. Above 1.5 μg of carbon, the standard deviation is one to two percent of the determined value, and is dependent on the nature of the sample and sampling substrate as shown in Table 2. Several factors could cause the observed variability:

- nonuniform distribution of carbon containing particles in the deposit;
- nonuniform distribution of the carbonaceous material throughout the filter medium; and
- variability in the size of the aliquot cut from the filter.

TABLE 2
Precision of Carbon Analyses Based on
Replicate and Repeat Determinations

Measurement	Area Analyzed cm^2	Standard Deviations		
		μg/Filter	μg/cm^2	μg/m$^{3(a)}$
TONC or TC[b]				
on TGF	0.07	76.0	5.4	3.4
on QAST	0.07	12.0	0.9	1.2
WOC				
on TGF	1.4	7.9	0.6	0.4
EC				
on QAST	0.07	16.9	1.2	1.7
OC				
on QAST	0.07	12.0	0.9	1.2
Solvent Extn.				
on TGF	1.4	7.7	0.6	0.5

(a) Applicable to the sampling volumes for data presented in this paper
(b) TC = total carbon and WOC = water soluble organic carbon.

The amounts of carbon measured on 1.4 cm^2 filter discs in the analysis of water soluble organic carbon (WOC) in the aqueous extracts, and of organic carbon in the solvent extracts of sample and blank filters are mostly below 1.5 μg. Therefore, for these samples the analytical precision is limited by the precision of the carbon analyzer.

The accuracy for measuring the carbon content of airborne particulate matter has not been assessed directly due to the lack of standardized materials for this purpose. However, the accuracy of the carbon determination has been assessed in two ways:

• Quartz fiber filters impregnated with organic compounds and graphitic materials were treated at 500° and 850°C and the relative recovery of carbon was compared for each material. Total recovery was expected for the organics at 500°C and for the graphitic materials at 850°C over the MnO$_2$ catalyst. Table 3 shows results from some typical tests. Some of the polymeric materials, i.e., OV-17 and Teflon, were not entirely converted to CO$_2$ at 500°C. Subsequent tests indicated this is a matter of reaction time as a function of the amount subjected to analysis. In addition, the fine carbon and graphite oxidized at 500°C represented only the adsorbed organic materials in the sample, since no further oxidation could be observed while these samples remained in the analyzer. The graphitic materials were converted primarily to CO$_2$ at 850°C. This is illustrated by the analyzer response to graphitic material in Fig. 2. In

TABLE 3
Relative Recoveries of Carbonaceous Materials

Material	Percent of Total Recovery at	
	500°C	850°C
Dioctylphthalate	100	0
OV-17	60(a)	40(a)
Polystyrene	99	1
Teflon FEP	48(a)	52(a)
Submicron Graphite Particles	11(b)	89
Fine Carbon Powder (N57202)	15(b)	85

(a) These materials require several instrument cycles for complete combustion at 500°C, see text.
(b) Includes organic impurities, see text.

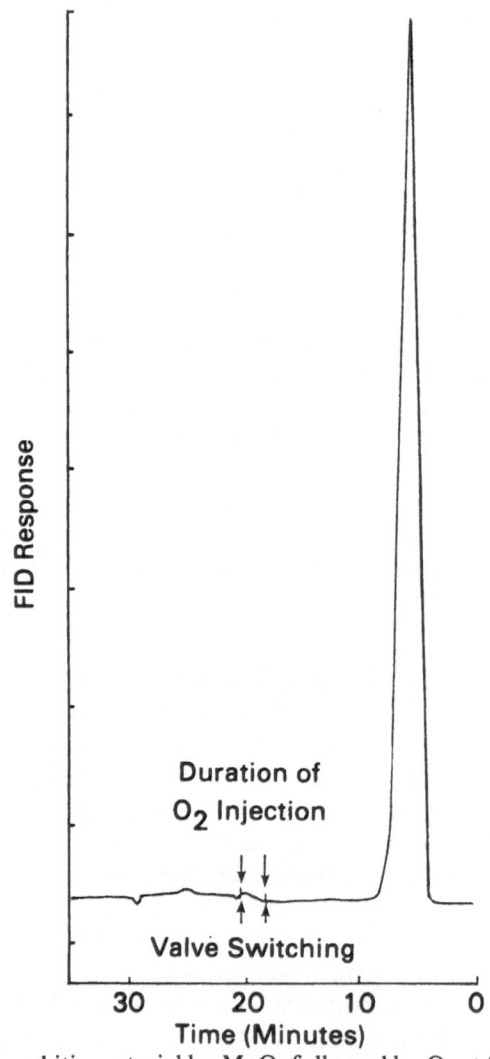

Fig. 2. Oxidation of graphitic material by MnO_2 followed by O_2 at 850°C.

this case, the submicron ARCO graphite was treated with MnO_2 at 850°C and then with O_2 at 850°C. No additional signal was obtained, demonstrating complete combustion with MnO_2 at 850°C. The linearity of response to increasing amounts of graphite deposited on quartz filters is shown in Fig. 3. The slope of the regression line was near unity with an intercept of 0.4 μg (or 5.7 μg/cm²), equivalent to the amount of carbon originally present as background impurity in the filter material. These results demonstrate that the instrument may be calibrated with an organic carbon standard to obtain reliable elemental carbon determinations, provided the MnO_2 is not exhausted.

• Results obtained by two different laboratories and methods were compared when analyzing samples of airborne particulate matter. Aliquots of samples obtained on quartz filters in Denver, CO, were distributed to the ERT and General Motors Research (GMR) laboratories. The paired comparison included 151 filters with an average loading of 208 μgC per filter. The regression of the data resulted in ERT = 0.908 (GMR) −2.9 μg/filter with a correlation coefficient of 0.99. Thus, there is substantial evidence that the reported carbon data are accurate within the limits of the reported precision of the measurements. The results of a comparison study involving GMR, ERT and other laboratories will be available shortly.

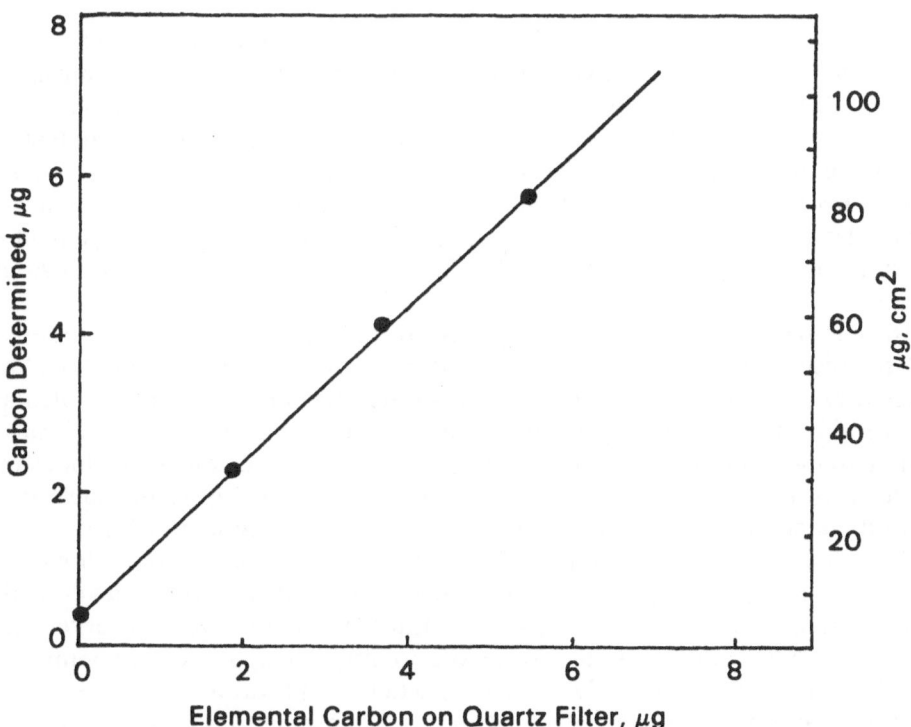

Fig. 3. Determination of elemental carbon by oxidation over MnO_2 at 850°C.

References pp. 367-368.

AMBIENT OBSERVATIONS

A variety of carbon measurements has been made as part of field studies we have conducted in several geographical areas of the United States, including urban areas in New England, nonurban areas of the northeastern United States, the Houston, TX area, the Denver, CO area, the California South Coast Air Basin (Los Angeles area), and the San Joaquin Valley in California. The types of carbon measurements performed in these studies are summarized in Table 4 and are discussed in the following paragraphs.

New England States — From the 19 sites studied, several 24-hour high volume filter (HIVOL) samples were selected for the analysis of total noncarbonate carbon (TONC), which includes elemental plus organic carbon and for volatile organic carbon (VOC). All samples had been analyzed also for mass, sulfate and nitrate, though the nitrate fraction is uncertain because of positive and negative filter and sampling artifacts. Samples from Connecticut were also analyzed for benzene soluble organics (BSO) and chloroform-methanol soluble organic (CMO). The results are summarized in Table 5. They indicate that the contribution of volatile organic components to the total carbon concentration measured in these samples was negligible.

Benzene-soluble organics and TONC were measured in samples from the Connecticut towns of Greenwich, Hartford, and Waterbury obtained during 1977. Arithmetic means of the observations are given in Table 6. Assuming the benzene soluble organics contained 70% by weight carbon, the average observed BSO was calculated and listed in Table 6 together with the percent of the TONC that was BSO. This value ranged from 34 to 54%. Thus, nearly half of the carbonaceous material was not recovered when extracting with benzene alone, in agreement with previous observations of Grosjean [12] for samples collected in the Los Angeles area.

The geometric means were calculated for the available data from the urban sites in Greenwich CT; Hartford CT; Waterbury CT; Westminster RI; and Arcadia RI. From these data, the percent composition of the particulate matter was calculated. Ten to 16% of the mass was accounted for without scaling the sulfate, nitrate and carbon to their compounds. The remainder of the mass presumably includes metal oxides or salts, soil dust, ammonium, as well as other components such as condensed water. The average composition of the particulate matter for these areas with respect to TONC ranged from 3.4% to 7.7% C by mass, except for Arcadia (10.7%). With the exception of one high value of 30.4 μgC/m^3, TONC levels in these samples ranged from 1.8 to 14.4 μgC/m^3. Large variations were observed within that range, from day to day and from site to site. Thus, TONC accounted for a significant fraction of the total suspended particles, and carbon was more abundant than sulfur or nitrogen in virtually all samples. If these distributions are considered typical for this area, urban samples will contain 4% to 10% TONC. This is similar to the mass fraction of sulfate, but is substantially larger than particulate nitrate in these samples.

TABLE 4

Summary of Ambient Particulate Carbon Data from Selected Field Studies

Parameter	New England [30]	Northeastern U.S. [27,32]	Houston TX [28]	Denver CO[23]	Los Angeles[29]	Rural CA [31]
Filter medium	GF	TFG and QAST	TGF	QAST	GF	QAST
Sampling flow rate ℓ/min	1130	120	120	30	1130	30
Sampling time (hrs)	24	3, 2	12	4	12	12
Total C	X	X	X	X	X	X
TONC				X	X	
Elemental C						
TOC plus CO_3^{-2}					X	
BSO	X					
CMO			X			
WOC		X			X	
VOC	X					
Seasonal variations		X				
Diurnal profiles		X	X	X	X	X
Size distribution						
Large (<10μmAD)	X	X	X	X	X	X
Refined (<2.5 μmAD)		X	X			X

TABLE 5

Relative Abundance of Volatile Organic Carbon (VOC) and Total Noncarbonate Carbon (TONC) in New England Particulate Samples [26, 30]

State	Number of Samples	Number of Sites	VOC*		TONC*	
			Range	Average	Range	Average
Connecticut (1977)	15	6	0-1.0	0.15	18.3-70.9	36.1
Rhode Island (1978)	18	3	0-1.0	1.10	8.2-154.3	36.0
New Hampshire (1975-77)	24	3	0-0.5	0.09	14.3-69.0	27.8
Vermont (1977-78)	39	6	0-2.0	0.39	6.2-55.7	24.8

*$\mu gC/cm^2$ of filter.

TABLE 6

Benzene Soluble Organic Carbon (BSO) and Total Noncarbonate Carbon (TONC) in Samples from Urban Areas in Connecticut in 1977 (Averages of 24-hour samples)

Town	N	Month	TONC $\mu g/m^3$	BSO* $\mu g/m^3$	BSOC/TONC %
Greenwich	1	Feb	4.4	1.5	34.1
Hartford	2	Feb	13.3	6.9	51.9
Waterbury	6	Feb to Oct	6.7	3.6	53.7

*Assumes 70% dry weight of benzene solubles was carbon.

Nonurban Northestern USA — Previous work has shown that the noncarbonate carbon in urban areas is mainly contained in small particles, less than a few micrometers in diameter [10]. This observation has been explained by the emissions of small carbon-containing particles from combustion sources and by atmospheric chemical reactions of hydrocarbon vapors. While small amounts (ng/m³) of organic carbon in atmospheric particles are biogenic [33], μg/m³ amounts of carbon would be ascribable as a first approximation to anthropogenic sources.

To assess the occurrence of carbonaceous material in nonurban areas, refined particulate matter samples (<2.5 μm AD), were collected at nine sites in the northeastern U.S. of which seven were essentially nonurban [27]. These samples were collected on Teflon coated glass fiber filters (TGF) every three hours every day for 30 days during August and October 1977 and during January, April, July and October 1978. Subsets of three days each month were selected for the determination of TONC and WOC. These days represented the evolution and decay of a meteorological situation leading to changes and widespread distribution in ambient sulfate concentrations.

To provide an overview of the findings the averages of the maximum 3-hour values from 6 months of data are plotted in Fig. 4 for TONC, WOC and SO₄⁻² in μg/m³ for

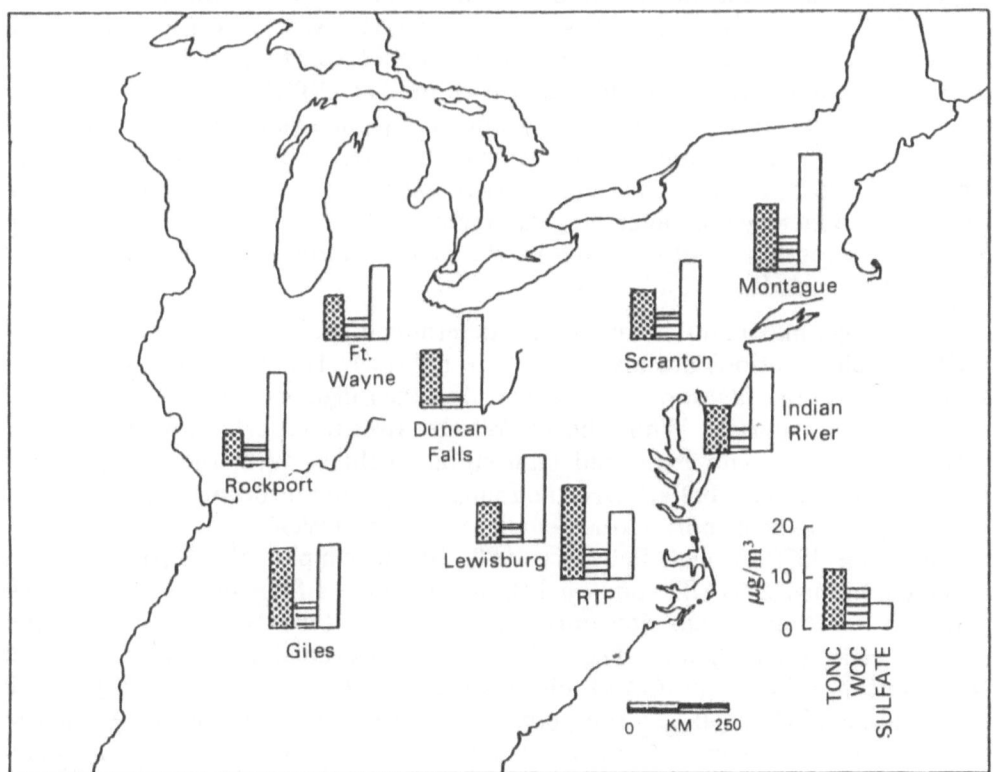

Fig. 4 Mean values of maxima for 3-hour concentrations of TONC, WOC and SO₄⁻² at nine nonurban sites in the eastern United States. Data were taken between August 1977 and October 1978 (data from ref. 27, and are in μg/m³).

References pp. 367-368.

each of the sampling sites. The range among the sites of maximum values observed in the TONC data varied almost by a factor of three. The highest concentrations were observed at a suburban-to-nonurban site at Research Triangle Park, NC, the second highest at a rural site in Giles County, TN and the lowest at a nonurban site at Rockport, IN, along the lower Ohio River Valley. The differences are believed to be related to the influences of sources near the individual sites. For example, elevated lead concentrations and diurnal NO_X and O_3 patterns at the North Carolina and Tennessee sites also indicate exposure to urban influences. It is interesting that relatively high average maximum values were also observed at Montague, MA, Scranton, PA, Indian River, DE and Duncan Falls, OH. Depending on the site, these values may have been due to the passage of urban plumes during events of regional pollution, or local combustion sources such as wood fires (see also ref. 27).

The average maximum WOC concentrations varied among the sites by a factor of about two and a half, nearly as much as the TONC. However, the single highest value was obtained for Scranton, PA and the lowest for Lewisburg, WV. The initial reasoning for determining WOC was based on the assumption that these materials represent particularly the organic material formed in the atmosphere by photochemically initiated reactions of hydrocarbons. The concentrations of WOC relative to TONC were thus hypothesized to serve as a surrogate for determining qualitatively the relative importance of photochemical reactions in the formation of sulfates from SO_2. However, the patterns of the WOC/TONC ratios are not unequivocally consistent with these expectations. The high values at Scranton could have been due to the possibility that that location is a receptor for such material. But they could also be explained by temporary fumigation of that site from the nearby emission of a factory manufacturing detergents and soaps. The other high values for WOC are consistent with the high TONC values at Montague, Giles County and Research Triangle Park.

The average maximum sulfate values determined similarly for the TONC and WOC sampling periods are also displayed in Fig. 4. The range of these values among the sites of a factor of 1.8 is smaller than the range of concentrations for the carbonaceous materials. While the Ft. Wayne, IN site reported low TONC and WOC values, this site exhibited high sulfate values. With the exception of Montague and Duncan Falls, TONC and sulfate appear to be negatively correlated. There appears to be no correlation between sulfate and WOC.

From June 1979 through February 1980, 2-hour samples of <2 μm AD were collected on quartz (QAST) and on Teflon coated glass fiber filters (TGF) at the Scranton and Duncan Falls sites during daylight hours (0800 to 1600) five days per week on a staggered schedule. These filters were analyzed for mass, SO_4^{-2}, NO_3^- and NH_4^+ concentrations. The QAST filters were analyzed for TONC and the light absorption coefficient, b_{abs}. The percent compositions, based on the geometric means for each of the measurements for the months of July and October 1979 and January 1980 are illustrated in the diagrams of Fig. 5. The percent of the total mass that was TONC ranged from 6.3 to 16.7%. At these two sites, the TONC percentages were similar or higher than those observed at the small urban New England sites. Its fraction of the total mass of fine particles collected is generally less than

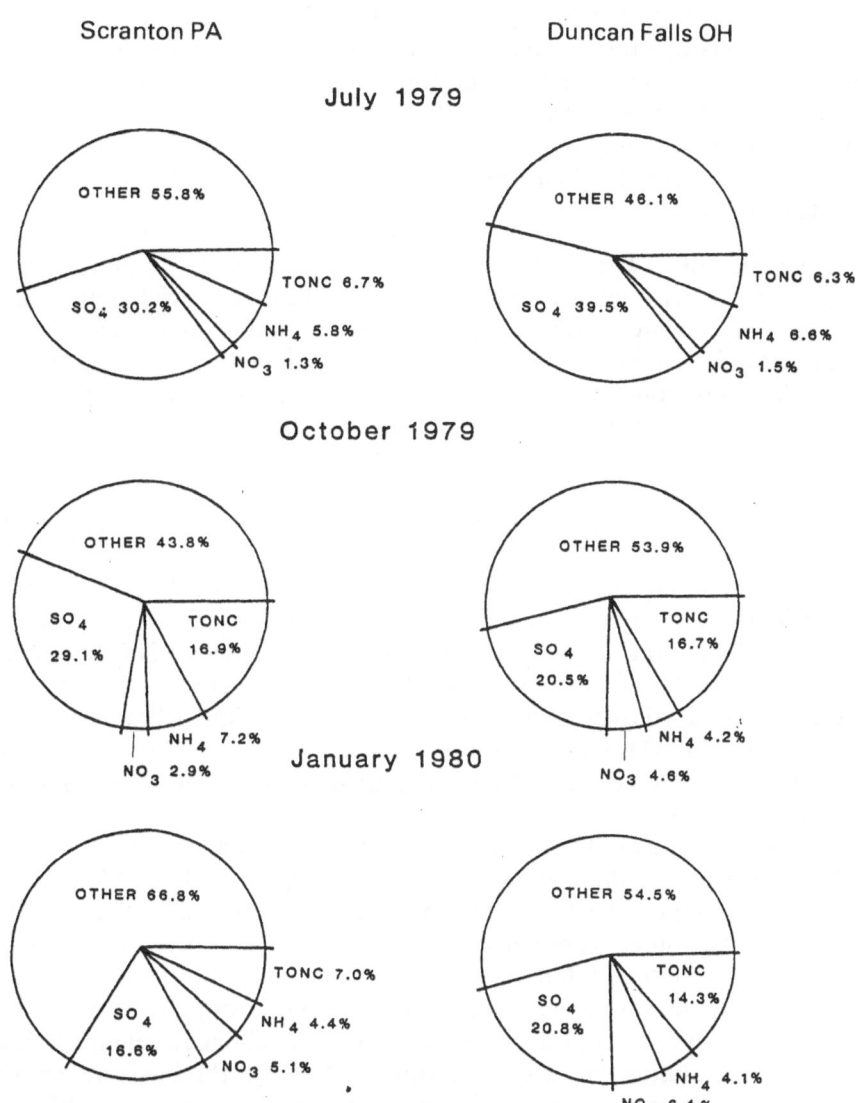

Fig. 5. Distribution of total noncarbonate carbon in small particles ($\leq 2.5 \mu$m) for two rural sites in the industrial northeast. Average composition based on the geometric means are shown for three seasonal months, October, July and January [27].

sulfate found at these sites, but is the largest amount of a single element other than oxygen found in the particles. The carbon content of the samples was found to be highest in the October samples and lowest in the July samples. The inverse is true for sulfate. The elevated concentrations of carbon at the Ohio and Pennsylvania sites may be related to wood or leaf burning in these areas during fall and winter. Thus, carbon is a regionally important, probably universal, component of the particulate matter in the nonurban areas of the eastern United States.

References pp. 367-368.

Comparison of the TONC with babs for these samples showed that there was no correlation between the amount of incident light absorbed by the samples and the amount of TONC deposited on the filters. This indicates that the variability in TONC at rural locations in the NE USA is more likely to be governed by the organic matter than by elemental carbon. Thus, further studies of the carbonaceous material from these areas should include determinations that will resolve the particulate organic compounds.

The Gulf Coast — The Gulf Coast area contains many large petroleum refining, and petrochemical facilities and paper production. An enrichment in the carbon content of particulate matter from these sources in this region might be expected. The Houston, TX area is an example of an urban-industrial complex in this region.

Samples were collected in 1977 at five locations as part of the Houston Area Oxidant Study (HAOS) [28]. These samples were collected on TFG. Both the small particles smaller than 2.5 μm AD and the large particles less than 10 μm AD were subjected to analysis for organic compounds soluble in a 1:1 mixture of dichloromethane and methanol. This solvent was adopted to extract as much organic material of a diverse nature as possible.

The small particle samples were collected simultaneously at each of five sites, located in the downtown area, in the northern suburban area near the airport, in the south suburban part of the city, to the east and to the west. The large particle and fine particle fractions were measured only at the downtown site. Sampling was generally conducted over 12-hour periods during daytime, from 0400 to 1600 hours or from 0500 to 1700 hours Central Standard Time (CST). Diurnal variations were investigated on selected days by collecting several consecutive 3-hour or 4-hour samples.

Dichloromethane-methanol soluble organic carbon levels in the small particles are listed in Table 7, along with the corresponding solvent-soluble organic carbon values for the large particles collected at the downtown site. The solvent-soluble organic carbon values from small particulate samples ranged from about 1 to 11 μgC/m^3, with significant, but nonuniform diurnal and site-to-site variations. From the mass, chloride, sulfate, nitrate and solvent soluble organic carbon data summarized in Fig. 6, it is evident that carbon is the most abundant element measured in small particles during this Houston study. Thus, on the basis of measurements conducted at five sites and involving approximately 220 samples, dichloromethane-methanol soluble organic carbon accounted for 30.2% of the mass concentration of the small particles in the Houston area. This is essentially the same as the sulfate concentration in this material. These results suggest that organic carbon material is substantially greater in the Houston area than in the urban areas of the New England states.

The distribution of the solvent soluble organic carbon with respect to particle size is illustrated with the samples collected at the downtown site (Crawford). The corresponding results and the small to large particle ratio are listed in Table 7 (3-hour and 4-hour samples). These results show that a large fraction of the total solvent soluble organic carbon occurred in the small particles. The lowest ratios are associated with low particle mass concentrations. In those cases, a few large carbon-

TABLE 7

Dichloromethane/Methanol-Soluble Organic Carbon in Inhalable (IPMC, ≤ 10 μm Diameter) and Respirable Particles (RSPC, ≤ 2.5 μm Diameter) in the Houston Area [28]

Date, 1977	Time (CST)	RSPC, μgC/m³					IPMC μgC/m³ Downtown	RSPC/ IPMC Downtown
		North	South	East	West	Downtown		
7/15	5-9	7.44	7.52	7.64	3.35	3.93	6.16	0.39
7/15	9-13	1.38	5.20	3.14	6.35	1.63	3.14	0.52
7/15	13-17	2.90	5.00	4.58	5.71	1.84	5.41	0.34
7/16	5-9	—	4.83	3.68	4.60	2.38	6.70	0.35
7/16	9-13	—	6.11	4.80	8.81	3.39	5.19	0.65
7/17	4-9	3.23	7.36	5.18	7.82	11.41	10.70	1.06
7/17	8-12	2.00	3.67	2.95	4.63	1.83	7.44	0.25
7/17	12-16	2.08	1.17	1.42	1.59	3.74	1.44	>1
7/18	4-8	8.35	5.93	9.59	7.51	9.16	13.62	0.67
7/18	8-12	2.67	7.16	1.41	1.88	0.58	3.47	0.17
7/18	12-16	3.64	1.77	3.09	2.50	0.58	4.83	0.12
9/21	5-8	9.40	—	14.99	7.50	15.44	16.75	0.92
9/21	8-11	8.13	3.58	11.48	6.39	6.48	—	—
9/21	11-14	15.38	—	12.96	5.57	11.05	—	—
9/21	14-17	17.89	5.43	5.91	3.55	6.43	8.28	0.78
9/22	5-8	8.41	6.77	6.83	18.86	18.83	—	—
9/22	8-11	—	5.60	10.40	21.80	11.51	10.85	1.06
9/22	11-14	9.44	6.55	5.91	2.07	12.37	9.55	1.29
9/22	14-17	9.40	2.64	9.56	6.08	4.08	6.93	0.59

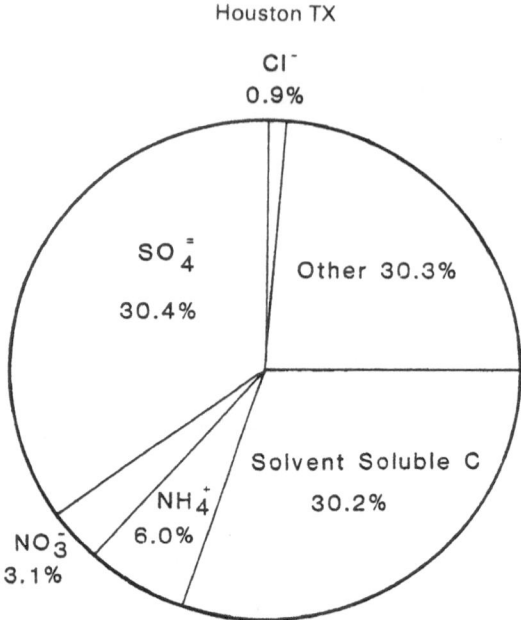

Fig. 6. The average distribution of particulate components in particles less than 2.5 μm aerodynamic mass median diameter. Data were taken in the summer and early fall of 1977 in the Houston, Texas area.

containing particles heavily influence the carbon mass distribution with respect to size.

The Western United States — The behavior of carbon in atmospheric aerosol found in the West appears to be as diverse as in the East. This tentative conclusion is derived from sketchy data available from sampling in the Denver area as well as in parts of California. The Denver data were obtained during the recent winter haze investigation [23]. The California results came from an observational program in the Los Angeles area [29] and from a project in the San Joaquin Valley near Fresno and Bakersfield [31].

The Denver study covered a period of November through December in 1978. This period was selected to be within a time when the Denver winter haze is intense and occurs at greatest frequency. Sampling was conducted at five different sites in the metropolitan area. The program included particle collection on two building tops at two heights, and at ground level north of the downtown area. The Cadillac-Fairview Building (CFB) was the highest elevation, <96 meters above street level; the downtown location was on the Denver Athletic Club (DAC) roof at approximately 23 meters above the street. The principal ground level site was the Denver Sewage Treatment Plant (STP). Samples were collected for inorganic analysis on Teflon membrane filters; quartz fiber filters were used for organic and elemental carbon analysis. Samples were obtained for fine particles $\lesssim 2$ μm AD and total suspended particles <30 μm AD. Daily samples were collected every four hours during the course of the study, beginning at midnight.

Some results of this study are shown in Fig. 7. They indicate that both elemental and organic carbon are significant constituents of the Denver winter urban aerosol. The average composition of the fine particle fraction and the coarse particle fraction (nominally 2.5 to 30 μm AD) for three sites at different elevations are shown in Fig. 7. The data indicate that most of the carbonaceous material was in the fine particles. This portion of the particle samples was actually larger than the sulfate component of the Denver particulate matter. The elemental carbon fraction represents up to half of the total carbon fraction in the particles. There evidently is a decrease in carbonaceous material with height as might be expected if urban ground level sources are prevalent. Like Houston in summer, the Denver winter aerosol contains more carbon than was observed in urban areas of New England.

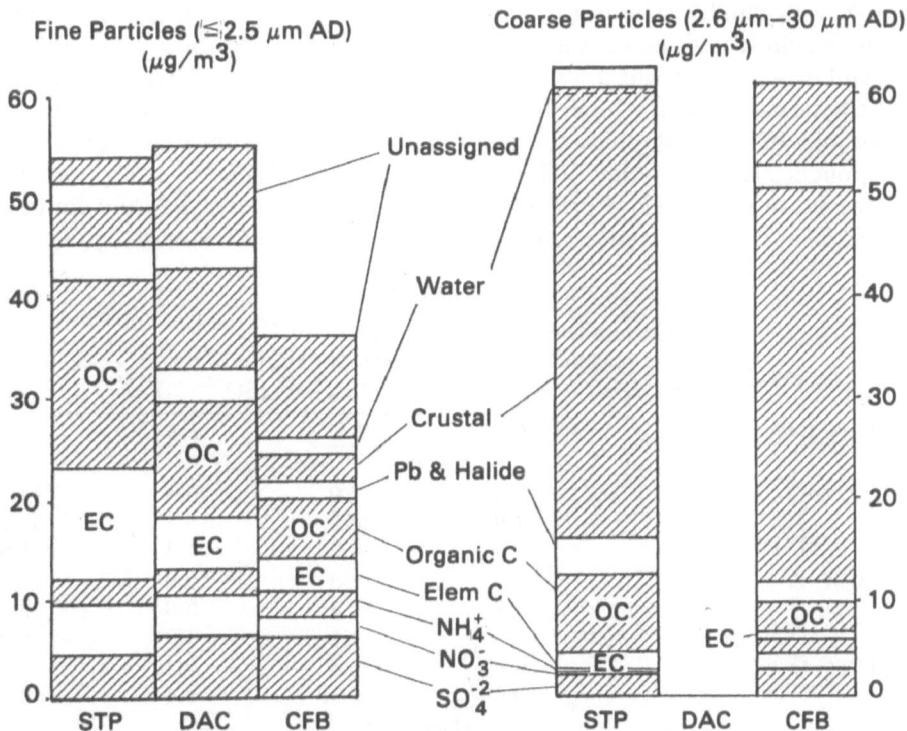

Fig. 7. Composition of Denver Winter Haze particle samples in November-December 1978. The STP site is at ground level north of the downtown area; the DAC site is a downtown building roof top site at 23 meters height and the CFB is a downtown building rooftop site at 96 meters height [23].

Early work in southern California documented the importance and the variability of carbonaceous material in the smog aerosol [11]. Subsequent work has reinforced this conclusion. However, the initial expectation of the importance of organic particle production by atmospheric chemical reactions has still not been thoroughly tested. The significance of photochemical processes was studied between August

1977 and April 1978 at two urban south coast air basin locations, Azusa and Anaheim, CA [29]. In this study, samples were obtained during the "day' (1000 to 2200 hours nominally) and "night" (2200 to 1000 hours nominally) for carbon analysis and other particle chemistry. The pollution intensity was determined in terms of combined levels of particulate sulfate over a 24-hour period and hourly maximum ozone reported at the stations. The data were stratified according to four regimes: (1) low sulfate (<10 $\mu g/m^3$) and low ozone (<10 pphm)(LL); (2) low sulfate and high ozone concentration (>20 pphm) (LH); (3) high sulfate (>20 $\mu g/m^3$) and low ozone (HL); and (4) high sulfate and high ozone concentrations (HH). An attempt was made to obtain samples for six days in each category. Sampling for noncarbonate carbon analysis was conducted on glass fiber media with a high-volume sampler; water soluble and total carbon were reported for the samples taken.

The results of the study were summarized in Table 8. The concentrations of WOC and TC were similar at both sites, with substantially more carbonaceous material on the high sulfate and high ozone days than during the other periods. Likewise, the low sulfate and low ozone days had the lowest carbon levels. There appears to be no systematic pattern in the other two categories. In general, there was more carbonaceous material in the daytime samples, with a weak preference for the water soluble component also in daytime. This is not necessarily indicative of photochemical processes, however, since primary carbon emissions increase in the daytime with industrial and transportation activity.

TABLE 8
Carbon Data Summary by Category Days for
Azusa and Anaheim in $\mu g/m^3$ and Ratios [29]

Parameter		LL	LH	HL	HH
		Category (SO_4^{-2}: O_3)			
		AZUSA			
TC	$\overline{D}$	11.0	17.9	23.5	34.6
	$\overline{N}$	11.9	20.0	19.2	33.8
	$\overline{D/N}$	1.3	0.88	1.3	1.0
WOC	$\overline{D}$	4.4	12.0	9.8	17.3
	$\overline{N}$	3.9	7.8	8.0	15.5
	$\overline{D/N}$	1.1	1.6	1.7	0.98
WOC/TC	$\overline{D}$	0.42	0.67	0.37	0.49
	$\overline{N}$	0.35	0.38	0.40	0.48
		ANAHEIM			
TC	$\overline{D}$	11.8	15.6	14.7	33.4
	$\overline{N}$	12.5	16.2	17.3	22.4
	$\overline{D/N}$	1.4	1.1	0.72	1.6
WOC	$\overline{D}$	3.6	7.7	6.7	17.3
	$\overline{N}$	3.6	9.1	8.0	13.1
	$\overline{D/N}$	1.3	1.6	0.84	1.6
WOC/TC	$\overline{D}$	0.30	0.52	0.51	0.58
	$\overline{N}$	0.34	0.46	0.49	0.58

$\overline{D}$, $\overline{N}$, $\overline{D/N}$ are averages of day and night values and of the ratio of day-to-night. TC is total carbon, and WOC is water soluble organic carbon.

The fraction of total carbon in the Los Angeles samples for particles less than ~10μm AD for different categories is shown in Fig. 8. The fractional contribution of total carbon to particulate matter is quite high, ranging from 22% to 33% in these samples based on 24-hour averges. If the solvent-soluble fraction in Houston is considered equivalent to the thermal volatilization method, then the Los Angeles and Houston samples have similar carbon contents. The two methods are not entirely equivalent since direct thermal volatilization will give elemental carbon, which presumably the solvent extraction does not. Thus, the total noncarbonate carbon content of particles in Houston may be greater than that in Los Angeles.

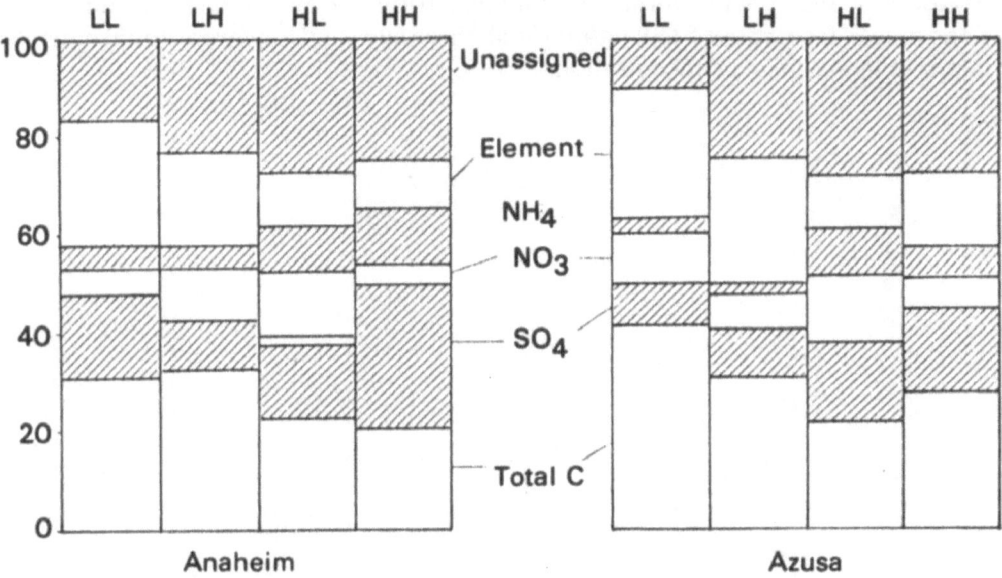

Fig. 8. Fraction (%) of total carbon in 24-hour average aerosol particle samples less than 10 μm for Azusa and Anaheim, CA [29]. Classes of days are low SO_4^{-2}, low O_3 (LL), low SO_4^{-2}, high O_3 (LH), etc.

As a contrast to the urban sites examined in the Los Angeles area, samples have been analyzed from rural sites in the California San Joaquin Valley. These were taken as part of a larger experiment to study the regional air quality in this enormous agricultural and oil-producing area. Samples were taken at essentially the same locations in November and December of 1978 and again in September of 1979. Samples were taken on Teflon membrane for subsequent analyses for inorganics and on QAST for measurements of carbon. About 30 sequential 12-hour samples were taken each season at sites south of Bakersfield and south of Fresno on the central-east side of the valley. A third site was located at Lost Hills northwest of Bakersfield.

References pp. 367-368.

The average concentration of major components of the San Joaquin Valley particles less than 20 μm AD are shown in Fig. 9. The data show interesting features, some of which are similar to other geographic areas and others which are not. The average total mass concentration was quite high, tending to exceed 100 μm/m³. More than 60% of the mass and about 40% of the carbon was in particles larger than 2.5 μm AD. Nitrate was a very substantial fraction of the collected particulate matter during the winter month. Silicon dominated the particulate concentration during the summer months. Noncarbonate carbon was present during both seasonal months. Its average concentration of about 25 μg/m³ was similar to nitrate during winter at Fresno. The carbon concentrations were lowest (5 to 7 μg/m³ in winter near Bakersfield. During the summer period, the average carbon concentrations ranged from about 5 to 9 μg/m³ for all three sites and they were about equal or larger than the sulfate concentrations.

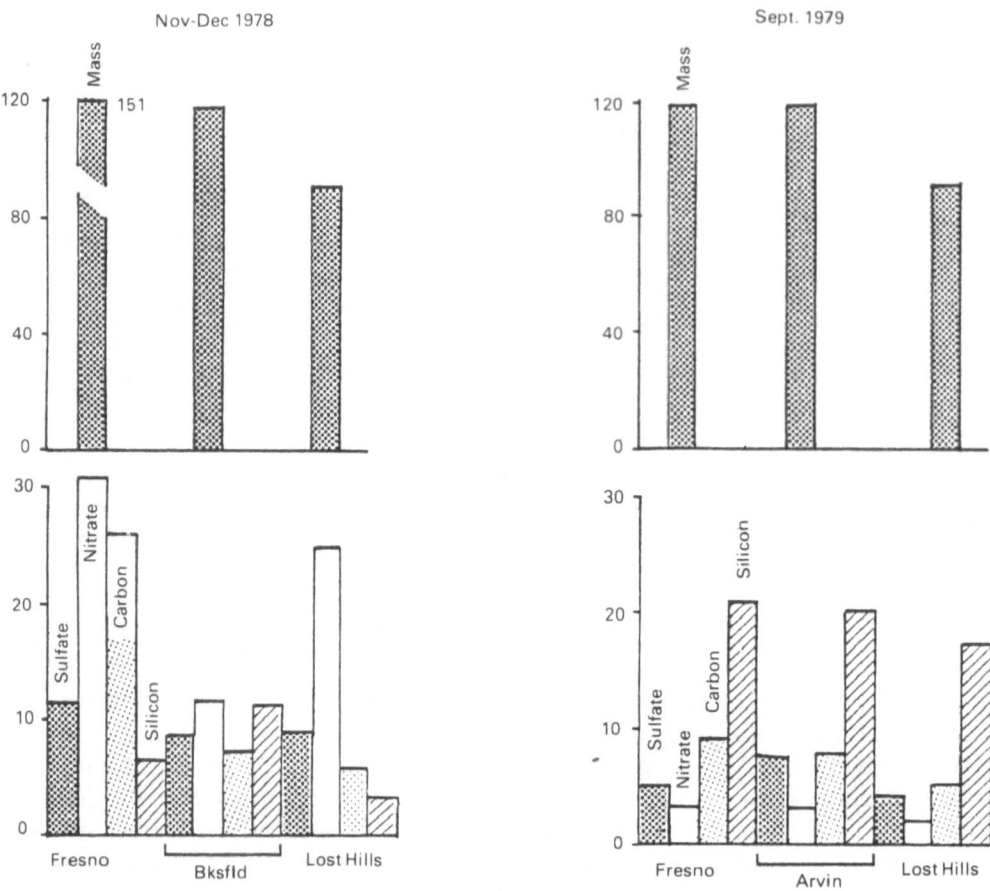

Fig. 9. Chemical components in μg/m³ of San Joaquin Valley Aerosol Particles ($\leq$20 μm AD) for winter and later summer [31].

The average relative composition of particles from these sites for carbonaceous carbon and the other major components are shown in Table 9. The carbon content of the particulate matter remained virtually constant for two of the three sites. The high averge value for the Fresno site indicates an important nearby source for carbon in winter. Thus, carbonaceous material occurred ubiquitously in the aerosol sampled in the rural San Joaquin Valley. The carbon fraction of the total particulate mass sampled in both the nonurban eastern and western states occurred in similar ranges. Carbon was one of the dominant elements present in the ambient particulate samples taken in California.

TABLE 9
Average Percent Composition of Particulate Matter Samples
at Nonurban Sites in the California San Joaquin Valley

Site Name	Component					
	C	SO_4^{-2}	NO_3^-	NH_4^+	Si	Other
November to December 1978						
Fresno	17.2	8.6	21	8.5	4.2	40.5
Bakersfield	6.2	7.3	10	7.7	9.8	59
Lost Hills	6.8	11	29	12	4.2	37
September 1979						
Fresno	7.8	4.1	2.9	1.2	18	66
Arvin	6.9	6.3	3.0	2.0	17	64.8
Lost Hills	6.1	4.8	2.7	1.3	19	69.1

SUMMARY AND CONCLUSIONS

Results have been presented concerning the concentration, size distribution and temporal variations of particulate carbon in various parts of the USA. Our studies have included the determination of total organic, total noncarbonate, benzene-soluble, chloroform-methanol-soluble, water-soluble and volatile organic carbon.

The results derived from field studies conducted in the New England states, the nonurban northeastern United States, nonurban locations in California and the Houston and Los Angeles urban areas are summarized in Table 10 for TONC. The average concentration of TONC generally increased from about 2 $\mu g/m^3$ at the nonurban to 7 $\mu g/m^3$ at the small urban locations in the northeastern USA. At rural sites in an agricultural and petroleum producing valley in California, the average concentration was about 9.5 $\mu g/m^3$. The highest TONC concentrations were observed in urban-industrial areas such as Houston, Denver and Los Angeles. Particulate matter samples from these areas contained typically 20 to 40% TONC as compared with 3 to 20% TONC for samples from areas of lower population density. Particulate carbon was found to be the most abundant element not only in Los Angeles aerosols but also in the other urban areas and regions investigated to date. Our findings underline the importance of carbonaceous materials in ambient particles.

References pp. 367-368.

TABLE 10
Total Noncarbonate Carbon in Atmospheric Particles
as Found at Various Locations in the USA

Location	Date	Sampling Time	No. of Samples	No. of Sites	TONC ($\mu g/m^3$)		Average % of Mass
					Range	Mean	
Non-urban NE USA	Jul 79–Jan 80	2 hr 24 hr	480	2	0.3 to 12	1.9	11.4
Small Urban NE USA							
CT	1977		15	6	3.5 to 13	6.8	3 to 8
RI	1978		18	3	1.6 to 10	6.8	3 to 19
NH	1975-77		24	3	2.7 to 13	5.3	—
VT	1977-78		39	6	1.2 to 11	4.7	—
Calif. Valley	1978-79	12 hr	137	9	5.6 to 26	9.5	9.2
Houston TX	1977	24 hr	70	5	11.2 to 21	16.6	30
Denver CO	Winter 1978	4 hr	360	6	1.2 to 89	17.5	36
Los Angeles CA	Fall 1977	12 hr	72	2	3.6 to 40	22.6	29

By virtue of its molecular complexity, particulate carbon represents the most likely candidate material of toxicological significance in the ambient aerosol, and elemental carbon contributes to visibility reduction. Resolution of the carbonaceous material by volatility and solvent solubility does not provide sufficient knowledge about the ambient particulate carbon. This applies especially to the organic fraction. The molecular composition of the organic fraction needs to be better characterized, not only for toxicological reasons, but also for the purpose of establishing the origins of this material. However, the thermal methods used in our studies will continue to be useful for monitoring purposes and as part of studies on visual air quality.

ACKNOWLEDGEMENT

The studies leading to this paper were sponsored by several institutions. They include: The Electric Power Research Institute, the Southern California Edison Co., the Motor Vehicle Manufacturers Association of the United States, Inc., the Houston Chamber of Commerce, USEPA Region I, and the California Air Resources Board. We are indebted to many individuals at ERT who carred out the field, laboratory, data processing and the manuscript production work.

REFERENCES

1. *J. Leiter, M. B. Shimkin and M. J. Shear, J. Nation Cancer Inst., Vol. 3 (1942), p. 155.*
2. *P. Kotin, H. L. Falk, P. Mader, M. Thomas, Arch. Ind. Hyg. Occup. Med., Vol. 9 (1954), p. 153.*
3. *National Academy of Sciences, Particulate Polycyclic Organic Matter, National Research Council, Washington, DC. (1972)*
4. *D. Grosjean, Aerosols, Chapt. 3 in Ozone and Other Photochemical Oxidants, National Academy of Sciences, Washington, D.C., (1977), pp. 45-125.*
5. *S. L. Heisler and S. K. Friedlander, Atmos. Environ., Vol. 11 (1977), p. 157.*
6. *P. H. McMurray and S. K. Friedlander, J. Colloid Interface Sci., Vol. 64 (1978), p. 248.*
7. *D. Schuetzle, A. L. Crittenden and R. J. Charlson, J. Air Pollut. Control Assoc., Vol. 23 (1973), p. 704.*
8. *D. Schuetzle, D. Cronn, A. L. Crittenden and R. J. Charlson, Environ. Sci. Technol., Vol. 9 (1975), p. 838.*
9. *D. Grosjean, K. Van Cauwenberghe, J. P. Schmid, P. E. Kelley and J. N. Pitts, Jr., Environ. Sci. Technol., Vol 12 (1978), p. 313.*
10. *P. K. Mueller, R. W. Mosley and L. B. Pierce, J. Colloid Interface Sci., Vol. 39 (1972), p. 235.*
11. *G. M. Hidy, P. K. Mueller, D. Grosjean, B. R. Appel and J. J. Weslowski (Ed.), The Character and Origins of Smog Aerosols, Wiley, New York, NY (1979).*
12. *D. Grosjean, Anal. Chem., Vol. 47 (1975), p. 797.*
13. *E. S. Macias, C. D. Radcliffe, C. W. Lewis and C. R. Sawicki, Anal. Chem., Vol. 50 (1978), p. 1120.*
14. *J. J. Huntzicker and R. L. Johnson, Paper 2, Proceedings of the Conference on Carbonaceous Particles in the Atmosphere, Lawrence Berkeley Labortory, Berkeley, CA. (1979).*

15. D. Grosjean and S. K. Friedlander, *Formation of Organic Aerosols from Cyclic Olefins and Diolefins in The Character and Origins of Smog Aerosols, G. M. Hidy, (Ed.). Wiley, New York, N.Y., (1979), pp. 435-473.*
16. T. Novakov, S. G. Chang and A. B. Harkins, *Science, Vol. 186 (1974), p. 159.*
17. H. Rosen and T. Novakov, *Nature, Vol. 266 (1977), p. 708.*
18. S. G. Chang and T. Novakov, *Atmos. Environ., Vol. 9 (1975), p. 495.*
19. J. N. Pitts, Jr., D. Grosjean, T. M. Mishke, V. F. Simmon and D. Poole, *Toxicol. Letters, Vol. 1 (1977), p. 65.*
20. J. N. Pitts, Jr., K. Van Cauwenberghe, D. Grosjean, J. P. Schmid, D. R. Fitz, W. L. Belser, G. B. Knudson and P. M. Hynds, *Science, Vol. 202 (1978), p. 515.*
21. D. Grosjean, K. Van Cauwenberghe, D. R. Fitz and J. N. Pitts, Jr., *Amer. Chem. Soc. Div. Environ. Chem. Preprints, Vol. 18 (1978), pp. 354-356.*
22. National Research Council, Diesel Impacts Study Committee, *Draft Report, Washington, DC. (1980).*
23. S. L. Heisler, R. C. Henry, J. G. Watson, and G. M. Hidy, *The 1978 Denver Winter Haze Study. ERT Project for the Motor Vehicle Manufacturers of the United States, Inc., Westlake Village, CA. (1980), p. 5417.*
24. G. T. Wolff, R. J. Countess, P. J. Groblicki, M. A. Ferman, S. H. Cadle and J. L. Muhlbaier, *Atmos. Environ., Vol. 15 (1981), p. 2485.*
25. Cautreels and K. Van Cauwenberghe, *Atmos. Environ., Vol. 12 (1978), pp. 1133-1142.*
26. D. Grosjean, K. Fung, P. K. Mueller, S. Heisler, and G. M. Hidy, *Paper No. 45G, Symposium on Sampling and Analysis of Particulate Matter, 72nd Annual AICHE Meeting, San Francisco, CA, November 27. AICHE Symposium Series AIR-1979, (1979).*
27. P. K. Mueller, G. M. Hidy, R. L. Baskett, K. K. Fung, R. C. Henry, T. F. Lavery, N. J. Lordi, A. C. Lloyd, J. W. Thrasher, K. K. Warren, and J. G. Watson, *The Sulfate Regional Experiment: Report of Findings, Report EA-1901, Electric Power Research Institute, Palo Alto, CA 94306, Vol. 2 (1981).*
28. S. L. Heisler, *Particulate Sampling and Analysis, Final Report ERT P-5190, for the Houston Area Oxidant Study (HAOS), Radian Corporation, Austin, TX (1979).*
29. S. L. Heisler, R. C. Henry, P. K. Mueller, G. M. Hidy and D. Grosjean, *Aerosol Behavior Patterns in the South Coast Air Basin with Emphasis on Airborne Sulfate. Final Report ERT Project P-A085, for the Southern California Edison Co. (1980).*
30. J. C. Chow, A. Flaherty, E. Moore and J. Watson, *Filter Analysis for TSP-SIP Development. EPA901/9-78-003, U.S. EPA Region I, Lexington, MA (1980).*
31. S. L. Heisler and R. Baskett, *Particle Sampling and Analysis in the California San Joaquin Valley. Final Report ERT P-5381-700 for the California Air Resources Board, Environmental Research & Technology, Inc., Westlake Village, CA (1980).*
32. J. G. Watson and P. K. Mueller, *The Eastern Regional Air Quality Study Report EA1914, Electric Power Research Institute, Palo Alto CA, in manuscript (1981).*
33. B. R. T. Simoneit, *Eolian Particulates from Oceanic and Rural Areas in Prog. Phys. and Chem. of the Earth. A. G. Douglas and J. R. Maxwell (editors), Pergamon Press Ltd., London, in press (1980).*

DISCUSSION

S. Arnold, *(State of Colorado)*

You indicated the temperature used for the organic determination is 550°C. Earlier papers indicate that there was a formation of pyrolitic carbon and some of the elemental carbon that would start to come off long before that 550°C temperature was reached. Could you comment on that?

P. Mueller

If you temperature program, continuously without MnO_2, that kind of chemical change can be demonstrated, but in our method this is not done. We load our sample onto a magnesium oxide boat which is then moved into the 550°C zone immediately. Under our conditions, only the organic material was converted to carbon dioxide, and the conversion was complete as indicated in the text of the paper.

B. Appel, *(California Department of Health)*

What do you do about the carbonate carbon?

P. Mueller

When carbonates are shown to be significant constituents of the sample (small particles from urban areas typically contain virtually no carbonates) then each sample is acidified first and the CO_2, is swept from the combustion apparatus automatically prior to the next step.

C. Spicer, *(Battelle Columbus)*

From some of the earlier papers, there seemed to be very little secondary organic aerosols. You seem to be drawing different conclusions. George Wolff and one or two of the earlier authors [*], [†], indicated that there was not very much room left for secondary organic aerosol as part of the total carbon burden in urban areas.

P. Mueller

When all of our papers are published, we should mutually study the magnitude of the differences you are referring to, and the assumptions underlying the respective methods. Then we can decide whether or not the apparent differences are real.

L. Gundel, *(Lawrence Berkeley Laboratories)*

Some work I did a couple of years ago showed the presence of water soluble organics in Berkeley particulates in the wintertime. I would not say that that is an indication of secondary material and I would be a little hesitant to generalize.

P. Mueller

I think that is a good point. Our allocation is based on the premise that water soluble material should be an indication of secondary material. We realize this is not necessarily true. Water soluble organics can come directly from sources so we have

---

* *Novakov, T., These Proceedings, p. 19.*
†*Wolff, G. T., Groblicki, P. J., Cadle, S. H. and Countess, R. J., These Proceedings . p. 297*

to be careful about interpreting data. The amounts of water soluble carbon that we found in Los Angeles were consistent with our expectations for that area based on photochemistry. There is an urgent need to obtain molecular characterization of organics emitted by various types of major sources.

R. Draftz, *(I.I.T. Research)*

I wonder if you had any experience seeing whether elemental carbon is coming through as water soluble carbon. Our Henderson Cloud* study suggested that it did.

P. Mueller

We tested for this possibility. We centrifuged some of our extracts at ultra high G and found no difference in the carbon content.

R. Draftz

Let me just add one other thing. We are filtering with a 0.05 micron nucleopore filter and we could not get rid of the turbidity. We did the analysis by a number of techniques and found that the organic was primarily elemental carbon.

P. Mueller

I would like to study your techniques and also the nature of the differences in our respective samples. A great deal of care is needed to avoid interference from colloidal elemental carbon when analyzing extracts.

Baker, D. H., Rush, T., Brookman, D. J. and Draftz, R. G., IIT Research Institute, Chicago, IL, 1980

371

MEASUREMENT OF LIGHT ABSORPTION AND ELEMENTAL CARBON IN ATMOSPHERIC AEROSOL SAMPLES FROM REMOTE LOCATIONS

J. HEINTZENBERG

University of Stockholm
Stockholm, Sweden

ABSTRACT

A new double beam, single detector integrating sphere photometer has been developed for light measurements on aerosol samples. The light measurements are completely digital (photon counting), controlled by a desk computer. The resolution of the digital photometer is such that it can detect a 10^{-5} difference between the sample and the reference signals. Taking a typical background aerosol sample (10^5 m length of air column swept by the filter), this results in a detection limit of 10^{-10} m^{-1} for the light absorption coefficient of the aerosol particles. This sensitivity is sufficient for the detection of elemental carbon in size segregated samples at ground level background locations or in stratospheric air.

Two independent light absorption parameters can be determined by the instrument:

a. the conventional absorption coefficient.
b. the pure light absorption of the particles independent of the filter they are sampled on and independent of their scattering properties.

In the second case the supporting filter is dissolved and the particles are analyzed as a hydrosol in the integrating sphere. By means of a calibration with hydrosols of known amounts of soot with known composition, the results can be expressed in terms of equivalent amounts of elemental carbon in the atmospheric samples.

The first results from size segregated aerosol samples in Arctic haze on Spitsbergen show light absorption coefficients similar to those measured in Northern Alaska. The dominating part of the elemental carbon in Artic haze is located in particles smaller than 0.1 μm radius.

References p. 376.

INTRODUCTION

Elemental or graphitic carbon or soot as it will be called here has been shown to be the one component in many atmospheric aerosols which dominates their light absorption [1]. Moreover, soot has a number of physical and chemical properties which stimulate the interest of the atmospheric scientist.

1. Most soot particles are primary particulate combustion products, i.e., they are emitted as particles into the atmosphere.
2. The black material in soot is chemically inert during its lifetime in the atmosphere. In contrast to most other trace substances in the air, it will not undergo chemical transformations, but it will maintain the characteristic features.
3. While being chemically inert, it carries reactive groups on its surface which are involved in chemical reactions in the atmosphere.
4. We suspect that the combustion processes which cause the major part of soot in the atmosphere generate particles predominantly in the Aitken nuclei range, i.e., below $0.1 \mu m$ radius. Through dry particle coagulation and physical-cloud processes, the primary particles will be distributed over all sizes of the atmospheric aerosol. Hence, the study of the size distribution of light absorbing or soot in atmospheric particles should yield information on the physical transformation or aging processes of the atmospheric aerosol.

THE STUDY OF SOOT AT REMOTE LOCATIONS

In 1978, a study of the light extinction properties and soot content in size-segregated atmospheric aerosol samples was started. We chose remote locations in the Arctic as field sites. This area is especially sensitive to anthropogenic influence on the atmosphere. It is covered by a thin film of pack ice causing a continental thermal character in the Arctic. This pseudo continent now has a high surface albedo. However, an albedo reduction from 70 to 60% would destroy the ice in eight to ten years. A summer air temperature anomaly of $+2°C$ would cause the same destruction in a few decades [2]. Global changes of climate are to be an expected consequence of such a drastic change. Light scattering and absorbing soot-containing aerosol particles can contribute to an air temperature anomaly. In deposited form, soot can cause surface albedo changes. Even embedded in snow, soot causes temperature changes within the ice layer.

At the same time the Arctic sink region provides a unique possibility for studying long distance transport and transformation processes of continental aerosols emitted in Eurasia and we intend to make use of the tracer properties of soot in investigations of long distance transport.

SAMPLING METHODS

The amount of suspended particulates in remote areas such as the Arctic is extremely low (on the order of $0.1 - 1.0 \mu g/m^3$). Therefore, soot particles had to be collected from large volumes of air. We chose latex fiber filters because of their low flow resistance in high-volume sampling and because of their high particle retention (> 90%) in the Aitken nuclei range. They can be dissolved in organic solvents, which is necessary for our method of soot analysis.

High volume sampling was done either through total filters (47 mm diameter) or through 1-stage high-volume impactors with a 50% radii cut-off in the 0.1 to 0.5 μm range following the design of Winkler [3]. The same type of filter was used to collect the non-impacted fraction of the aerosol. Sample volumes of 300 to 800 m^3 covered time periods of 1 to 6 days.

One of the major problems in atmospheric sampling in remote regions is that of local contamination by practically unavoidable nearby combustion sources. We made a special effort to reduce contamination from local sources by controlling the pumps of the samplers with the signals of Aitken nuclei counters [2].

SAMPLE ANALYSIS

For the analysis of our samples, a new double beam, single detector integrating sphere photometer has been developed for light measurements on aerosol samples. The light measurements are completely digital (photon counting), and are controlled by a desk computer. The resolution of the digital photometer is such that it can detect a 10^{-5} difference between the sample and the reference signals. For a typical background aerosol sample (10^5 m length of air column swept by the filter), there is a detection limit of 10^{-10} m^{-1} for the light absorption coefficient of the aerosol particles at about a 550 nm wavelength. This sensitivity is sufficient for the detection of elemental carbon in size segregated samples at ground level background locations or in stratospheric air.

Two independent light absorption parameters can be determined by the instrument: the conventional absorption coefficient [4], and the pure light absorption of the particles which is independent of the filter media and independent of their scattering properties. In the second case the supporting filter is dissolved. The sample is dispersed by an ultrasonic bath and the particles are analyzed as a hydrosol in an integrating sphere [5]. The integrating sphere brings all the light scattered by the soot particles to the optical detector. Fig. 1 shows a schematic drawing of the soot photometer plus the data recording equipment.

CALIBRATION OF THE SOOT PHOTOMETER

Measuring the light absorption of the particles and inferring the soot content from the results require that the instrument be calibrated with known amounts of soot and that the results be compared with measurements which are specific to graphitic carbon, e.g., Raman scattering [1]. A calibration with artificial soot has been performed. We chose a channel type soot (Monarch M71 of the Cabot Corp.) which exhibits a narrow size range near 80Å radius according to the manufacturer. It is about 95% nonvolatile carbon plus 5% volatile material. Its specific light

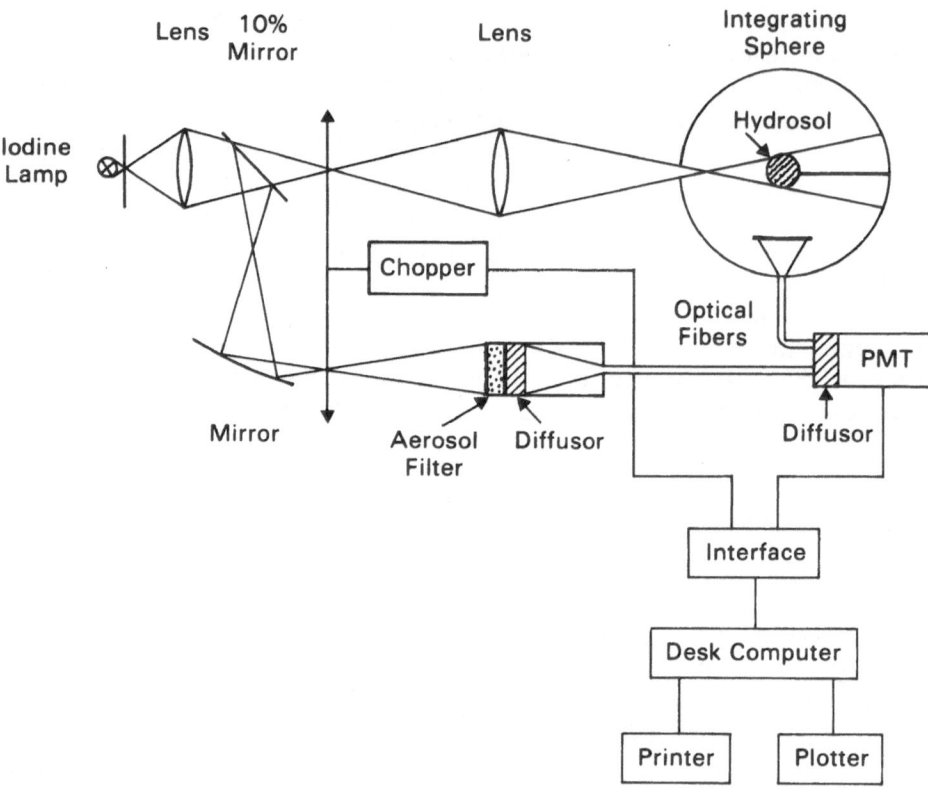

Fig. 1. Double beam single detector integrating sphere photometer for the analysis of soot aerosols.

absorption of 9.68 m²/g [6] is very similar to values found for atmospheric aerosols. With a microbalance, samples containing 100-1000 μg of soot were weighed and volume-diluted in hydrosol form down to about 500 ng/L.

Fig. 2 gives the resulting calibration curve relating the mass of soot in the hydrosol to the optical density, relative to that of the sample flask filled with a dissolved blank filter. From this calibration we determined a lower detection limit of about 500 ng and an upper of about 100 μg because of the highly nonlinear response to large amounts of soot. The signal for a sample flask filled completely with soot is marked with an infinity symbol in Fig. 2 to show that the optical method can only see the shell of such a large volume of soot. From a smaller number of tests we give 10 % as the accuracy in terms of M71-soot within the usable mass range.

AMBIENT MEASUREMENTS IN REMOTE AREAS

In two field experiments aerosol samples for soot analysis were taken in remote locations. In the late winter of 1979, direct optical measurements and aerosol sampling for physical and chemical analysis was done at Ny-Ålesund, Spitsbergen. The high winter levels of air pollution first found by Rahn [7] in Northern Alaska

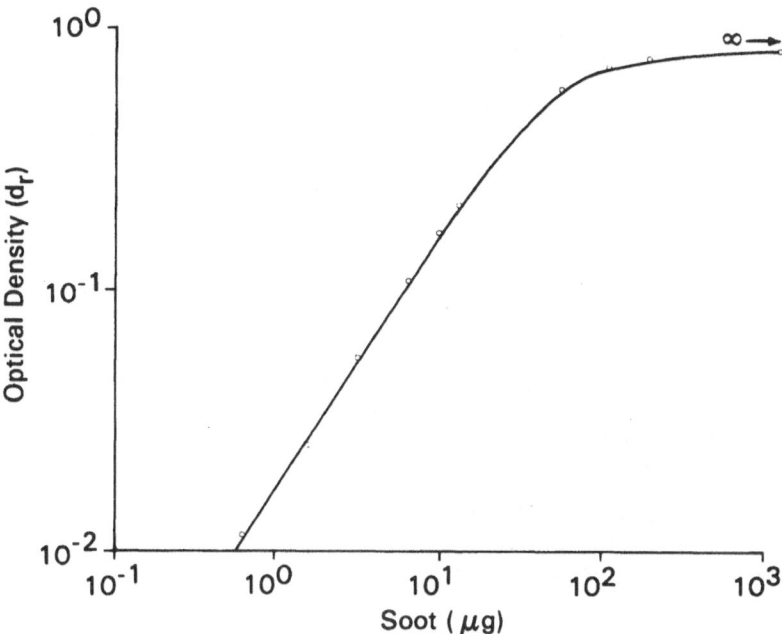

Fig. 2. Calibration curve of the soot photometer.

were confirmed by the results of our experiment [8]. The observed size distribution was as expected from our knowledge of aerosol residence times versus particle size [9] in a highly aged continental aerosol. The number, surface and mass concentration peaks in the accumulation mode were at about 1.0 μm radius. The total suspended mass was about 4000 ng/m^3 and about 1 % was found to be soot. Of these 50 ng/m^3 of soot, about 60% was in the Aitken nuclei range less than 0.1 μm radius. This roughly agrees with the total mass size distribution which we found at Ny-Ålesund. The light absorption coefficients as measured on the aerosol in the hydrosol state were about 2.2x10^{-6} m^{-1} or about 15 % of the light scattering coefficient we measured with an integrating nephelometer.

The few existing results obtained during the summer show that the aerosol concentration decreases by a factor of 10-20 compared to the winter levels. The Swedish icebreaker expedition, Ymer-80, gave a unique chance to study this clean Arctic summer atmosphere. During July, August and September 1980, it explored the area North of Greenland to North of Franz-Josefsland. The University of Stockholm had a complete air chemistry lab-container on board for the study of the arctic aerosol. The formidable contamination problem, arising from the fact that samples at levels down to 1 particle per cm^3 were taken near a chimney emitting particle concentrations on the order of 10^6 cm^{-3}, again was solved by controlling our samplers with an Aitken nuclei counter (TSI 3020) which essentially has no lower concentration limit.

Our results show that the Arctic summer aerosol is indeed in much lower concentration levels than in the winter. So far only one sample has been analyzed for its soot content. The results are listed in Table 1 together with winter results from our

TABLE 1
Physical Characteristics of the Arctic Aerosol

Location	Time	CNC cm^{-3}	SCP m^{-1}	ACT m^{-1}	ACN m^{-1}	MST ngm^{-3}	MSN ngm^{-3}
79°N, 12°E	790/05	309	1.5 E-5	2.2 E-6	1.4 E-6	54	32
4 MER-80/II	8008/09	103	1.6 E-6	1.4 E-7	6.3 E-8	2.4	1.1

Note added in print

Since the symposium, a large number of samples from the YMER-expedition has been analyzed and the better blank values have been established. Table 1 has been updated accordingly. There are, however, some calibrations which are still incomplete, so the results are somewhat preliminary.

CNC = *Aitken nuclei concentration*
SCP = *scattering coefficient of the dry aerosol particles at 550 nm wavelength*
ACT = *absorption coefficient of a total aerosol sample*
ACN = *absorption coefficient of an aerosol sample < 0.1 μm radius*
MST = *mass of soot in a total aerosol sample*
MSN = *mass of soot in an aerosol sample < 0.1 μm radius*

Spitsbergen experiment. The total amount of soot, 2.4 ng m^{-3}, is about 20 times lower than in winter, and about 50% of it is found in the Aitken nuclei range. The light absorption coefficients measured in the hydrosol state amount to about 10% of the total light scattering.

Since our samples have not yet been analyzed for the total suspended mass, summer results from Northern Greenland [10] are used to relate our soot values to the total mass. As in late winter, we find about 1% of the mass to be soot. It should be noted that the results from the one sample analyzed so far, do not represent the cleanest air we sampled during the experiment. Stable situations with scattering coefficients down to 4 x 10^{-7} m^{-1} and total particle concentrations down to 3 cm^{-3} occurred and aerosol samples were taken during these time periods. These results will be reported later.

REFERENCES

1. H. Rosen, A. D. A. Hansen, L. Gundel, T. Novakov, Appl. Opt., Vol. 17 (1978), p. 3859.
2. J. Heintzenberg, H.-C. Hansson, H. Lannefors, Tellus, Vol. 33, in print (1981).
3. P. Winkler, Geophys. Res. Lett., Vol. 2 (1975), p. 45.
4. C.-I. Lin, M. B. Baker, R. J. Charlson, Appl. Opt., Vol. 12 (1973), p. 1356.
5. R. Ulbricht, E.T.Z., Vol. 21 (1900), p. 595.
6. H. C. Donian, A. I. Medalia, J. Paint Techn., Vol. 39 (1967), p. 716.
7. K. A. Rahn, Sources of trace elements in aerosols – an approach to clean air. Ph.D. thesis, University of Michigan, Department of Atmospheric and Oceanic Sciences, 1971.
8. J. Heintzenberg, Tellus, Vol. 32 (1980), p. 251.
9. R. Jaenicke, Phys. Chem., Vol. 82 (1978), p. 1198.
10. H. Flyger and N. Z. Heidam, J. Aerosol Sci., Vol. 9 (1978), p. 157.

DISCUSSION

W. Richards; *Meteorology Research, Inc.*

The commercial grades of carbon have a range of a factor of ten in specific absorption coefficient. Could you say something about the calibration material you used?

J. Heintzenberg

I do not claim to be able to determine the light absorption coefficient in the aerosol particles. I did it on the filters but that is not a direct measurement so I cannot compare them. The direct measurements of light absorption measured with the Lin* method and those which I calculate with the artificial soot are incompatible. Consequently, either the method or the soot I used has an unrealistic light absorption per mass.

W. Richards

Which soot did you use?

J. Heintzenberg

We used Cabot No. 71 which is a channel soot with a rather narrow size distribution.

K. Rahn, *(University of Rhode Island)*

I'm puzzled by some of your data, particularly your coefficient of absorption. Do you trust the optical measurement or the mass measurement?

J. Heintzenberg

The optical measurement is measured directly on the filters with the Lin method.* But these are fiber filters and I don't trust the method because I suspect there are multiple scattering problems within the fibrous filter material. However, I trust the soot mass measurements which indicate soot is 1 % of the total mass. They seem realistic for an aerosol in a very remote area.

Note added in print concerning the apparent discrepancy between absorption coefficients determined in the integrating sphere and with the Lin* method: after elimination of a calculating error in the integrating sphere results and after evaluating the first sample filters with individual blank values, agreement within 20 % is reached between the two methods.

*Lin, C. I., Baker, M. and Charlson, R. J., Appl. Optics, Vol. 12, p. 1356 (1973).

DEPOSITION OF PARTICULATE ELEMENTAL CARBON FROM THE ATMOSPHERE

J. A. OGREN

University of Washington
Seattle, Washington

ABSTRACT

Elemental carbon is removed from the atmosphere by both precipitation scavenging and dry deposition at the Earth's surface. Measurements of the wet removal flux may be obtained from optical or chemical analyses of the insoluble material in precipitation samples. Sampling considerations include losses to the walls of the sampling system, contamination from fugitive dust, pollen, or biological growth in the sample, and extraction of the carbon on to a suitable filter medium. An estimate of the dry deposition flux may be obtained by exposing a clean surface to the atmosphere but not to precipitation, washing with water subsequent to collection, and then analyzing in the same manner as precipitation samples. This may not be representative of the true flux because of the wide variety of surfaces to which deposition may occur, but does provide an estimate of the contribution of dry deposition to the precipitation samples.

Preliminary samples collected in Washington State suggest that precipitation scavenging is an important removal mechanism, and that atmospheric lifetimes for elemental carbon are comparable to those reported for sulfate aerosols. This implies that the removal mechanisms for the two aerosol types are similar, and is consistent with the hypothesis that sulfates and elemental carbon are mixed within the same particles in the atmosphere.

INTRODUCTION

It seems appropriate that removal of particulate elemental carbon (PEC) from the atmosphere is the closing topic for this symposium, because from the point of view of atmospheric chemistry and physics, this is the end of the cycle. However, the atmospheric deposition flux is just one source for the geochemical cycle of PEC through the hydrosphere and lithosphere to its "final" resting place in lake and ocean sediments. Separation of the atmospheric portion from the remainder of the

References p. 388.

geochemical cycle is justifiable because the geochemical sinks are only weakly coupled to the atmosphere through the weathering of exposed sediments. Furthermore, the particles produced by weathering processes typically are large enough to have high settling velocities and hence short atmospheric residence times. Although the effects of this and other sources of the coarse size fraction of PEC (diameters larger than a few micrometers) may be important within a few kilometers of the source, the primary interactions of PEC with atmospheric processes are due to the fine size fraction (diameters smaller than about one micrometer). Accordingly, the emphasis of this paper is the regional-scale deposition of fine, particulate, elemental carbon.

That deposition of PEC occurs is evidenced by the fact that windows get dirty (and black!), and that samples of filtered rain and snow are frequently gray or black. While this doesn't prove that the black deposits are PEC, the identification of PEC as a dominant absorbing substance in submicron aerosols [1, 2] suggests that this may well be the case.

Deposition is a particularly important part of the atmospheric cycle of PEC because, unlike other species (e.g., sulfur dioxide), its chemical stability dictates that the removal mechanisms are physical rather than chemical. As a result, deposition processes may control the atmospheric lifetime of PEC and thereby the spatial scales of importance to the cycle.

In addition to the importance of deposition to the overall cycling of PEC through the atmosphere, deposition taken by itself has several important effects. As suggested by Brosset [3], the catalytic activity of PEC is transferred to raindrops during the washout process, and this may enhance the washout of sulfur species. Deposition of PEC also results in a flux of adsorbed species, some of which might have important health effects [4]. Moreover, the deposition of a material such as PEC with a high efficiency for absorption of solar radiation causes an attenuation of radiation reaching or leaving the surfaces on which deposition occurs. Two examples of this effect are the soiling of windows and buildings, and the reduction of the albedo of snow [5] with consequent effects on radiative climate.

This paper begins with a discussion of the physical and chemical processes that are important for deposition, and relates the physical and chemical properties of PEC to these processes. The emphasis is to describe the deposition process qualitatively rather than to review theoretical treatments of deposition available in the literature. Following this, several considerations for measurement of deposition fluxes are discussed, the experimental approaches are described, and the results of preliminary measurements are presented.

THEORETICAL ASPECTS OF DEPOSITION

A review of the theoretical approach to predicting wet and dry deposition fluxes is far beyond the scope of this paper; the interested reader is referred to the work of Slinn et al. [6] for such a review. The approach taken here is to discuss in primarily qualitative terms the parameterizations generally used in calculating deposition rates, the physical and chemical processes underlying these parameterizations, and the physical and chemical properties of PEC that dictate which of these processes may predominate.

Dry Deposition — Transfer through any one of several layers in the atmosphere may be the rate limiting step for dry deposition, although the major barrier is usually the layer within a few centimeters of the surface. The rate at which submicron particles are delivered to this layer from aloft is determined by atmospheric dynamical processes such as downdrafts, subsidence, and large-scale turbulent eddies. These processes are controlled by meteorological factors such as temperature lapse rate and wind speed rather than the physical or chemical properties of the particles. However, the vertical distribution of the particles is important in this context, and particles with sources near the surface would be expected to be removed faster than particles emitted from elevated sources. In addition, it is possible that high concentrations of PEC may absorb enough solar radiation to increase the static stability of the atmosphere, thereby inhibiting vertical mixing [7]. This would tend to increase the lifetime of existing particles or those emitted aloft, but would decrease the lifetime of particles emitted close to the surface. This positive feedback is probably limited to highly polluted areas representing a limited portion of the cycle, so that transport of submicron PEC to within a few centimeters of the surface is essentially independent of the properties of the particles.

Turbulence within a few centimeters of the surface is strongly damped, causing Brownian diffusion and gravitational settling to be the most efficient mechanisms for transport through this layer. A commonly used parameterization for the flux through this layer assumes that it is proportional to the atmospheric concentration at some reference level, generally 1 - 10 m above the surface. The dry removal flux, R_d $(ML^{-2}T^{-1})$, is written as

$$R_d = V_d \chi \qquad (1)$$

where χ is the atmospheric concentration (ML^{-3}) and V_d is called the deposition velocity (LT^{-1}). There is a minimum in the deposition velocity for particles with both low Brownian diffusivities and low settling velocities, typically in the 0.1 - 1 μm diameter interval.

Deposition velocities are also influenced by the nature of the surface to which deposition occurs. Not only does the surface have a strong effect on micrometeorological conditions near it, but it also determines the efficiency of retention of particles that collide with it. Accordingly, use of deposition velocities in regional flux calculations requires consideration of the relative abundance of surfaces such as forests, grasslands, and bodies of water. Furthermore, this dependence on the nature of the surface makes it difficult to relate real deposition velocities to those measured by artificial collectors.

Beyond the first-order effects of vertical distribution, atmospheric stability, wind speed, Brownian diffusivity, settling velocity, and texture of the surface, other processes might be important in special circumstances. Examples of these proc-

References p. 388.

esses are thermophoresis, as anyone who has cleaned interior walls above heating registers will confirm, and diffusiophoresis, which over the ocean might terminate the dry deposition of some small particles under conditions of high evaporation rates [6].

Although measurements of PEC deposition velocities have not been reported, an estimate of the range of values expected can be obtained from theoretical calculations and from measurements of deposition velocities for other species in the same size range as PEC. For dry deposition to a smooth surface in moderate winds (5-10 ms^{-1}), Slinn [8] calculated deposition velocities of $0.002 - 0.03$ cm s^{-1} for $0.02 - 0.2\,\mu$m diameter particles. Electrical and phoretic effects were neglected for the calculation, under the assumption that Brownian diffusion is the dominant mechanism for particles smaller than $0.2\,\mu$m diameter. Measurements of deposition velocities to a water surface reported by Sehmel and Sutter [9] and Möller and Schumann [10] yielded results similar to those calculated by Slinn [8].

One element which might exhibit dry deposition behavior similar to that expected for PEC is lead. Both have strong sources in vehicular exhaust and similar initial size distributions [11]. Slinn et al. [6] reported an average deposition velocity for lead of 0.31 cm s^{-1} which is at least an order of magnitude larger than that expected for sub-$0.2\,\mu$m diameter particles. This suggests that either the theoretical estimates are too low, or that the lead particles sampled were dominated by large particles with correspondingly large settling velocities.

The small size and near-surface source of vehicular PEC emissions suggest that dry deposition is a more important removal mechanism for them than for particles emitted from elevated stacks or jet aircraft. This, coupled with the larger concentrations in urban areas, indicates that dry deposition fluxes in cities are probably larger than in rural areas. Mixing in the lower few kilometers of the atmosphere will tend to decrease the dry deposition rate of near-surface emissions, and coagulation will cause the small particles to grow into the 0.1-$1\,\mu$m diameter "accumulation mode" where both Brownian diffusivities and settling velocities are low. Accordingly, deposition velocities for PEC are expected to be in the range $10^{-3} - 10^{-1}$ cm s^{-1}, with the highest values occurring for freshly emitted particles.

Wet Deposition — Mechanisms for incorporating PEC into precipitation may be divided into two categories: those in which the particle acts as a nucleating agent in the formation of cloud droplets or ice crystals, and those in which the particle collides with an existing droplet or crystal. The nucleating action of a particle depends strongly on its physical and chemical properties, while collisions with existing cloud particles may be thought of as a special case of dry deposition. Parameterizations used for calculating wet deposition rates generally neglect nucleation mechanisms, or else are based on empirical relationships that are independent of the mechanisms.

The ability of PEC to act as a cloud condensation nucleus (CCN) is dependent on the size and the chemical nature of the surface of the particle. The small size and hygrophobic nature of freshly-emitted PEC suggest that it is not likely to act as a CCN. After the PEC has had adequate time to coagulate with other particles, its cloud nucleating properties may be controlled by the chemical composition of the coagulated mixture. For example, collisions of PEC with sulfuric acid or ammonium sulfate particles might result in a mixed aerosol with cloud nucleating properties similar to those of sulfate aerosols alone. The catalytic action of PEC in oxidizing SO_2 to sulfate in aqueous solution [12] is likely to enhance the cloud nucleating ability of aged aerosols containing PEC. At present, however, little is known about the effects of PEC on the nucleation of cloud droplets.

Even less is known about the ability of PEC to act as an ice nucleus. About all that can be said is that particles with hexagonal crystal structures, similar to that of ice, appear to be more effective than other structures. Crystalline graphite has a hexagonal structure, which suggests that PEC has the potential, as yet unconfirmed, to act as an ice nucleus.

Collisions of PEC with existing cloud or rain drops are controlled by many of the same factors controlling dry deposition. Slinn's [8] calculated collision rates display a minimum at about 0.2μm diameter for the same reasons that deposition velocities have a similar minimum: both Brownian diffusion and inertial effects are weak near this size. Electrical attractions between charged particles and cloud (or rain) droplets may play a role in determining collision rates, but this effect has not been quantified.

A frequently reported parameter for studies of wet deposition is the washout ratio, defined as a particular species' concentration in precipitation divided by its average near-surface concentration in air*.

Wet deposition rates are influenced by the physical properties of the precipitating clouds, e.g., droplet number and size distribution, rainfall intensity and duration, and entrainment rate of ambient air into the cloud. Although the washout ratio parameterization does not explicitly account for these properties, it does provide a means for comparing removal rates of different species by the same storm.

An estimate of the washout ratio for PEC may be obtained by considering the similarity between its physical and chemical properties and those of lead. Both are relatively insoluble, and have comparable initial sizes and common vehicular sources. This suggests that they also may have common mechanisms leading to their wet removal. On this basis, the washout ratio for PEC is expected to be similar to the values for lead of $6 \times 10^4 - 2 \times 10^5$ tabulated in Slinn et al. [6].

MEASUREMENT OF DEPOSITION FLUXES

Measurements of regional-scale wet deposition fluxes of PEC may be obtained using techniques similar to those used for sulfur fluxes [13], with appropriate modifications incorporated where necessary to account for particular physical and chemical properties of PEC. Measurements of regional dry deposition fluxes for other species have not been reported, presumably due to the difficulties associated with measuring dry deposition rates to natural surfaces. Resolution of these problems is beyond the scope of this paper, which will rely on measurements of dry deposition to a smooth surface for initial estimates of dry deposition rates. These

---

Some authors report dimensional washout ratios; others report washout ratios with respect to mass mixing ratios. This paper uses the notation of Slinn et al. [6] in reporting washout ratios as the ratio of mass concentrations per unit volume in precipitation and air.

References p. 388.

measurements also will allow calculation of the contribution of dry deposition to the wet deposition measurements.

In addition to design parameters common to regional-scale measurements of wet deposition fluxes of any species (e.g., number of samplers, distance between samplers, site selection, sampling frequency), the physical and chemical properties of PEC impose special considerations for obtaining quantitative sample recovery. These complications result from the chemical nature of the surface of PEC, its insolubility, and the need to exclude coarse PEC from the measurement. Three basic study areas have been identified during the development of experimental methods for measuring PEC concentrations in precipitation: the relationship between PEC in actual precipitation and in the sample, quantitative extraction of PEC from water, and constraints imposed by the analytical technique used to determine the amount of PEC on a filter.

During and after collection of precipitation samples, several factors may combine to alter the apparent concentration of PEC. During sampling, simultaneous dry deposition of PEC, evaporation of water, and losses of PEC to the walls of the sampling apparatus may occur. Samples collected or stored over long periods of time may be altered by coagulation, and algal growth in such samples may create logistical problems in extracting PEC from the water.

Simultaneous dry deposition and evaporation may be avoided by sampling only during precipitation events, either by manual control or with an automatic sampler which keeps the collector tightly covered during dry periods. Alternatively, a co-located rain gauge may be used to determine the extent of evaporation, and a duplicate collector open to the atmosphere but protected from rainfall can be used to estimate the contribution of dry deposition to the sample. Because sedimentation is not important for fine particles, the dry collector is most easily deployed upside down. This will not collect coarse particles which might fall into the wet collector, but those particles may be removed by mechanical means prior to extraction of the fine PEC. This approach is not appropriate for soluble species such as sulfate, because of the inability to distinguish coarse from fine particles once they have entered solution.

PEC demonstrates strong affinity for the walls of some sample containers. This is presumably due to the chemical nature of the surface of PEC, which often is coated with condensed hydrocarbons [14]; these hydrocarbons adhere strongly to vessels made from or coated with similar materials. Measurements on conventional polyethylene buckets indicate that about half of the PEC is retained on the walls, even after vigorous scrubbing with distilled water and a surfactant (sodium lauryl sulfate). The remainder may be removed with a non-polar solvent (1, 1, 1-trichloroethane was used in the tests in our laboratory). Similar wall losses were observed for a stainless steel beaker. Tests performed on lgass containers indicate that they are preferable for sample collection, as only 5 percent of the total could not be recovered with distilled water and surfactant washes. Limited tests on linear polyethylene sample bottles showed similar behavior to glass containers, suggesting that some plastics might prove to be acceptable. Whatever the sampling system, tests of this nature are needed if quantitative measurements of PEC are to be made.

An estimate of the effects of coagulation on a stored sample may be obtained using the similarity solution for Brownian coagulation described by Friedlander [15]. For initial PEC number concentrations in rainwater greater than 10^6 cm^{-3}, the number concentration after 30 days is essentially independent of the initial number concentration. An initially monodisperse 0.1 μm dia PEC aerosol with a concentration in air of 1 μg/m^3 will have a number concentration in water of 10^9 cm^{-3}, assuming a washout ratio of 10^6. After 30 days of coagulation (neglecting sedimentation and wall losses), the number concentration will about 7×10^4 cm^{-3}, resulting in a volume mean diameter of 2.4 μm. This is probably an upper bound because the calculation neglects sedimentation and losses to the wall of the container. Repeating the calculation for a washout ratio of 10^5 and an air concentration of 0.1 μg/m^3 results in a final volume mean diameter of 0.52 μm. These calculations suggest that PEC in rainwater does not grow by coagulation to more than few microns. However, it appears that PEC which behaves optically in air as a volume absorber [16] may not demonstrate the same behavior after incorporation into rainwater, potentially limiting the usefulness of optical absorption techniques for measuring PEC.

Samples which are collected or stored over extended periods of time are subject to algal growth, particularly in warm climates. This biological carbon hinders sample extraction by rapidly clogging filters, and poses potential interference in measurement systems that remove non-graphitic carbon by volatilization in inert atmospheres. These problems can be avoided with a suitable biocide, although care must be taken to ensure that the biocide does not interfere with the PEC measurement. Two potential candidates are mercuric chloride and pentachlorophenol; preliminary tests indicate that both are effective biocides, although one rain sample using mercuric chloride contained a white precipitate after collection. Further experiments are planned to determine the suitability of these and other biocides.

Quantitative extraction of fine PEC from water requires separation of coarse PEC from the sample, documentation of losses, and obtaining an extract that is representative of the entire sample. Removal of coarse PEC and other large particles (e.g., pollens) is effected by mechanical separation with a nylon mesh*. A mesh opening of 5 microns has been selected as being large enough to minimize losses of fine particles (even after coagulation times of one month) while still removing the undesirable larger material. Sample losses due to collection of fine PEC by the nylon mesh prefilters were estimated by comparing the optical transmission change of the prefilters with that of glass-fiber final filters. For two samples which had little biological growth clogging the prefilter, the fractions of total absorption removed by the prefilter were 0.5 and 0.32. These values, which may be considered as upper limits because they include absorption due to large particles as well as multiple scattering and absorption effects of the nylon mesh, indicate that sample losses on the prefilters are minor. Additional evaluation of these losses will be included as part of continuing studies of recovery efficiencies.

Measurements of the recovery of absorbing material from solution were performed using Nuclepore filter samples of ambient aerosols. These were measured for optical transmission, and then dissolved in concentrated ammonium hydroxide.

Tetco Inc., Monterey Park, CA 91754.

References p. 388.

Hydrochloric acid was used to neutralize the solution, which subsequently was filtered through another Nuclepore filter. The fraction of the initial filter absorption which was recovered ranged from 0.29 to 0.85 for eight filters, with a median value of 0.51. This indicates that loss of absorbing material to the walls of the glass beaker, or changes or absorption efficiency due to coagulation in the solution, introduced uncertainties of the order of a factor of two. More tests of this nature, using a controlled source of graphitic carbon aerosols, are planned to validate this approach more completely.

An additional investigation of the behavior of absorbing aerosols in solution examined the variability of light absorption among extracts taken from the same precipitation sample. Both Nuclepore and glass-fiber filters were included in these tests, and the variabilities reported below include differences between the two filter types. An empirical multiplier of 0.35 was used to correct for multiple scattering and absorption in the glass-fiber filters [17]. The variability typically was less than 50 percent, and in some cases was less than 10 percent. Additional tests are needed, but these preliminary results indicate that the absorbing material is distributed uniformly in the solution and that the absorption measurement may be performed on either Nuclepore or glass-fiber filters.

Special considerations in sample processing are imposed by the analytical method used to determine the amount of PEC on a filter. If combustion to CO_2 is employed [18], then special care must be taken to exlude large carbonaceous particles (e.g., pollens) which might pyrolyze to form PEC in the analyzer; the nylon mesh prefilter should accomplish this. For this technique, high filter loadings are desirable to minimize any variations of the filter blanks.

An alternative to combustion techniques is to measure the optical absorption on the filter using the integrating plate method [19], and interpret the data under the assumption that PEC is the dominant absorber [20]. Calculations by Bergstrom [21] and measurements on acetylene smoke by Roessler and Faxvog [22] suggest the use of an absorption efficiency for fine PEC of about 10 m²/gm. For this approach to work, the filter must not be heavily loaded because of the non-linearities this introduces. A limited comparison of absorption and combustion determinations of PEC extracted from water has been performed as part of the present work. Unfortunately, most of the filters were too heavily loaded to obtain reliable absorption measurements, and the resulting data showed poor agreement between the two techniques. Measurements on lightly loaded filters are planned as part of ongoing method validation tests.

PRELIMINARY EXPERIMENTAL RESULTS

With the development and preliminary verification of methods for quantitative extraction of absorbing material from precipitation, it is possible to estimate the atmospheric lifetimes of PEC. Between 25 April and 12 May 1980, four different types of samples were obtained simultaneously on the University of Washington campus: submicron aerosol, wet deposition, dry deposition, and total deposition. The deposition samples were collected in one-liter glass beakers, containing about

40 mg of mercuric chloride as a biocide, and were passed through a 5 micron nylon mesh and onto 25 mm glass-fiber filters. The wet and dry deposition samples were collected in an automatic sampler which exposed only one sample at a time, depending on whether or not it was raining. The dry deposition beaker was washed repeatedly with distilled water and the resultant wash water processed as a wet deposition sample. The aerosol sample was collected on a 47 mm quartz-fiber filter; coarse particles were removed from the sample by a cyclone ahead of the filter.

The measurements reported below are based on optical measurements, and hence are not specific for PEC. However, the identification of PEC as the dominant absorbing species in urban aerosols by [1] suggests that the optical effects may be attributed to PEC. For clarity of presentation, the data will be reported in terms of PEC mass, based on an absorption efficiency of 10 m^2/gm and a multiple scattering/ absorption correction factor of 0.35. It should be noted that these two assumed factors do not affect the calculated residence times, deposition velocity, or washout ratio.

During the sampling period, the average PEC concentration in air was 0.5 $\mu g/m^3$, with measured deposition fluxes of 9.0, 7.0, and 0.36 $\mu g/m^2$/hr for total, wet, and dry removal, respectively. Assuming a mixed layer depth of 1.5 km, the column burden of PEC was 750 $\mu g/m^2$. Defining the residence time as the ratio of column burden to flux, the residence times were caluclated to be 3.5, 4.5, and 87 days for total, wet, and dry deposition, respectively. These values are comparable to those for sulfate aerosols [23] and support the hypothesis that the removal mechanisms for the two aerosol types are the same.

The concentration of PEC in the wet deposition sample was 240 $\mu g/\ell$, leading to a washout ratio of 5 x 10^5. The dry deposition velocity was 0.02 cm s^{-1}. Both these parameters are near the expected range of values discussed earlier, lending credence to the experimental methodology and encouraging further refinement and applications.

Returning to the effects discussed earlier, it is instructive to examine the wet and dry deposition fluxes in terms of the rate of increase of absorption optical depths of the deposition surfaces. For the case of dry deposition to a window, unit absorption optical depth is reached after 35 years; the corresponding time for wet deposition is only 2.3 years. These results indicate that the contribution of dry deposition to the total deposition measurement was negligible for the site studied.

CONCLUSIONS

Particulate elemental carbon (PEC) may be removed from the atmosphere by both wet and dry processes. Preliminary measurements indicate that wet deposition is an important removal mechanism. Dry deposition to natural surfaces may also be significant, although dry removal on a glass collector was negligible compared to wet removal. The limited data obtained to date suggest that lifetimes for PEC are similar to those for sulfur aerosols, which implies that the removal mechanisms for the two aerosol types may be similar. This is consistent with the hypothesis that sulfates and PEC are mixed within the same particles in the atmosphere.

References p. 388.

ACKNOWLEDGEMENT

Special thanks are due to Dr. John Miller of NOAA for the loan of the automated wet/dry deposition sampler, and to Dr. Peter Groblicki of General Motors Research Laboratories for performing carbon analyses. This work was supported in part by the National Science Foundation Graduate Fellowship Program.

REFERENCES

1. H. Rosen, A. D. A. Hansen, L. Gundel and T. Novakov, App. Optics, Vol. 24 (1978), p. 3859.
2. R. E. Weiss, A. P. Waggoner, R. J. Charlson, D. L. Thorsell, J. S. Hall and L. A. Riley, "Studies of the Optical, Physical, and Chemical Properties of Light Absorbing Aerosols," Proc. Conf. on Carbonaceous Particles in the Atmosphere, Lawrence Berkeley Laboratory, Berkeley, California, 1979.
3. C. Brosset, Ambio, Vol. 5 (1976), p. 157.
4. B. D. Crittenden and R. Long, "The Mechanisms of Formation of Polynuclear Aromatic Compounds in Combustion Systems," in Carcinogenesis, Vol. 1. Polynuclear Aromatic Hydrocarbons: Chemistry, Metabolism, and Carcinogenesis, edited by R. I. Freudenthal and P. W. Jones, Raven Press, New York, 1976.
5. T. C. Grenfell, D. K. Perovich and J. A. Ogren (1981) "Spectral Albedos of an Alpine Snowpack," Cold Regions Sci. Technol. (In press).
6. W. G. N. Slinn, L. Hasse, B. B. Hicks, A. W. Hogan, D. Lal, P. S. Liss, K. O. Munnich, G. A. Sehmel and O. Vittori, Atmos. Environ., Vol. 12 (1978), p. 2055.
7. R. J. Charlson and M. J. Pilat, J. App. Met., Vol. 8 (1969), p. 1001.
8. W. G. N. Slinn, J. Wat. Air Soil Poll., Vol. 7 (1977), p. 513.
9. G. A. Sehmel and S. L. Sutter, J. Rechs. Atmos., Vol. 3 (1974), p. 911.
10. U. Möller and G. Shumann, J. Geophys. Res., Vol. 75 (1970), p. 3013.
11. G. L. Ter Haar, D. L. Lenane, J. N. Hu and M. Brandt, Air Pollut. Control Assoc., Vol. 22 (1972), p. 39
12. S.-G. Chang, R. Brodzinsky, R. Toossi, R. P. Markowitz and T. Novakov, "Catalytic Oxidation of SO_2 on Carbon in Aqueous Solution," Proc. Conf. on Carbonaceous Particles in the Atmosphere, Lawrence Berkeley Laboratory, Berkeley, California, 1979.
13. L. Granat, Atmos. Environ., Vol. 12 (1978), p. 413.
14. W. R. Pierson, "Particulate Organic Matter and Total Carbon from Vehicles on the Raod," Proc. Conf. on Carbonaceous Particles in the Atmosphere, Lawrence Berkeley Laboratory, Berkeley, California, 1979.
15. S. K. Friedlander, Smoke, Dust and Haze, Wiley, New York, 1977.
16. A. P. Waggoner, M. B. Baker and R. J. Charlson, App. Optics, Vol. 12 (1973), p. 896.
17. M. Sadler, R. J. Charlson, H. Rosen and T. Novakov, Atmos. Environ., (In press).
18. S. H. Cadle, P. J. Groblicki and D. P. Stroup, Anal. Chem., Vol. 52 (1980), p. 2201.
19. C. Lin, M. Baker and R. J. Charlson, App. Optics, Vol. 12 (1973), p. 1356.
20. R. E. Weiss and A. P. Waggoner, These Proceedings.
21. R. W. Bergstrom, Beitr. Phys. Atm., Vol. 46 (1973), p. 223.
22. D. M. Roessler and F. R. Faxvog, J. Opt. Soc. Am., Vol. 70 (1980), p. 230.
23. H. Rodhe, Atmos. Environ., Vol. 12 (1978), p. 671.

DISCUSSION

J. Heicklen, *(Penn State University)*

Where did you get the 1-1/2 kilometer scale factor?

J. Ogren

I estimated it from George Holzworth's* study of mixing heights based on rawinsonde data. The springtime average is about 1.5 kilometers for our part of the country.

J. Heicklen

I have taken a scale height of three kilometers and calculated a lifetime of eleven days. If I increase the scale height, the lifetime increase to the square of that. That is a factor of four which would be the difference between the 3-1/2 days you get and the 11 days I get. Was it particularly rainy in Seattle during this time?

J. Ogren

It was one of our highly advertised periods when it drizzles constantly.

J. Heicklen

Do you think that because it rains frequently in Seattle, that you are going to get shorter lifetimes than other parts of the country.

J. Ogren

It would be interesting to make these measurements under different general synoptic conditions. However, our average annual rainfall is only about 85 cm, less than that in many areas of the country.

D. Stedman, (University of Michigan)

There is a great deal of conventional wisdom that has arisen from people who live in London about the wet and dry deposition of soot particles. It is certainly very clear that your conclusion is that wet deposition is much more important. If you put out some laundry in London to dry and it rains, the problem is not that it comes in wet, but that it comes in dirty.

J. Heicklen

That is not a fair argument because precipitation is intermittent. So you have to average the wet deposition over periods when it is not raining.

---

*Holzworth, G.C. "Mixing Heights, Wind Speeds, and Potential for Urban Air Pollution Throughout the Contiguous United States," office of Air Programs Publication AP-11, U.S. Environmental Protection Agency, Research Triangle Park, North Carolina, 1972.

L. Newman, *(Brookhaven National Laboratory)*

I guess I am not clear on what your wet deposition fluxes are. The actual number depends on the fraction of time that it is raining and I assume that is included in your wet deposition flux.

J.Ogren

The flux which I reported, 7 micrograms per square meter per hour, is the flux over this entire seventeen day period. That is averaged over several periods of rain, but I did not measure the percentage of the time that it rained.

L. Newman

I would like to know what the X_c term is and how it was measured.

J. Ogren

That is the concentration of sub-micron absorbing aerosol as inferred from the absorption measurement, using a specific absorption efficiency of $10m^2/g$.

J. Muhlbaier, *(General Motors Research Laboratories)*

Since you collected the dry deposition on a glass slide, how did you eliminate the coarse material?

J. Ogren

The dry deposition sample was collected in a glass beaker and then washed and processed in the same manner as a wet deposition sample. I filtered it through a 5 micron mesh nylon screen after the sample was collected. So at the time of collection, it contained both the coarse and fine deposition fractions. In the process of handling it, I used the nylon mesh to mechanically separate it.

K. Rahn

One of the points that I was thinking about in preparation for this conference was what I call the large scale lifetime of black carbon in the atmosphere. I did not have a chance to go into that but on one of my slides, the aging diagram,* you could see evidence that carbon decreased with distance or age at roughly the same rate as submicron vanadium and manganese. I take this as an indication that, even over the long haul, carbon is behaving quite similar to these submicron species.

---

*Rahn, K., Brosset, C., Ottar, B. and Patterson, E. M. , These Proceedings, p. 327.

J. Ogren

One final point on that, comparing Heitzenberg's* ice breaker cruise data with what you find in Point Barrow clearly shows that there must be a major difference in the sinks for air flowing over the Arctic and over the ocean. In the more rainy areas which Heintzenberg studied the concentrations are 3 orders of magnitude less the 1 μg/m^3 which you reported.

---

*Heintzenberg, J., These Proceedings, p. 371.

PARTICIPANTS

Ackerman, Thomas P.
  NASA Ames Research Center
  Moffett Field, CA

Agnew, William G.
  General Motors Research Laboratories
  Warren, MI

Albers, W. A. Jr.
  General Motors Research Laboratories
  Warren, MI

Amann, Charles A.
  General Motors Research Laboratories
  Warren, MI

Ancker-Johnson, Betsy
  GM, Environmental Activities Staff
  Warren, MI

Anderson, Larry G.
  General Motors Research Laboratories
  Warren, MI

Ang, Carolina C.
  General Motors Research Laboratories
  Warren, MI

Appel, Bruce R.
  California Department of Health
  Berkeley, CA

Arnold, Steve
  Colorado Department of Health
  Denver, CO

Atkinson, Dwight
  USEPA
  Ann Arbor, MI

Bauer, Mark H.
  GM, Industry-Government Relations
  Detroit, MI

Bidwell, Joseph B.
  General Motors Research Laboratories
  Warren, MI

Bove, John L.
  The Cooper Union
  New York, NY

Bradow, Ronald L.
  U. S. Environmental Protection Agency
  Research Triangle Park, NC

Brosset, Cyrill
  Swedish Water & Air Pollution
    Research Laboratory
  Gothenburg, Sweden

Budiansky, Stephen P.
  Environmental Science & Technology
  Washington, DC

Buzan, Leroy R.
  General Motors Research Laboratories
  Warren, MI

Cadle, Steven H.
  General Motors Research Laboratories
  Warren, MI

Caplan, John D.
General Motors Research Laboratories
Warren, MI

Carey, Penny
USEPA
Ann Arbor, MI

Carrigan, Richard A.
National Science Foundation
Washington, DC

Cass, Glen R.
California Institute of Technology
Pasadena, CA

Cermak, Gregory W.
General Motors Research Laboratories
Warren, MI

Chan, Tai L.
General Motors Research Laboratories
Warren, MI

Chang, S.-G.
University of California
Berkeley, CA

Charlson, Robert J.
University of Washington
Seattle, WA

Chenea, Paul F.
General Motors Research Laboratories
Warren, MI

Chock, David P.
General Motors Research Laboratories
Warren, MI

Coffey, Peter E.
New York State
Dept. of Environmental Conservation
Albany, NY

Colucci, Joseph M.
General Motors Research Laboratories
Warren, MI

Countess, Richard J.
General Motors Research Laboratories
Warren, MI

Currie, Lloyd A.
National Bureau of Standards
Washington, DC

Daisey, Joan M.
New York University Medical Center
Tuxedo, NY

Dattner, Stuart
Texas Air Control Board
Austin, TX

Dod, Raymond L.
University of California
Berkeley, CA

Draftz, Ronald G.
I.I.T. Research Institute
Chicago, IL

Dunker, Alan M.
General Motors Research Laboratories
Warren, MI

Elder, Charles J.
GM, Environmental Activities Staff
Warren, MI

Everett, Robert L.
GM, Environmental Activities Staff
Warren, MI

Fagley, Walter S.
Chrysler Corporation
Highland Park, MI

Faix, Louis J.
GM, Chevrolet Motor Division
Warren, MI

Ferek, Ronald
Nat'l Center for Atmospheric Research
Boulder, CO

Ferman, Martin A.
  General Motors Research Laboratories
  Warren, MI

Gardels, Keith D.
  General Motors Research Laboratories
  Warren, MI

Gerber, Hermann E.
  Naval Research Laboratory
  Washington, DC

Gibson, Thomas L.
  General Motors Research Laboratories
  Warren, MI

Gorse, Robert A.
  Ford Motor Company
  Dearborn, MI

Groblicki, Peter J.
  General Motors Research Laboratories
  Warren, MI

Gundel, Lara
  University of California
  Berkeley, CA

Haberstadt, Marcel
  Motor Vehicle Manufacturers Assoc.
  Detroit, MI

Hansen, Anthony D. A.
  University of California
  Berkeley, CA

Heicklen, Julian P.
  Pennsylvania State University
  University Park, PA

Heintzenberg, Jost
  Intern. Meteorological Institute
  Stockholm, Sweden

Heuss, Jon M.
  General Motors Research Laboratories
  Warren, MI

Hickling, Robert
  General Motors Research Laboratories
  Warren, MI

Hidy, George M.
  Environ. Research & Technology, Inc.
  Westlake Village, CA

Hilden, David L.
  General Motors Research Laboratories
  Warren, MI

Hilst, Glen R.
  Electric Power Research Institute
  Palo Alto, CA

Holzwarth, James C.
  General Motors Research Laboratories
  Warren, MI

Hummel, John R.
  General Motors Research Laboratories
  Warren, MI

Huntzicker, James J.
  Oregon Graduate Center
  Beaverton, OR

Jakelvic, Joseph
  University of California
  Berkeley, CA

Jamerson, Frank E.
  General Motors Research Laboratories
  Warren, MI

Japar, Steven
  Ford Motor Company
  Dearborn, MI

Kamal, Mounir M.
  General Motors Research Laboratories
  Warren, MI

Kelly, Nelson A.
  General Motors Research Laboratories
  Warren, MI

Klimisch, Richard L.
General Motors Research Laboratories
Warren, MI

Kloian, John Jr.
GM, Detroit Diesel Allison
Detroit, MI

Kumar, Sudarshan
General Motors Research Laboratories
Warren, MI

Kummler, Ralph
Wayne State University
Detroit, MI

Larson, John G.
General Motors Research Laboratories
Warren, MI

Lee, William R.
General Motors Research Laboratories
Warren, MI

Lewis, Lynn L.
General Motors Research Laboratories
Warren, MI

Lipari, Frank
General Motors Research Laboratories
Warren, MI

Macias, Edward S.
Washington University
St. Louis, MO

Marks, Craig
GM, Environmental Activities Staff
Warren, MI

Martens, Stuart J.
GM, Environmental Activities Staff
Warren, MI

Mason, Conrad
University of Michigan
Ann Arbor, MI

Mason, Mark
Northrup Services
Research Triangle Park, NC

McDonald, Richard J.
General Motors Research Laboratories
Warren, MI

Mohan, Phillip V.
GM, Engineering Staff
Milford, MI

Mohnen, Volker A.
State University of New York
Albany, NY

Mueller, Peter K.
Electric Power Research Institute
Palo Alto, CA

Muench, Nils L.
General Motors Research Laboratories
Warren, MI

Muhlbaier, Jean L.
General Motors Research Laboratories
Warren, MI

Mulawa, Patricia A.
General Motors Research Laboratories
Warren, MI

Nebel, George J.
General Motors Research Laboratories
Warren, MI

Newman, Leonard
Brookhaven National Laboratory
Upton, NY

Novakov, T.
University of California
Berkeley, CA

Ogren, John
University of Washington
Seattle, WA

Palmer, R. A.
  Duke University
  Durham, NC

Patty, R. R.
  North Carolina State University
  Raleigh, NC

Phillips, Robert J.
  GM, Environmental Activities Staff
  Warren, MI

Pierrard, J. M.
  E. I. DuPont
  Wilmington, DE

Pimenta, James A.
  Ontario Ministry of the Environment
  Rexdale, Ontario, Canada

Rahn, Kenneth A.
  University of Rhode Island
  Kingston, RI

Reck, Ruth A.
  General Motors Research Laboratories
  Warren, MI

Richards, L. Willard
  Meteorology Research, Inc.
  Santa Rosa, CA

Rifkin, Ellis
  Ethyl Corporation
  Ferndale, MI

Roessler, David M.
  General Motors Research Laboratories
  Warren, MI

Rosen, Hal
  University of California
  Berkeley, CA

Russell, Philip A.
  Denver Research Institute
  Denver, CO

Ruthkosky, Martin J.
  General Motors Research Laboratories
  Warren, MI

Samson, Perry
  University of Michigan
  Ann Arbor, MI

Schreck, Richard M.
  General Motors Research Laboratories
  Warren, MI

Schroeder, John
  Michigan Dept. of Natural Resources
  Lansing, MI

Shelef, Mordecai
  Ford Motor Company
  Dearborn, MI

Siegla, Donald M.
  General Motors Research Laboratories
  Warren, MI

Sloane, Christine S.
  General Motors Research Laboratories
  Warren, MI

Smith, George W.
  General Motors Research Laboratories
  Warren, MI

Soderholm, Sidney C.
  General Motors Research Laboratories
  Warren, MI

Spengler, John D.
  Harvard School of Public Health
  Boston, MA

Spicer, Chester W.
  Battelle Memorial Institute
  Columbus, OH

Stedman, Donald
  University of Michigan
  Ann Arbor, MI

Stevens, Robert K.
U. S. Environmental Protection Agency
Research Triangle Park, NC

Stroup, David P.
General Motors Research Laboratories
Warren, MI

Sverdrup, George
Battelle Memorial Institute
Columbus, OH

Swarin, Steven J.
General Motors Research Laboratories
Warren, MI

Taylor, Kathleen C.
General Motors Research Laboratories
Warren, MI

Tracy, J. C.
General Motors Research Laboratories
Warren, MI

Vostal, Jaroslav J.
General Motors Research Laboratories
Warren, MI

Waggoner, Alan P.
University of Washington
Seattle, WA

Weaver, Harry B.
Motor Vehicle Manufacturers Assoc.
Detroit, MI

Whitby, Robert A.
New York State, Department
of Environmental Conservation
Albany, NY

White, Warren H.
Washington University
St. Louis, MO

Williams, Ronald L.
General Motors Research Laboratories
Warren, MI

Wolff, George T.
General Motors Research Laboratories
Warren, MI

AUTHOR AND CONTRIBUTOR INDEX

This index contains the names of the authors who contributed to this volume, the authors who have been cited and the names of the individuals who participated in the discussions.

Numbers in parenthesis indicate the number of the references cited in the text without the names of the authors.

Numbers set in boldface refer to the first page of a contributor's paper.

Burnstein, R., 167(30)

Cadle, S.H., 80(12), **89**, 89(10), 90(20),
    90(23), 91(10), 95(30), 95(30), 95(31),
    98, 99(10), 112(4), 185(5), 188(5),
    188(9), 207(10), 214(10), 216(10),
    240(10), 252(28), 254(28), 262(2),
    297, 297(6), 298(8), 301(13), 302(6),
    302(16), 304(20), 305(16), 306(20),
    307(20), 323(10), 336(22), 341,
    344(24), 369, 386(18)
Cahill, T.A., 131(10)
Calder, J.A., 246(11)
Campbell, P.P., 10(17)
Carlson, T.H., 147, 148, 150
Carr, R.C., 10(18)
Cary, R.A., **79**, 131(1), 214(34), 215(34),
    232(34), 234(34), 236(34)
Cass, G.R., 131(11), 131(12), **207**,
    207(5), 208, 208(5), 210(26), 214(33),
    216(24), 218(50), 219(5), 220(33),
    226, 227, 230, 241
Cassate, W.A., 279(11)
Cautreels, 345(25)
Caverly, R.S., 147, 148, 150
Chan, T., 302(15)
Chandrasekhar, S., 59(23)
Chang, S.G., 14(28), 20(2), **159**, 159(1),
    160(1), 166(21), 166(23), 167(26),
    168, 168(34), 168(36), 173(35),
    173(37), 174(37), 174(49), 185(2),
    297(5), 344(16), 344(18), 383(12)
Charlson, R.J., **3**, 7(4), 8(13), 13(24),
    54(5), 69, 89(14), 89(15), 91(14),
    113(9), 113(10), 113(12), 114(9),
    114(12), 117(9), 188(9), 146, 147(6),
    147(7), 147(30), 149(30), 150(5),
    150(6), 150(7), 151(6), 151(7),
    151(54), 152(5), 152(6), 153(5),
    153(6), 154(59), 155(63), 174(63),
    208(16), 219(54), 245(1), 273(3),
    274(5), 276(7), 278(3), 278(8), 281(7),
    297(1), 302(17), 302(18), 309(17),
    318(12), 318(4), 323(8), 325, 344(7),
    344(8), 345(8), 373(4), 377, 380(2),
    381(7), 385(16), 386(17), 386(19)
Chmutok, K., 167(27)
Chock, D.P., 92(29)
Chow, J.C., 348(30), 353(30), 354(30)

Chu, L.-C., 80(14), 89(16), **131**, 131(5),
    131(6), 131(7), 131(8), 151(47),
    207(5), 207(6), 208(5), 219(5), 219(6),
    241, 336(24)
Chylek, P., 54(6), 59(24)
Ciaccio, L., 207(1)
Clark, A., 171(47), 172(47)
Clark, R.L., 174(51)
Clarke, A.G., 168
Clemenson, M.S., 166(21)
Coakley, J.A., 54(6), 59(24)
Cofer, III, W.R., 168
Coffey, P., 71
Colodny, P., 80(3), 83, 89(2), 92(2),
    112(3), 207(3)
Conklin, M.H., 131(12), 207(5), 208,
    219, 241
Conway, T.J., 329(15), 334(15), 339(15)
Cooper, J.A., 245(4), 249(24), 252(24),
    253(24)
Cotten, F.A., 12(21)
Coughlin, R.W., 160(6), 168(33)
Coulson, K.L., 54(7)
Countess, R.J., 39, 90(20), 91(24),
    95(30), 95(31), 96(24), 207(10),
    208(10), 214(10), 216(10), 240(10),
    252(28), 254(28), **297**, 297(3), 297(6),
    298(3), 298(10), 302(6), 302(16),
    303(3), 304(20), 305, 306(20),
    307(20), 310(3), 311(20), 323(10),
    336(22), 341, 344(24), 369
Courtney, W.J., **111**, 245(5), 247(19),
    254(19)
Covert, D.A., 245(1), 278(8)
Cox, S.T., 147(28), 149(28)
Craig, H., 246(10)
Crittenden, A.L., 344(7), 344(8), 345(8)
Crittenden, B.D., 380(4)
Cronn, D., 344(8)
Cukor, P., 207(1)
Cullis, C.F., 8(7), 160(16), 185(6)
Currie, L.A., 38, 143, **245**, 247(13),
    247(14), 247(17), 249(17), 249(21),
    249(22), 249(24), 250(21), 250(21),
    251(14), 251(25), 252(24), 252(30),
    252(31), 253(30), 254(24), 255(31),
    256(31), 262(6)
Cushing, K.M., 147(18), 148(18),
    150(18)
Czeplak, G., 56(19)

SUBJECT INDEX

For subjects that appear on 2 or more consecutive pages, only the first page is listed.